ENCYCLOPEDIA OF MATHEMATICS AND ITS APPLICATIONS

Edited by G.-C. ROTA

Volume 50

T0234376

Handbook of Categorical Algebra 1

ENCYCLOPEDIA OF MATHEMATICS AND ITS APPLICATIONS

ENCYCLOPEDIA OF MATHEMATICS AND ITS APPLICATIONS

Handbook of Categorical Algebra 1

Basic Category Theory

Francis Borceux

Département de Mathématique
Université Catholique de Louvain

CAMBRIDGE
UNIVERSITY PRESS

CAMBRIDGE UNIVERSITY PRESS
Cambridge, New York, Melbourne, Madrid, Cape Town, Singapore, São Paulo

Cambridge University Press
The Edinburgh Building, Cambridge CB2 8RU, UK

Published in the United States of America by Cambridge University Press, New York

www.cambridge.org
Information on this title: www.cambridge.org/9780521441780

First published 1994
This digitally printed version 2008

A catalogue record for this publication is available from the British Library

ISBN 978-0-521-44178-0 hardback
ISBN 978-0-521-06119-3 paperback

to Sammy Eilenberg,
mathematician and friend

Contents

Preface to volume 1

Volume 1 of the *Handbook of category theory* is concerned with those notions and techniques which turn out to appear quite naturally in most developments of category theory, independently of additional structures or properties one requires from the categories involved in the study.

Any book on category theory must say a word on the non-obvious problems concerned with the logical foundations of the theory. We mention both the axiom system of *classes* and that of *universes*, and later we freely use the presentation which fits best the problem we study. We chose not to dwell on foundational questions as long as the development of the theory does not really depend on them.

Another general principle in this volume is to develop the general notions from more accessible special cases, for which we have given a large supply of examples. This is by no means the most economical way of developing the theory, but we hope inexperienced readers will appreciate our pedagogical choice.

Of course we start with the basic vocabulary of categories, functors, natural transformations, monomorphisms, epimorphisms, isomorphisms. The analogies between monomorphisms and epimorphisms, covariant and contravariant functors, lead to the famous *duality principle* which is, with the Yoneda lemma, one of the key results of the first chapter.

Starting with the notions of products, coproducts, equalizers, coequalizers, pullbacks, and so on, we reach the general notion of limit of a functor and prove the corresponding existence theorem. We devote special attention to some classes of colimits, like absolute ones, filtered ones and universal ones. We also study limits in categories of functors and dwell on the case of colimits of set-valued functors.

Adjoint functors are the next fundamental notion of categorical algebra. We prove the classical general and special existence theorems in

terms of limits and particularize our study to the case of reflective subcategories. We treat separately the case of Kan extensions, which is a
fruitful example of an adjunction which cannot be deduced in its general
form from the previous existence theorems.

The special adjoint functor theorems refer explicitly to notions like
subobject (equivalence class of monomorphisms), generator and cogenerator. We decided to group in a separate chapter a specific study of
the various kinds of monomorphisms, epimorphisms (plain, extremal,
strong, regular) and correspondingly for generators, cogenerators (plain,
strong, regular, dense). This is also the opportunity for studying epi-
mono factorizations as well as notions related to monomorphisms or
epimorphisms, like injective and projective objects. We pay special attention to the case of injective cogenerators, whose existence is very often
related to rather deep specific theorems.

Chapter 5 is probably the most difficult one in this first volume. It is
essentially concerned with rather sophisticated methods for constructing
reflective subcategories. We start with the problem of "formally inverting
an arrow $g\colon A \longrightarrow C$ of a category" (categories of fractions) and relate
this with the problem of "uniquely extending a morphism $f\colon A \longrightarrow B$
along the arrow $g\colon A \longrightarrow C$" (the orthogonal subcategory problem). Under good assumptions, we prove by transfinite induction the reflectivity
of the subcategory of those objects "orthogonal to a set of mappings".
We relate this to the general notion of $(\mathcal{E}, \mathcal{M})$-systems of factorization.
We discuss the special case where the reflection is left exact (the "localizations") and relate it to the notions of universal closure operation and
bidense morphism.

The rest of the book is essentially devoted to some generalizations or
special cases of the basic notions of chapters 1, 2 and 3.

When $\mathscr{C}$ is a category with finite limits, set-valued functors on $\mathscr{C}$ which
preserve finite limits (the "left exact functors") admit interesting properties and characterizations: in particular, they are exactly the filtered
colimits of representable functors. When $\mathscr{C}$ is an arbitrary small category, this characterization serves as a definition for the notion of "flat
functor". Flat functors share most properties of left exact functors and
will turn out to play a key role in a categorical approach of model theory
(see chapter 5, volume 2, accessible categories).

Another elementary notion which turns out to have unexpectedly rich
applications is that of "splitting idempotents"; this happens to be equivalent to a rather weak completeness property: having just those limits
or colimits which are preserved by all functors. This is called the *Cauchy*

completeness of the category, for reasons which will become apparent in chapter 6 of volume 2. Replacing "completeness" by "Cauchy completeness" and "preserving limits" by "being α-flat, for every regular cardinal α" yields amazingly enough a generalization of the adjoint functor theorem; this generalization now contains as a special case the Kan extension theorem of chapter 3.

While a category has just objects and arrows, the category of categories and functors can be provided with additional devices, namely natural transformations between functors. This leads to the richer notion of a 2-category, where besides objects and arrows one gives also "2-cells" between the arrows. There are corresponding enrichments of the notions of functor, natural transformation, limit, adjoint functors, and so on.

Now in category theory, many constructions are defined uniquely... up to isomorphism! This results in very often obtaining isomorphisms where one would have expected equalities. So it is not unusual to reach a situation where a composite $f \circ (g \circ h)$ (whatever f, g, h and the composition are) is just isomorphic to $(f \circ g) \circ h$, not necessarily equal. Taking this seriously, one gets the notion of a *bicategory*: one has objects, arrows and 2-cells, but in various axioms "equalities" are replaced by "isomorphic 2-cells". A basic example of a bicategory is that of small categories, distributors and natural transformations between them: in category theory, a distributor is to a functor what, in set theory, a relation is to a function. Every functor turns out to be a distributor with a right adjoint, and the converse holds when working with Cauchy complete categories. In the same spirit as bicategories, one can "relax" the notions of functor, natural transformation, limit, colimit, working now "up to an isomorphic 2-cell" or even "up to an arbitrary 2-cell".

We conclude this first volume with an elementary study of internal categories. While a small category has a *set* of objects and a *set* of arrows, together with some operations "source", "target", "unit", "composition" given by *mappings*, one is now interested in replacing *set* by *object of a category* $\mathscr{C}$ and *mapping* by *arrow of the category* $\mathscr{C}$. This is the notion of "category internal to $\mathscr{C}$". We study the corresponding notions of internal functors and internal limits or colimits.

Introduction to this handbook

My concern in writing the three volumes of this *Handbook of categorical algebra* has been to propose a directly accessible account of what – in my opinion – a Ph.D. student should ideally know of category theory before starting research on one precise topic in this domain. Of course, there are already many good books on category theory: general accounts of the state of the art as it was in the late sixties, or specialized books on more specific recent topics. If you add to this several famous original papers not covered by any book and some important but never published works, you get a mass of material which gives probably a deeper insight in the field than this *Handbook* can do. But the great number and the diversity of those excellent sources just act to convince me that an integrated presentation of the most relevant aspects of them remains a useful service to the mathematical community. This is the objective of these three volumes.

The first volume presents those basic aspects of category theory which are present as such in almost every topic of categorical algebra. This includes the general theory of limits, adjoint functors and Kan extensions, but also quite sophisticated methods (like categories of fractions or orthogonal subcategories) for constructing adjoint functors. Special attention is also devoted to some refinements of the standard notions, like Cauchy completeness, flat functors, distributors, 2-categories, bicategories, lax-functors, and so on.

The second volume presents a selection of the most famous classes of "structured categories", with the exception of toposes which appear in volume 3. The first historical example is that of abelian categories, which we follow by its natural non-additive generalizations: the regular and exact categories. Next we study various approaches to "categories of models of a theory": algebraic categories, monadic categories, locally presentable and accessible categories. We introduce also enriched category theory and devote some attention to topological categories. The volume ends with the theory of fibred categories "à la Bénabou".

The third volume is entirely devoted to the study of categories of sheaves: sheaves on a space, a locale, a site. This is the opportunity for developing the essential aspects of the theory of locales and introducing Grothendieck toposes. We relate this with the algebraic aspects of volume 2 by proving in this context the existence of a classifying topos for coherent theories. All these considerations lead naturally to the notion of an elementary topos. We study quite extensively the internal logic of toposes, including the law of excluded middle and the axiom of infinity. We conclude by showing how toposes are a natural context for defining sheaves.

Besides a technical development of the theory, many people appreciate historical notes explaining how the ideas appeared and grew. Let me tell you a story about that.

It was in July, I don't remember the year. I was participating in a summer meeting on category theory at the Isles of Thorns, in Sussex. Somebody was actually giving a talk on the history of Eilenberg and Mac Lane's collaboration in the forties, making clear what the exact contribution of the two authors was. At some point, somebody in the audience started to complain about the speaker giving credit to Eilenberg and Mac Lane for some basic aspect of their work which – he claimed – they borrowed from somebody else. A very sophisticated and animated discussion followed, which I was too ignorant to follow properly. The only things I can remember are the names of the two opponents: the speaker was Saunders Mac Lane and his opponent was Samuel Eilenberg. I was not born when they invented category theory. With my little story in mind, maybe you will forgive me for not having tried to give credit to anybody for the notions and results presented in this *Handbook*.

Let me conclude this introduction by thanking the various typists for their excellent job and my colleagues of the Louvain-la-Neuve category seminar for the fruitful discussions we had on various points of this *Handbook*. I want especially to acknowledge the numerous suggestions Enrico Vitale has made for improving the quality of my work.

Handbook of categorical algebra

Contents of the three volumes

1

The language of categories

1.1 Logical foundations of the theory

It is a common practice, when developing mathematics, to consider a statement involving "all groups" or "all topological spaces" .... For example we say that an abelian group A is projective when, for every surjective homomorphism of abelian groups $f\colon B \longrightarrow C$ and every group homomorphism $g\colon A \longrightarrow C$, g factors through f (see diagram 1.1). This definition of "A being projective" starts thus with a list of universal quantifiers

$$\forall B \ \forall C \ \forall f \ \forall g \ \ldots$$

This formula, from the point of view of set theory, creates a problem: the variables B and C are "running through something ($=$ the collection of all abelian groups) which is not a set". This last fact is an easy consequence of the following well-known paradox.

Proposition 1.1.1 *There exists no set S such that*

$$x \in S \Leftrightarrow x \text{ is a set.}$$

Proof In other (bad) words: "the set of sets does not exist"! To prove this, let us assume such an S exists. Since $x \notin x$ is a formula of set theory

$$T = \{x \mid x \in S \text{ and } x \notin x\}$$

defines a subset T of S, thus in particular a set T. The law of excluded middle tells us that

$$T \in T \text{ or } T \notin T.$$

But from the definition of T itself we conclude that

$$T \in T \Rightarrow T \notin T,$$

1

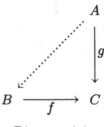

Diagram 1.1

$$T \notin T \Rightarrow T \in T,$$

thus in both cases a contradiction. □

Category theory will in fact be handling all the time "the collection of all groups", "the collection of all sets", "the collection of all topological spaces", and so on .... Therefore it is useful to pay some attention to these questions of "size" at the very beginning of this book.

A first way to handle, in category theory, problems of this type is to assume the axiom on the existence of "universes".

Definition 1.1.2 *A universe is a set $\mathcal{U}$ with the following properties*

(1) $x \in y$ and $y \in \mathcal{U} \Rightarrow x \in \mathcal{U}$,

(2) $I \in \mathcal{U}$ and $\forall i \in I \ x_i \in \mathcal{U} \Rightarrow \bigcup_{i \in I} x_i \in \mathcal{U}$,

(3) $x \in \mathcal{U} \Rightarrow \mathcal{P}(x) \in \mathcal{U}$,

(4) $x \in \mathcal{U}$ and $f : x \longrightarrow y$ surjective function $\Rightarrow y \in \mathcal{U}$,

(5) $\mathbb{N} \in \mathcal{U}$,

where $\mathbb{N}$ denotes the set of finite ordinals and $\mathcal{P}(x)$ denotes the set of subsets of x.

Notice some easy consequences of the definition.

Proposition 1.1.3

(1) $x \in \mathcal{U}$ and $y \subseteq x \Rightarrow y \in \mathcal{U}$,

(2) $x \in \mathcal{U}$ and $y \in \mathcal{U} \Rightarrow \{x, y\} \in \mathcal{U}$,

(3) $x \in \mathcal{U}$ and $y \in \mathcal{U} \Rightarrow x \times y \in \mathcal{U}$,

(4) $x \in \mathcal{U}$ and $y \in \mathcal{U} \Rightarrow x^y \in \mathcal{U}$.

Proof We prove (1) and leave the rest as an easy exercise. First of all $\emptyset \in \mathbb{N}$ and $\mathbb{N} \in \mathcal{U}$, thus $\emptyset \in \mathcal{U}$. Now if $x \in \mathcal{U}$ and $y \subseteq x$ with $y \neq \emptyset$, choose $z \in y$. Define $f : x \longrightarrow y$ to be

$$f(t) = t \text{ if } t \in y,$$
$$f(t) = z \text{ if } t \notin y.$$

Obviously f is surjective and therefore $y \in \mathcal{U}$. □

It should be noticed that – assuming the axiom of choice in our set theory – condition (4) in definition 1.1.2 could have been replaced precisely by

$$x \in \mathcal{U} \text{ and } y \subseteq x \Rightarrow y \in \mathcal{U}.$$

Now the axiom on the existence of universes is just

Axiom 1.1.4 *Every set belongs to some universe.*

Not much is known about this axiom from the point of view of set theory. Because of the property

$$x \in \mathcal{U} \text{ and } y \subseteq x \Rightarrow y \in \mathcal{U},$$

it sounds reasonable to think of the elements of a universe as being "sufficiently small sets". If you choose to use the theory of universes as a foundation for category theory, the following convention has to remain valid throughout this book.

Convention 1.1.5 *We fix a universe $\mathcal{U}$ and call "small sets" the elements of $\mathcal{U}$.*

Obviously we now have the following

Proposition 1.1.6 *There exists a set S with the property $x \in S \Leftrightarrow x$ is a small set.*

Proof Just choose $S = \mathcal{U}$. □

An analogous statement is valid for small abelian groups, small topological spaces, and so on .... For example a small group is a pair $(G, +)$ where G is a small set (and there is just a set of them) and $+$ is a suitable mapping $G \times G \longrightarrow G$ (and there is just a set of them); so we can draw the conclusion by proposition 1.1.3.

An alternative way to handle these problems of size is to use the Gödel–Bernays theory of sets and classes. In the Zermelo–Fränkel theory, the primitive notions are "set" and "membership relation". In the Gödel–Bernays theory, there is one more primitive notion called "class" (think of it as "a big set"); that primitive notion is related to the other two by the property that every set is a class and, more precisely:

Axiom 1.1.7 *A class is a set if and only if it belongs to some (other) class.*

The axioms concerning classes imply in particular the following "comprehension scheme" for constructing classes.

Comprehension scheme 1.1.8 *If $\varphi(x_1,\ldots,x_n)$ is a formula where quantification just occurs on set variables, there exists a class A such that*

$$(x_1,\ldots,x_n) \in A \text{ if and only if } \varphi(x_1,\ldots,x_n).$$

For example, there exists a class A with the property

$$(G,+) \in A \text{ if and only if } (G,+) \text{ "is a group"}$$

(where "is a group" is an abbreviation for the group axioms); in other words, this defines the "class of all groups". In the same way we deduce the existence of the class of sets, the class of topological spaces, the class of projective abelian groups, and so on.

When the axiom of universes is assumed and a universe $\mathcal{U}$ is fixed, one gets a model of the Gödel–Bernays theory by choosing as "sets" the elements of $\mathcal{U}$ and as "classes" the subsets of $\mathcal{U}$. It makes no relevant difference whether we base category theory on the axiom of universes or on the Gödel-Bernays theory of classes. We shall use the terminology of the latter, thus using the words "set" and "class"; the reader who prefers the terminology of the former should thus read "small set" when we write "set" and should read "set" when we write "class".

1.2 Categories and functors

With every mathematical structure on a set is generally associated a notion of "mapping compatible with that structure": a group homomorphism between groups, a linear mapping between vector spaces, a continuous mapping between topological spaces, and so on .... The basic examples of a category are designed in precisely that way: those sets provided with a prescribed structure and, between them, those mappings which are compatible with the given strucure.

Definition 1.2.1 *A category $\mathscr{C}$ consists of the following:*

(1) a class $|\mathscr{C}|$, whose elements will be called "objects of the category";

(2) for every pair A, B of objects, a set $\mathscr{C}(A,B)$, whose elements will be called "morphisms" or "arrows" from A to B;

(3) for every triple A, B, C of objects, a composition law

$$\mathscr{C}(A,B) \times \mathscr{C}(B,C) \longrightarrow \mathscr{C}(A,C);$$

the composite of the pair (f,g) will be written $g \circ f$ or just gf;

(4) for every object A, a morphism $1_A \in \mathscr{C}(A,A)$, called the identity on A.

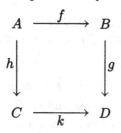

Diagram 1.2

These data are subject to the following axioms.

(1) Associativity axiom: *given morphisms* $f \in \mathscr{C}(A, B)$, $g \in \mathscr{C}(B, C)$, $h \in \mathscr{C}(C, D)$ *the following equality holds*:

$$h \circ (g \circ f) = (h \circ g) \circ f.$$

(2) Identity axiom: *given morphisms* $f \in \mathscr{C}(A, B)$, $g \in \mathscr{C}(B, C)$ *the following equalities hold*:

$$1_B \circ f = f, \quad g \circ 1_B = g.$$

A morphism $f \in \mathscr{C}(A, B)$ will often be represented by the notation $f \colon A \longrightarrow B$; A is called the "domain" or the "source" of f and B is called the "codomain" or the "target" of f. In the situation of diagram 1.2, we say that the given square is "commutative" when the equality $g \circ f = k \circ h$ holds between the two possible composites; an analogous terminology holds for diagrams of arbitrary shape.

As usual 1_A is the only morphism from A to A which plays the role of an identity for the composition law. Indeed if $i_A \in \mathscr{C}(A, A)$ is another such morphism

$$1_A = 1_A \circ i_A = i_A.$$

Let us now define a "homomorphism of categories".

Definition 1.2.2 *A functor F from a category $\mathscr{A}$ to a category $\mathscr{B}$ consists of the following:*

(1) *a mapping*

$$|\mathscr{A}| \longrightarrow |\mathscr{B}|$$

between the classes of objects of $\mathscr{A}$ and $\mathscr{B}$; the image of $A \in \mathscr{A}$ is written $F(A)$ or just FA;

(2) *for every pair of objects A, A' of $\mathscr{A}$, a mapping*

$$\mathscr{A}(A, A') \longrightarrow \mathscr{B}(FA, FA');$$

the image of $f \in \mathscr{A}(A, A')$ is written $F(f)$ or just Ff.

These data are subject to the following axioms:

(1) for every pair of morphisms $f \in \mathcal{A}(A, A')$, $g \in \mathcal{A}(A', A'')$

$$F(g \circ f) = F(g) \circ F(f);$$

(2) for every object $A \in \mathcal{A}$

$$F(1_A) = 1_{FA}.$$

Given two functors $F \colon \mathcal{A} \longrightarrow \mathcal{B}$ and $G \colon \mathcal{B} \longrightarrow \mathcal{C}$, a pointwise composition immediately produces a new functor $G \circ F \colon \mathcal{A} \longrightarrow \mathcal{C}$. This composition law is obviously associative. The identity functor on the category $\mathcal{A}$ (i.e. choose every mapping in definition 1.2.2 to be the identity) is clearly an identity for that composition law. A careless argument could thus lead to the conclusion that categories and functors constitute a new category ... but this can easily be reduced to a contradiction using proposition 1.1.1! The point is that, in the axioms for a category, it is required to have a *set* of morphisms between any two objects. And when the categories $\mathcal{A}$ and $\mathcal{B}$ merely have a class of objects, there is no way to force the functors from $\mathcal{A}$ to $\mathcal{B}$ to constitute a *set*. All along in this book we shall realize how crucial it is, in category theory, to distinguish all the time between sets and classes. To facilitate the language, we particularize definition 1.2.1.

Definition 1.2.3 *A category $\mathcal{C}$ is called a small category when its class $|\mathcal{C}|$ of objects is a set.*

The next result is then obvious (see 1.1.8).

Proposition 1.2.4 *Small categories and functors between them constitute a category.*   □

Examples 1.2.5

Let us first list some obvious examples of categories and the corresponding notation, when it is classical.

1.2.5.a   Sets and mappings: Set.

1.2.5.b   Topological spaces and continuous mappings: Top.

1.2.5.c   Groups and group homomorphisms: Gr.

1.2.5.d   Commutative rings with unit and ring homomorphisms: Rng.

1.2.5.e   Real vector spaces and linear mappings: $\text{Vect}_{\mathbb{R}}$.

1.2.5.f   Real Banach spaces and bounded linear mappings: Ban_{∞}.

1.2.5.g   Sets and injective mappings.

1.2.5.h   Real Banach spaces and linear contractions: Ban_1.

And so on.

Examples 1.2.6

Here is a list of some mathematical devices which can also be seen as categories.

1.2.6.a Choose as objects the natural numbers and as arrows from n to m the matrices with n rows and m columns; the composition is the usual product of matrices.

1.2.6.b A poset $(X, \leq)$ can be viewed as a category $\mathscr{X}$ whose objects are the elements of X; the set $\mathscr{X}(x, y)$ of morphisms is a singleton when $x \leq y$ and is empty otherwise. The possibility of defining a (unique) composition law is just the transitivity axiom of the partial order; the existence of identities is just the reflexivity axiom.

1.2.6.c Every set X can be viewed as a category $\mathscr{X}$ whose objects are the elements of X and the only morphisms are identities. ($\mathscr{X}(x, y)$ is a singleton when $x = y$ and is empty otherwise). A category whose only morphisms are the identities is called a discrete category.

1.2.6.d A monoid $(M, \cdot)$ can be seen as a category $\mathscr{M}$ with a single object $*$ and $M = \mathscr{M}(*, *)$ as a set of morphisms; the composition law is just the multiplication of the monoid. As a special case, we can view any group as a category. When a ring with unit is considered as a special case of a category, the composition law of that category is generally that induced by the multiplication of the ring.

Examples 1.2.7

From a given category $\mathscr{C}$, one very often constructs new categories of "diagrams in $\mathscr{C}$". Here are some basic contructions.

1.2.7.a Let us fix an object $I \in \mathscr{C}$. The category $\mathscr{C}/I$ of "arrows over I" is defined by the following.

- Objects: the arrows of $\mathscr{C}$ with codomain I.
- Morphisms from the object $(f \colon A \longrightarrow I)$ to the object $(g \colon B \longrightarrow I)$: the morphisms $h \colon A \longrightarrow B$ in $\mathscr{C}$ such that $g \circ h = f$ (the "commutative triangles over I"); see diagram 1.3.

The composition law is that induced by the composition of $\mathscr{C}$. Notice that when $\mathscr{C}$ is the category of sets and mappings, a mapping $f \colon A \longrightarrow I$ can be identified with the I-indexed family of sets $\left(f^{-1}(i)\right)_{i \in I}$ so that the previous category is just that of I-indexed families of sets and I-indexed families of mappings.

1.2.7.b Again fixing an object $I \in \mathscr{C}$, we define the category $I/\mathscr{C}$ of "arrows under I".

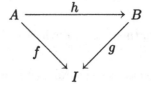

Diagram 1.3

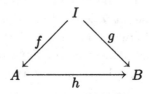

Diagram 1.4

- Objects: the arrows of $\mathscr{C}$ with domain I.
- Morphisms from the object $f\colon I \longrightarrow A$ to the object $g\colon I \longrightarrow B$: the morphisms $h\colon A \longrightarrow B$ in $\mathscr{C}$ such that $h \circ f = g$ (the "commutative triangles under I"); see diagram 1.4.

The composition law is induced by that of $\mathscr{C}$.

1.2.7.c   The category $\mathsf{Ar}(\mathscr{C})$ of arrows of $\mathscr{C}$ has for objects all the arrows of $\mathscr{C}$; a morphism from the object $(f\colon A \longrightarrow B)$ to the object $(g\colon C \longrightarrow D)$ is a pair $(h\colon A \longrightarrow C, k\colon B \longrightarrow D)$ of morphisms of $\mathscr{C}$, with the property $k \circ f = g \circ h$ ("a commutative square"); see diagram 1.5. Again, the composition law is that induced pointwise by the composition in $\mathscr{C}$.

In examples 1.2.7.a,b,c, it is easy to check that when $\mathscr{C}$ is small, so are the three categories $\mathscr{C}/I$, $I/\mathscr{C}$, $\mathsf{Ar}(\mathscr{C})$.

Examples 1.2.8

Let us finally mention some first examples of functors.

1.2.8.a   The "forgetful functor" $U\colon \mathsf{Ab} \longrightarrow \mathsf{Set}$ from the category Ab of abelian groups to the category Set of sets maps a group $(G,+)$ to the underlying set G and a group homomorphism f to the corresponding mapping f.

1.2.8.b   If R is a commutative ring, let us write Mod_R for the category of R-modules and R-linear mappings. Tensoring with R produces a functor from the category Ab of abelian groups to Mod_R:

$$- \otimes R\colon \mathsf{Ab} \longrightarrow \mathsf{Mod}_R.$$

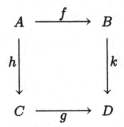

Diagram 1.5

An abelian group A is mapped to the group $A \otimes_{\mathbb{Z}} R$ provided with the scalar multiplication induced by the formula

$$(a \otimes r)r' = a \otimes (rr').$$

A group homomorphism $f: A \longrightarrow B$ is mapped to the R-linear mapping $f \otimes \mathrm{id}_R$.

1.2.8.c We obtain a functor $\mathcal{P}: \mathrm{Set} \longrightarrow \mathrm{Set}$ from the category of sets to itself by mapping a set A to its power set $\mathcal{P}(A)$ and a mapping $f: A \longrightarrow B$ to the "direct image mapping" from $\mathcal{P}(A)$ to $\mathcal{P}(B)$.

1.2.8.d Given a category $\mathscr{C}$ and a fixed object $C \in \mathscr{C}$, we define a functor

$$\mathscr{C}(C, -): \mathscr{C} \longrightarrow \mathrm{Set}$$

from $\mathscr{C}$ to the category of sets by first putting

$$\mathscr{C}(C, -)(A) = \mathscr{C}(C, A).$$

Now if $f: A \longrightarrow B$ is a morphism of $\mathscr{C}$, the corresponding mapping

$$\mathscr{C}(C, -)(f) \equiv \mathscr{C}(C, f): \mathscr{C}(C, A) \longrightarrow \mathscr{C}(C, B)$$

is defined by the formula

$$\mathscr{C}(C, f)(g) = f \circ g$$

for an arrow $g \in \mathscr{C}(C, A)$. Such a functor is called a "representable functor" (the functor is "represented" by the object C).

1.2.8.e Given two categories $\mathscr{A}, \mathscr{B}$ and a fixed object $B \in \mathscr{B}$, we define the "constant functor to B"

$$\Delta_B: \mathscr{A} \longrightarrow \mathscr{B}$$

by

$$\Delta_B(A) = B, \quad \Delta_B(f) = 1_B$$

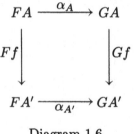

$$FA \xrightarrow{\quad \alpha_A \quad} GA$$

Diagram 1.6

for every object $A \in \mathcal{A}$ and every morphism f of $\mathcal{A}$.

1.3 Natural transformations

General topology studies, in particular, topological spaces and continuous functions between them. But given two continuous functions from a space to another one, there exists also the notion of a "homotopy" between those two continuous functions, which allows you to "pass" from one function to the other one. A similar situation exists for categories and functors.

Definition 1.3.1 *Consider two functors $F, G: \mathcal{A} \rightrightarrows \mathcal{B}$ from a category $\mathcal{A}$ to a category $\mathcal{B}$. A natural transformation $\alpha: F \Rightarrow G$ from F to G is a class of morphisms $(\alpha_A: FA \longrightarrow GA)_{A \in \mathcal{A}}$ of $\mathcal{B}$ indexed by the objects of $\mathcal{A}$ and such that for every morphism $f: A \longrightarrow A'$ in $\mathcal{A}$, $\alpha_{A'} \circ F(f) = G(f) \circ \alpha_A$. (see diagram 1.6)*

It is an obvious matter to notice that, when F, G, H are functors from $\mathcal{A}$ to $\mathcal{B}$ and $\alpha: F \Rightarrow G$, $\beta: G \Rightarrow H$ are natural tranformations, the formula

$$(\beta \circ \alpha)_A = \beta_A \circ \alpha_A$$

defines a new natural transformation $\beta \circ \alpha: F \Rightarrow H$. That composition law is clearly associative and possesses a unit at each functor F: this is just the natural transformation 1_F whose A-component is 1_{FA}. Again a careless argument would deduce the existence of a category whose objects are the functors from $\mathcal{A}$ to $\mathcal{B}$ and whose morphisms are the natural transformations between them. But since $\mathcal{A}$ and $\mathcal{B}$ have merely classes of objects, there is in general no way to prove the existence of a set of natural transformations between two functors! But when $\mathcal{A}$ is small, that problem disappears and we get the following result.

Proposition 1.3.2 *Let $\mathscr{A}$ be a small category and $\mathscr{B}$ an arbitrary category. The functors from $\mathscr{A}$ to $\mathscr{B}$ and the natural transformations between them constitute a category; that category is small as long as $\mathscr{B}$ is small.* □

We prove now the first important theorem of this book. We refer to example 1.2.8.d for the description of the representable functors.

Theorem 1.3.3 (The Yoneda lemma)
Consider a functor $F: \mathscr{A} \longrightarrow \mathsf{Set}$ from an arbitrary category $\mathscr{A}$ to the category of sets and mappings, an object $A \in \mathscr{A}$ and the corresponding representable functor $\mathscr{A}(A, -): \mathscr{A} \longrightarrow \mathsf{Set}$. There exists a bijective correspondence

$$\theta_{F,A}: \mathsf{Nat}\big(\mathscr{A}(A, -), F\big) \xrightarrow{\;\;\cong\;\;} FA$$

between the natural transformations from $\mathscr{A}(A, -)$ to F and the elements of the set FA; in particular those natural transformations constitute a set. The bijections $\theta_{F,A}$ constitute a natural transformation in the variable A; when $\mathscr{A}$ is a small category, the bijections $\theta_{F,A}$ also constitute a natural transformation in the variable F.

Proof   For a given natural transformation $\alpha: \mathscr{A}(A, -) \Rightarrow F$, we define $\theta_{F,A}(\alpha) = \alpha_A(1_A)$. With a given element $a \in FA$ we associate, for every object $B \in \mathscr{A}$, a mapping

$$\tau(a)_B: \mathscr{A}(A, B) \longrightarrow FB$$

defined by $\tau(a)_B(f) = F(f)(a)$. This class of mappings defines a natural transformation

$$\tau(a): \mathscr{A}(A, -) \Rightarrow F$$

since, for every morphism $g: B \longrightarrow C$ in $\mathscr{A}$, the relation

$$Fg \circ \tau(a)_B = \tau(a)_C \circ \mathscr{A}(A, g)$$

(see diagram 1.7) reduces to the equality

$$\forall f \in \mathscr{A}(A, B) \quad F(g \circ f)(a) = Fg((Ff)(a)),$$

which follows from the functoriality of F.

$\theta_{F,A}$ and τ are inverse to each other. Indeed, starting from $a \in FA$ we have

$$\theta_{F,A}\big(\tau(a)\big) = \tau(a)_A(1_A) = F(1_A)(a) = 1_{FA}(a) = a.$$

$$\mathscr{A}(A, B) \xrightarrow{\tau(a)_B} FB$$

$$\mathscr{A}(A, g) \Big\downarrow \qquad\qquad \Big\downarrow Fg$$

$$\mathscr{A}(A, C) \xrightarrow[\tau(a)_C]{} FC$$

Diagram 1.7

On the other hand, starting from $\alpha \colon \mathscr{A}(A, -) \Rightarrow F$ and choosing a morphism $f \colon A \longrightarrow B$ in $\mathscr{A}$,

$$\begin{aligned}
\tau\big(\theta_{F,A}(\alpha)\big)_B(f) &= \tau\big(\alpha_A(1_A)\big)_B(f) \\
&= F(f)\big(\alpha_A(1_A)\big) \\
&= \alpha_B\big(\mathscr{A}(A, f)(1_A)\big) \\
&= \alpha_B(f \circ 1_A) \\
&= \alpha_B(f),
\end{aligned}$$

where the third equality follows from the naturality of α. This proves the first part of the theorem.

To prove the naturality of the bijections, let us consider the functor $N \colon \mathscr{A} \longrightarrow \mathsf{Set}$ defined by

$$N(A) = \mathsf{Nat}\big(\mathscr{A}(A, -), F\big).$$

and for every morphism $f \colon A \longrightarrow B$ in $\mathscr{A}$

$$N(f) \colon \mathsf{Nat}\big(\mathscr{A}(A, -), F\big) \longrightarrow \mathsf{Nat}\big(\mathscr{A}(B, -), F\big)$$
$$N(f)(\alpha) = \alpha \circ \mathscr{A}(f, -)$$

(see example 1.3.6.c for the definition of $\mathscr{A}(f, -)$). We are claiming the existence of a natural transformation $\eta \colon N \Rightarrow F$ defined by $\eta_A = \theta_{F,A}$. Indeed, with the previous notation,

$$\begin{aligned}
\big(\theta_{F,B} \circ N(f)\big)(\alpha) &= \theta_{F,B}\big(\alpha \circ \mathscr{A}(f, -)\big) \\
&= \big(\alpha \circ \mathscr{A}(f, -)\big)_B(1_B) \\
&= \alpha_B(f), \\
\big(F(f) \circ \theta_{F,A}\big)(\alpha) &= F(f)\big(\alpha_A(1_A)\big) \\
&= \big(\alpha_B \circ \mathscr{A}(A, f)\big)(1_A) \\
&= \alpha_B(f).
\end{aligned}$$

Moreover, when $\mathscr{A}$ is a small category, it makes sense to consider the category $\mathsf{Fun}(\mathscr{A}, \mathsf{Set})$ of functors from $\mathscr{A}$ to Set and natural transformations between them. For a fixed object $A \in \mathscr{A}$ we consider this time the functor $M \colon \mathsf{Fun}(\mathscr{A}, \mathsf{Set}) \longrightarrow \mathsf{Set}$ defined by

$$M(F) = \mathsf{Nat}\big(\mathscr{A}(A, -), F\big);$$

for a functor $G \colon \mathscr{A} \longrightarrow \mathsf{Set}$ and a natural transformation $\gamma \colon F \Rightarrow G$,

$$M(\gamma) \colon \mathsf{Nat}\big(\mathscr{A}(A, -), F\big) \longrightarrow \mathsf{Nat}\big(\mathscr{A}(A, -), G\big)$$

is defined by $M(\gamma)(\alpha) = \gamma \circ \alpha$. On the other hand we consider the functor "evaluation in A" $\mathrm{ev}_A \colon \mathsf{Fun}(\mathscr{A}, \mathsf{Set}) \longrightarrow \mathsf{Set}$ defined by

$$\mathrm{ev}_A(F) = FA, \ \ \mathrm{ev}_A(\gamma) = \gamma_A.$$

We claim to have a natural transformation $\mu \colon M \Rightarrow \mathrm{ev}_A$ defined by $\mu_F = \theta_{F,A}$. Indeed, with the previous notation,

$$\big(\theta_{G,A} \circ M(\gamma)\big)(\alpha) = \theta_{G,A}(\gamma \circ \alpha)$$
$$= (\gamma \circ \alpha)_A(1_A),$$
$$\big(\mathrm{ev}_A(\gamma) \circ \theta_{F,A}\big)(\alpha) = \gamma_A\big(\alpha_A(1_A)\big). \qquad \square$$

In proposition 1.3.2 we have used a first composition law for natural transformations. In fact, there exists another possible type of composition for natural transformations.

Proposition 1.3.4 *Consider the following situation:*

$$\mathscr{A} \ \overset{F}{\underset{G}{\rightrightarrows}} \Downarrow\alpha \ \mathscr{B} \ \overset{H}{\underset{K}{\rightrightarrows}} \Downarrow\beta \ \mathscr{C}$$

where $\mathscr{A}$, $\mathscr{B}$, $\mathscr{C}$ *are categories,* F, G, H, K *are functors and* α, β *are natural transformations. The formula, for every* $A \in \mathscr{A}$,

$$(\beta * \alpha)_A = \beta_{GA} \circ H(\alpha_A) = K(\alpha_A) \circ \beta_{FA}$$

defines a natural transformation

$$\beta * \alpha \colon H \circ F \Rightarrow K \circ G.$$

called the "Godement product" of the two natural transformations α *and* β.

Proof $(\beta * \alpha)_A$ is thus defined considering diagram 1.8 which is indeed commutative by naturality of β. The proposition asserts, for every morphism $f \colon A \longrightarrow A'$ in $\mathscr{A}$, the commutativity of the outer rectangle

$$HFA \xrightarrow{H(\alpha_A)} HGA$$

$$\beta_{FA} \downarrow \qquad \qquad \downarrow \beta_{GA}$$

$$KFA \xrightarrow[K(\alpha_A)]{} KGA$$

Diagram 1.8

$$HFA \xrightarrow{H(\alpha_A)} HGA \xrightarrow{\beta_{GA}} KGA$$

$$HFf \downarrow \qquad HGf \downarrow \qquad \qquad \downarrow KGf$$

$$HFA' \xrightarrow[H(\alpha_{A'})]{} HGA' \xrightarrow[\beta_{GA'}]{} KGA'$$

Diagram 1.9

in diagram 1.9. It holds since the first square commutes by naturality of α and functoriality of H and the second square commutes by naturality of β. □

The proof of the next proposition is a straightforward exercise left to the reader.

Proposition 1.3.5 *Consider the situation*

$$\mathcal{A} \begin{array}{c} \xrightarrow{F} \\ \xrightarrow{H \; \Downarrow \alpha} \\ \xrightarrow{L \; \Downarrow \gamma} \end{array} \mathcal{B} \begin{array}{c} \xrightarrow{G} \\ \xrightarrow{K \; \Downarrow \beta} \\ \xrightarrow{M \; \Downarrow \delta} \end{array} \mathcal{C}$$

where $\mathcal{A}$, $\mathcal{B}$, $\mathcal{C}$ are categories, F, G, H, K, L, M are functors and α, β, γ, δ are natural transformations. The following equality holds:

$$(\delta * \gamma) \circ (\beta * \alpha) = (\delta \circ \beta) * (\gamma \circ \alpha). \qquad \square$$

For the sake of brevity and with the notations of the previous propositions, we shall often write $\beta * F$ instead of $\beta * 1_F$ or $G * \alpha$ instead of $1_G * \alpha$.

Examples 1.3.6

1.3.6.a Consider the power set functor $\mathcal{P} \colon \mathsf{Set} \longrightarrow \mathsf{Set}$ defined in 1.2.8.c and the identity functor $\mathrm{id} \colon \mathsf{Set} \longrightarrow \mathsf{Set}$. The mappings "singleton"

$$\sigma_E \colon E \longrightarrow \mathcal{P}(E)$$

which map an element $x \in E$ to the singleton $\{x\}$ constitute a natural transformation $\sigma: \mathrm{id} \Rightarrow \mathcal{P}$.

1.3.6.b Consider the category $\mathsf{Vect}_{\mathbb{R}}$ of real vector spaces and the bidual functor

$$(\ )^{**}: \mathsf{Vect}_{\mathbb{R}} \longrightarrow \mathsf{Vect}_{\mathbb{R}}.$$

The canonical morphisms

$$\sigma_V: V \longrightarrow V^{**}, \quad v \mapsto v^{**}$$

for every vector space V, define a natural transformation from the identity functor to the bidual functor.

1.3.6.c Consider a category $\mathscr{A}$ and a morphism $f: A \longrightarrow B$ of $\mathscr{A}$. We obtain a natural transformation

$$\mathscr{A}(f, -): \mathscr{A}(B, -) \Rightarrow \mathscr{A}(A, -)$$

between the functors represented by A and B (see 1.2.8.d) by putting, for every object $C \in \mathscr{A}$ and every morphism $g \in \mathscr{A}(B, C)$,

$$\mathscr{A}(f, -)_C(g) = g \circ f.$$

Generally we shall write $\mathscr{A}(f, C)$ for the mapping $\mathscr{A}(f, -)_C$.

1.3.6.d Given two categories $\mathscr{A}$, $\mathscr{B}$ and a fixed morphism $b: B \longrightarrow B'$, we define the "constant natural transformation on b" $\Delta_b: \Delta_B \Rightarrow \Delta_{B'}$ by $(\Delta_b)_A = b$ for every object $A \in \mathscr{A}$ (see 1.2.8.e for the definition of Δ_B, $\Delta_{B'}$).

1.4 Contravariant functors

If $\mathscr{A}$ is a small category, we know it makes sense to consider the category of functors from $\mathscr{A}$ to Set and natural transformations between them (see 1.3.2). In examples 1.2.8.d and 1.3.6.c we have defined a mapping

$$Y^*: \mathscr{A} \longrightarrow \mathsf{Fun}(\mathscr{A}, \mathsf{Set}),$$
$$Y^*(A) = \mathscr{A}(A, -), \quad Y^*(f) = \mathscr{A}(f, -),$$

where $A \in |\mathscr{A}|$ is an object of $\mathscr{A}$ and f is a morphism of $\mathscr{A}$. It is rather obvious that, given morphisms

$$A \xrightarrow{\ f\ } B \xrightarrow{\ g\ } C$$

in $\mathscr{A}$, we obtain the following equalities:

$$Y^*(g \circ f) = Y^*(f) \circ Y^*(g), \quad Y^*(1_B) = 1_{Y^*B}.$$

So Y^* is a mapping which "reverses the direction of every morphism",

$$f: A \longrightarrow B, \quad Y^*(f): Y^*(B) \longrightarrow Y^*(A),$$

and – up to this reversing process – preserves the composition law and identities. This is what we shall call a "contravariant functor".

Definition 1.4.1 *A contravariant functor F from a category $\mathscr{A}$ to a category $\mathscr{B}$ consists of the following:*

(1) a mapping

$$|\mathscr{A}| \longrightarrow |\mathscr{B}|$$

between the classes of objects; the image of $A \in \mathscr{A}$ is written $F(A)$ or just FA;

(2) for every pair of objects $A, A' \in \mathscr{A}$, a mapping

$$\mathscr{A}(A, A') \longrightarrow \mathscr{B}(FA', FA);$$

the image of $f \in \mathscr{A}(A, A')$ is written $F(f)$ or just Ff.

These data are subject to the following axioms:

(1) for every pair of morphisms $f \in \mathscr{A}(A, A')$, $g \in \mathscr{A}(A', A'')$,

$$F(g \circ f) = F(f) \circ F(g);$$

(2) for every object $A \in \mathscr{A}$,

$$F(1_A) = 1_{FA}.$$

When confusion could be possible, we shall emphasize the fact that we are definitely working with a functor in the sense of definition 1.2.2 by calling it a *covariant* functor.

The notion of a natural transformation can easily be carried over to the contravariant case.

Definition 1.4.2 *Consider two contravariant functors $F, G: \mathscr{A} \rightrightarrows \mathscr{B}$ from a category $\mathscr{A}$ to a category $\mathscr{B}$. A natural transformation $\alpha: F \Rightarrow G$ from F to G is a class of morphisms $(\alpha_A: FA \longrightarrow GA)_{A \in \mathscr{A}}$ of $\mathscr{B}$ indexed by the objects of $\mathscr{A}$ and such that for every morphism $f: A \longrightarrow A'$ in $\mathscr{A}$, $G(f) \circ \alpha_{A'} = \alpha_A \circ F(f)$ (see diagram 1.10).*

All the results of sections 1.2 and 1.3 can be transposed to the contravariant case; this is a straightforward exercise left to the reader. Moreover, we should mention at this point that the validity of this transposition can also be obtained as an application of the duality principle of section 1.10.

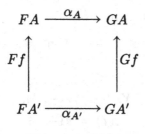

Diagram 1.10

Examples 1.4.3

1.4.3.a We started this section with the example of the "contravariant Yoneda embedding"

$$Y^*: \mathscr{A} \longrightarrow \mathsf{Fun}(\mathscr{A}, \mathsf{Set})$$

for a small category $\mathscr{A}$.

1.4.3.b Example 1.2.8.d can be "dualized"; given a category $\mathscr{A}$ and an object $A \in \mathscr{A}$ we define a contravariant functor

$$\mathscr{A}(-, A): \mathscr{A} \longrightarrow \mathsf{Set}$$

by the formulas

$$\mathscr{A}(-, A)(B) = \mathscr{A}(B, A)$$

for every object $B \in \mathscr{A}$, and

$$\mathscr{A}(-, A)(f): \mathscr{A}(C, A) \longrightarrow \mathscr{A}(B, A),$$
$$\mathscr{A}(-, A)(f)(g) = g \circ f$$

for all morphisms $f: B \longrightarrow C$ and $g: C \longrightarrow A$ in $\mathscr{A}$.

1.4.3.c Example 1.3.6.c can be "dualized" as well. With the previous notation we obtain a natural transformation

$$\mathscr{A}(-, f): \mathscr{A}(-, B) \longrightarrow \mathscr{A}(-, C)$$

for $f: B \longrightarrow C$, by putting

$$\mathscr{A}(-, f)_D(h) = f \circ h$$

for every object D and every morphism $h: D \longrightarrow B$. Generally, we shall write $\mathscr{A}(D, f)$ for the mapping $\mathscr{A}(-, f)_D$.

1.4.3.d Again using the previous notation, example 1.4.3.a itself can be "dualized". Let us write $\mathsf{Fun}^*(\mathscr{A}, \mathsf{Set})$ for the category of contravariant functors from a small category $\mathscr{A}$ to Set. The "covariant Yoneda embedding" is the covariant functor

$$Y_*: \mathscr{A} \longrightarrow \mathsf{Fun}^*(\mathscr{A}, \mathsf{Set})$$

defined by the formulas

$$Y_*(A) = \mathscr{A}(-, A),$$
$$Y_*(f) = \mathscr{A}(-, f)$$

for every object $A \in \mathscr{A}$ and every morphism f of $\mathscr{A}$.

1.4.3.e   Consider the category Rng of commutative rings with unit and the category Top of topological spaces and continuous mappings. The construction of the Zariski spectrum of a ring gives rise to a contravariant functor

$$\mathsf{Sp}: \mathsf{Rng} \longrightarrow \mathsf{Top}.$$

For a given ring A, $\mathsf{Sp}(A)$ is the Zariski spectrum of A, that is the set of prime ideals of A provided with the topology generated by the fundamental open subsets

$$\mathcal{O}_a = \{ P \in \mathsf{Sp}(A) \,|\, a \notin P \}$$

for every element $a \in A$. For a given ring homomorphism $f: A \longrightarrow B$, the inverse image process maps a prime ideal of B to a prime ideal of A; therefore we get a mapping

$$\mathsf{Sp}(f): \mathsf{Sp}(B) \longrightarrow \mathsf{Sp}(A),$$
$$\mathsf{Sp}(f)(P) = f^{-1}(P),$$

which is easily proved to be continuous.

1.4.3.f   The last example in this section is that of a contravariant functor $\mathcal{P}^*: \mathsf{Set} \longrightarrow \mathsf{Set}$ which coincides on the objects with the covariant functor $\mathcal{P}: \mathsf{Set} \longrightarrow \mathsf{Set}$ defined in 1.2.8.c. Thus $\mathcal{P}^*(X)$ is the power set of X and for a given mapping $f: X \longrightarrow Y$,

$$\mathcal{P}^*(f): \mathcal{P}^*(Y) \longrightarrow \mathcal{P}^*(X), \quad \mathcal{P}^*(f)(U) = f^{-1}(U)$$

is the inverse image mapping.

1.5 Full and faithful functors

An abelian group is a set provided with some additional structure; a group homomorphism is a mapping which satisfies some additional property. So, in some vague sense, the category of abelian groups is "included" in the category of sets... the expected "inclusion" being the functor described in example 1.2.8.a. But this functor is by no means injective since on the same set G, there exist in general many different abelian group structures. In fact this functor is what we shall call a "faithful functor".

Definition 1.5.1 *Consider a functor $F: \mathscr{A} \longrightarrow \mathscr{B}$ and for every pair of objects $A, A' \in \mathscr{A}$, the mapping*

$$\mathscr{A}(A, A') \longrightarrow \mathscr{B}(FA, FA'), \quad f \mapsto Ff.$$

(1) The functor F is faithful when the abovementioned mappings are injective for all A, A'.

(2) The functor F is full when the abovementioned mappings are surjective for all A, A'.

(3) The functor F is full and faithful when the abovementioned mappings are bijective for all A, A'.

(4) The functor F is an isomorphism of categories when it is full and faithful and induces a bijection $|\mathscr{A}| \longrightarrow |\mathscr{B}|$ on the classes of objects.

The reader will easily adapt definition 1.5.1 to the case of contravariant functors. Definiton 1.5.1.4 is a special instance, in the category of small categories and functors, of the general notion of isomorphism in a category.

Proposition 1.5.2 *The Yoneda embedding functors described in examples 1.4.3.a,d are full and faithful functors.*

Proof In the case of the contravariant Yoneda embedding, we have to prove that given two objects A, B in a small category A, the canonical mapping

$$\mathscr{A}(A, B) \longrightarrow \mathsf{Nat}\big(\mathscr{A}(B, -), \mathscr{A}(A, -)\big), \quad f \mapsto \mathscr{A}(f, -)$$

is bijective. This is a special case of the Yoneda lemma (see 1.3.3) applied to the functor $\mathscr{A}(A, -)$ and the object B.

The case of the covariant embedding is proved in a "dual" way. $\square$

Let us conclude with some terminology concerning subcategories.

Definition 1.5.3 *A subcategory $\mathscr{B}$ of a category $\mathscr{A}$ consists of:*

(1) a subclass $|\mathscr{B}| \subseteq |\mathscr{A}|$ of the class of objects,

(2) for every pair of objects $A, A' \in \mathscr{A}$, a subset $\mathscr{B}(A, A') \subseteq \mathscr{A}(A, A')$,

in such a way that

(1) $f \in \mathscr{B}(A, A')$ and $g \in \mathscr{B}(A', A'') \Rightarrow g \circ f \in \mathscr{B}(A, A'')$,

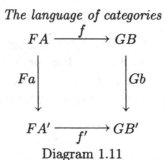

Diagram 1.11

(2) $\forall A \in \mathscr{B}$, $1_A \in \mathscr{B}(A, A)$.

A subcategory $\mathscr{B}$ of $\mathscr{A}$ thus gives rise to an injective (and therefore faithful) inclusion functor $\mathscr{B} \longrightarrow \mathscr{A}$.

Definition 1.5.4 *A subcategory $\mathscr{B}$ of a category $\mathscr{A}$ is called a full subcategory when the inclusion functor $\mathscr{B} \longrightarrow \mathscr{A}$ is also a full functor.*

$\mathscr{B}$ is thus full in $\mathscr{A}$ when

$$A, A' \in \mathscr{B} \Rightarrow \mathscr{B}(A, A') = \mathscr{A}(A, A').$$

The category of sets and injections between them is a (non-full) subcategory of the category of sets and mappings. The category of finite sets and mappings between them is a full subcategory of the category of sets and mappings. A full subcategory can clearly be defined by just giving its class of objects.

1.6 Comma categories

We indicate now a quite general process for constructing new categories from given ones. This type of construction will be used very often in this book.

Definition 1.6.1 *Consider two functors $F: \mathscr{A} \longrightarrow \mathscr{C}$ and $G: \mathscr{B} \longrightarrow \mathscr{C}$. The "comma category" (F, G) is defined in the following way.*

(1) *The objects of (F, G) are the triples (A, f, B) where $A \in \mathscr{A}, B \in \mathscr{B}$ are objects and $f: FA \longrightarrow GB$ is a morphism of $\mathscr{C}$.*

(2) *A morphism of (F, G) from (A, f, B) to (A', f', B') is a pair (a, b), where $a: A \longrightarrow A'$ is a morphism of $\mathscr{A}$, $b: B \longrightarrow B$ is a morphism of $\mathscr{B}$, and $f' \circ F(a) = G(b) \circ f$ (see diagram 1.11).*

(3) *The composition law in (F, G) is that induced by the composition laws of $\mathscr{A}$ and $\mathscr{B}$, thus*

$$(a', b') \circ (a, b) = (a' \circ a, b' \circ b).$$

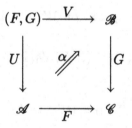

Diagram 1.12

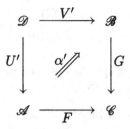

Diagram 1.13

Proposition 1.6.2 *Consider functors* $F: \mathscr{A} \longrightarrow \mathscr{C}$, $G: \mathscr{B} \longrightarrow \mathscr{C}$ *and their corresponding comma-category* (F, G). *There are two functors* $U: (F, G) \longrightarrow \mathscr{A}$, $V: (F, G) \longrightarrow \mathscr{B}$ *(see diagram 1.12); moreover there exists a canonical natural transformation*

$$\alpha: F \circ U \Rightarrow G \circ V.$$

Proof With the notation of 1.6.1 it suffices to define

$$U(A, f, B) = A, \quad V(A, f, B) = B,$$
$$U(a, b) = a, \quad V(a, b) = b.$$

The equality $F \circ U = G \circ V$ has no reason at all to hold in general. The natural transformation α is easily defined by $\alpha_{(A, f, B)} = f$; the fact that it is a natural transformation is just condition 1.6.1.(2). □

Proposition 1.6.3 *In the situation and with the notations of 1.6.2, consider a category* $\mathscr{D}$, *two functors* $U': \mathscr{D} \longrightarrow \mathscr{A}$, $V': \mathscr{D} \longrightarrow \mathscr{B}$ *(see diagram 1.13) and a natural transformation*

$$\alpha': F \circ U' \Rightarrow G \circ V'.$$

In that case there exists a unique functor $W: \mathscr{D} \longrightarrow (F, G)$ *such that*

$$U \circ W = U', \quad V \circ W = V', \quad \alpha * W = \alpha'.$$

Proof   The conditions imposed on W indicate immediately what it should be:

$$W(D) = (U'D, \alpha'_D, V'D)$$

for an object $D \in \mathscr{D}$ and

$$W(d) = (U'd, V'd)$$

for a morphism d of $\mathscr{D}$, which already proves the uniqueness of such a W. To prove the existence, it suffices to observe that the previous formulas indeed define a functor $W \colon \mathscr{D} \longrightarrow (F, G)$.   □

We shall refer to proposition 1.6.3 as the "universal property" of the comma category.

A special but very important case of a comma category is the "category of elements" of a functor $F \colon \mathscr{A} \longrightarrow \mathsf{Set}$.

Definition 1.6.4 *Consider a functor $F \colon \mathscr{A} \longrightarrow \mathsf{Set}$ from a category $\mathscr{A}$ to the category of sets. The category $\mathsf{Elts}(F)$ of "elements of F" is defined in the following way.*

(1) *The objects of $\mathsf{Elts}(F)$ are the pairs (A, a) where $A \in |\mathscr{A}|$ is an object and $a \in FA$.*

(2) *A morphism $f \colon (A, a) \longrightarrow (B, b)$ of $\mathsf{Elts}(F)$ is an arrow $f \colon A \longrightarrow B$ of A such that $Ff(a) = b$.*

(3) *The composition of $\mathsf{Elts}(F)$ is that induced by the composition of $\mathscr{A}$.*

Let us write $\mathbf{1}$ for the discrete category with a single object $\star$;

$$\mathbf{1} \colon \mathbf{1} \longrightarrow \mathsf{Set}, \quad \star \mapsto \{*\}$$

is the functor which maps the unique object $\star$ of $\mathbf{1}$ to the singleton $\{*\}$. In other words, we view $\mathbf{1}$ as the full subcategory of Set generated by a singleton set. Since an element $a \in FA$ can be seen as a morphism from a singleton to FA, thus as a morphism of the type $\mathbf{1}(A) \longrightarrow F(A)$ in Set, the category $\mathsf{Elts}(F)$ is exactly the comma category $(\mathbf{1}, F)$. Notice that the forgetful functor $\phi_F \colon \mathsf{Elts}(F) \longrightarrow \mathscr{A}$ is defined by $\phi_F(A, a) = A$ on the objects and by $\phi_F(f) = f$ on the morphisms.

Another interesting example of a comma category is the "product" of two categories.

Definition 1.6.5 *The product of two categories $\mathscr{A}$ and $\mathscr{B}$ is the category $\mathscr{A} \times \mathscr{B}$ defined in the following way.*

(1) *The objects of $\mathscr{A} \times \mathscr{B}$ are the pairs (A, B) with $A \in |\mathscr{A}|$, $B \in |\mathscr{B}|$ objects of $\mathscr{A}$, $\mathscr{B}$.*

(2) *The morphisms* $(A, B) \longrightarrow (A', B')$ *of* $\mathscr{A} \times \mathscr{B}$ *are the pairs* (a, b) *where* $a: A \longrightarrow A'$ *is a morphism of* $\mathscr{A}$ *and* $b: B \longrightarrow B'$ *is a morphism of* $\mathscr{B}$.

(3) *The composition in* $\mathscr{A} \times \mathscr{B}$ *is that induced by the compositions of* $\mathscr{A}$ *and* $\mathscr{B}$, *namely*

$$(a', b') \circ (a, b) = (a' \circ a, b' \circ b).$$

With the product $\mathscr{A} \times \mathscr{B}$ are associated the two "projection" functors

$$p_A: \mathscr{A} \times \mathscr{B} \longrightarrow \mathscr{A}, \quad p_B: \mathscr{A} \times \mathscr{B} \longrightarrow \mathscr{B}$$

defined by the formulas

$$p_{\mathscr{A}}(A, B) = A, \quad p_{\mathscr{B}}(A, B) = B,$$
$$p_{\mathscr{A}}(a, b) = a, \quad p_{\mathscr{B}}(a, b) = b.$$

These data satisfy the following "universal property".

Proposition 1.6.6 *Consider two categories* $\mathscr{A}$ *and* $\mathscr{B}$. *For every category* $\mathscr{D}$ *and every pair of functors* $F: \mathscr{D} \longrightarrow \mathscr{A}$, $G: \mathscr{D} \longrightarrow \mathscr{B}$, *there exists a unique functor* $H: \mathscr{D} \longrightarrow \mathscr{A} \times \mathscr{B}$ *such that* $p_{\mathscr{A}} \circ H = F$, $p_{\mathscr{B}} \circ H = G$.

Proof H is the functor defined by

$$H(D) = (FD, GD) \text{ for an object } D \text{ of } \mathscr{D},$$
$$H(d) = (Fd, Gd) \text{ for a morphism } d \text{ of } \mathscr{D}. \qquad \square$$

Let us now observe the existence of a unique functor $\Delta_{\mathscr{A}}: \mathscr{A} \longrightarrow \mathbf{1}$: this is the "constant functor" to the unique object of $\mathbf{1}$ (see 1.2.8.e). Since $\mathbf{1}$ has just one single mapping, the comma category $(\Delta_{\mathscr{A}}, \Delta_{\mathscr{B}})$ is isomorphic to the product category $\mathscr{A} \times \mathscr{B}$. Proposition 1.6.6 is then a particularization of proposition 1.6.3.

A point of terminology: a functor $F: \mathscr{A} \times \mathscr{B} \longrightarrow \mathscr{C}$ defined on the product of two categories is generally called a "bifunctor" (a functor of two "variables").

1.7 Monomorphisms

When a composition law appears in some mathematical structure, special attention is always paid to those elements which are "cancellable" or "invertible" for that composition. This section is devoted to the study of left cancellable morphisms in a category.

Definition 1.7.1 *A morphism* $f: A \longrightarrow B$ *in a category* $\mathscr{C}$ *is called a monomorphism when, for every object* $C \in \mathscr{C}$ *and every pair of morphisms* $g, h: C \underset{\longrightarrow}{\overset{\longrightarrow}{}} A$, *the following property holds:*

$$(f \circ g = f \circ h) \Rightarrow (g = h).$$

We shall generally use the symbol $f: A \rightarrowtail B$ to emphasize the fact that f is a monomorphism.

Proposition 1.7.2 *In a category* $\mathscr{C}$,

(1) every identity morphism is a monomorphism,

(2) the composite of two monomorphisms is a monomorphism,

(3) if the composite $k \circ f$ of two morphisms is a monomorphism, then f is a monomorphism.

Proof   We use the notation of 1.7.1 and consider another morphism $k: B \longrightarrow D$.

(1) is obvious.

(2) If f and k are monomorphisms,

$$k \circ f \circ h = k \circ f \circ g \Rightarrow f \circ h = f \circ g \Rightarrow h = g.$$

(3) If $k \circ f$ is a monomorphism,

$$f \circ g = f \circ h \Rightarrow k \circ f \circ g = k \circ f \circ h \Rightarrow g = h. \qquad \square$$

The following terminology is rather classical.

Definition 1.7.3 *Consider two morphisms $f: A \longrightarrow B$ and $g: B \longrightarrow A$ in a category. When $g \circ f = 1_A$, f is called a section of g, g is called a retraction of f and A is called a retract of B.*

Proposition 1.7.4 *In a category, every section is a monomorphism.*

Proof By 1.7.2.(1,3). $\qquad \square$

Let us now say a word about the effect of a functor on a monomorphism.

Definition 1.7.5 *Consider a functor $F: \mathscr{A} \longrightarrow \mathscr{B}$.*

(1) F preserves monomorphisms when, for every morphism f of $\mathscr{A}$,

$$f \text{ monomorphism} \Rightarrow Ff \text{ monomorphism}.$$

(2) F reflects monomorphisms when, for every morphism f of $\mathscr{A}$,

$$Ff \text{ monomorphism} \Rightarrow f \text{ monomorphism}.$$

Proposition 1.7.6 *A faithful functor reflects monomorphisms.*

Proof Consider a faithful functor $F: \mathscr{A} \longrightarrow \mathscr{B}$, a morphism $f: A \longrightarrow A'$ in $\mathscr{A}$, and suppose Ff is a monomorphism in $\mathscr{B}$. Choose another object $A'' \in \mathscr{A}$ and two morphisms $g, h: A'' \longrightarrow A$ in $\mathscr{A}$.

$$f \circ g = f \circ h \Rightarrow Ff \circ Fg = Ff \circ Fh$$
$$\Rightarrow Fg = Fh$$
$$\Rightarrow g = h$$

where the second implication holds since Ff is a monomorphism and the last one follows from the faithfulness of F. $\qquad\square$

Examples 1.7.7

1.7.7.a In the category Set of sets and mappings, the monomorphisms are exactly the injections. Indeed, an element $a \in A$ can be viewed as a mapping $\bar{a}: \{\star\} \longrightarrow A$ from the singleton to A; therefore, given a monomorphism $f: A \longrightarrow B$ and elements $a, a' \in A$,

$$f(a) = f(a') \Rightarrow f \circ \bar{a} = f \circ \bar{a'}$$
$$\Rightarrow \bar{a} = \bar{a'}$$
$$\Rightarrow a = a'.$$

Conversely, if $f: A \longrightarrow B$ is injective and $g, h: C \rightrightarrows A$ are mappings such that $f \circ g = f \circ h$, then for every element $c \in C$

$$f \circ g = f \circ h \Rightarrow f\big(g(c)\big) = f\big(h(c)\big)$$
$$\Rightarrow g(c) = h(c)$$

and therefore $g = h$.

1.7.7.b In the category Top of topological spaces and continuous mappings or its full subcategory Comp of compact Hausdorff spaces, the monomorphisms are exactly the continuous injections. Indeed, an element of a space A corresponds to a continuous mapping $\{\star\} \longrightarrow A$ from the singleton to A; therefore the argument of 1.7.7.a can be carried over.

1.7.7.c In the categories Gr of groups and Ab of abelian groups, the monomorphisms are exactly the injective group homomorphisms. The argument is again analogous, using now the bijective correspondence between the elements $a \in G$ of a group and the group homomorphisms $\bar{a}: \mathbb{Z} \longrightarrow G$ from the group of integers to G; we recall the correspondence:

$$\bar{a}(z) = z \cdot a, \quad a = \bar{a}(1).$$

1.7.7.d In the category **Rng** of commutative rings with a unit, the monomorphisms are exactly the injective ring homomorphisms. Repeat the argument using now the ring homomorphisms with domain the ring $\mathbb{Z}[X]$ of polynomials with integral coefficients: an element $r \in R$ of a ring R corresponds to the ring homomorphism $\bar{r} \colon \mathbb{Z}[X] \longrightarrow R$ mapping the polynomial $p(X)$ to $p(r)$; conversely $r = \bar{r}(X)$.

1.7.7.e In the category Mod_R of right modules on a ring R with unit, the monomorphisms are exactly the injective R-linear mappings. Use again the same argument using the R-linear mapping with domain the ring R itself: an element $m \in M$ of a R-module M corresponds to the linear mapping $\bar{m} \colon R \longrightarrow M$ mapping r to mr; conversely $m = \bar{m}(1)$.

1.7.7.f In the category Ban_1 of real Banach spaces and linear contractions, the monomorphisms are exactly the injective linear contractions. The elements of the unit ball of a Banach space B are in bijective correspondence with the linear contractions $\bar{a} \colon \mathbb{R} \longrightarrow B$; just put

$$\bar{a}(r) = ra \ , \ a = \bar{a}(1).$$

Therefore a monomorphism $f \colon B \longrightarrow B'$ is such that the implication

$$f(a) = f(a') \Rightarrow a = a'$$

holds for elements a, a' in the unit ball of B; by linearity of f, this fact extends to arbitrary elements a, $a' \in B$. The converse is once more obvious.

1.7.7.g The previous examples could give the wrong impression that, in "concrete" examples, a monomorphism is always exactly an injective morphism. This is false as shown by the following counterexamples. We give first an "algebraic" counterexample.

Consider the category Div of divisible abelian groups and group homomorphisms between them. The quotient morphism $q \colon \mathbb{Q} \longrightarrow \mathbb{Q}/\mathbb{Z}$ of the additive group of rational numbers by the group of integers is definitely not an injection, but it is a monomorphism in Div. Indeed, choose G a divisible group and $f, g \colon G \rightrightarrows \mathbb{Q}$ two group homomorphisms such that $q \circ f = q \circ g$. Putting $h = f - g$ we have $q \circ h = 0$ and the thesis becomes $h = 0$. Given an element $x \in G$, $h(x)$ is an integer since $q \circ h = 0$. If $h(x) \neq 0$ note that

$$h\left(\frac{x}{2h(x)}\right) = \frac{1}{2}$$

and therefore

$$(q \circ h) \left(\frac{x}{2h(x)} \right) \neq 0$$

which is a contradiction.

1.7.7.h Let us give now a "topological" counterexample. We consider the category whose objects are the pairs (X, x) where X is a connected topological space and $x \in X$ is a base point; in this category, a morphism $f\colon (X, x) \longrightarrow (Y, y)$ is a continuous mapping $f\colon X \longrightarrow Y$ which preserves the base points, i.e. such that $f(x) = y$. Let us consider the projection π of the circular helix $\mathcal{H}$ on the circle S^1,

$$\pi\colon (\mathcal{H}, h) \longrightarrow (S^1, s),$$

with $h \in \mathcal{H}$ and $s = \pi(h)$. If $f\colon (X, x) \longrightarrow (S^1, s)$ is a morphism in our category which admits a "lifting"

$$g\colon (X, x) \longrightarrow (\mathcal{H}, h)$$

through the projection π, that lifting is necessarily unique (see **Spanier**, page 67). But this expresses exactly the fact that π is a monomorphism.

1.8 Epimorphisms

We now turn our attention to right cancellable morphisms in a category.

Definition 1.8.1 *A morphism $f\colon B \longrightarrow A$ in a category $\mathscr{C}$ is called an epimorphism when, for every object $C \in \mathscr{C}$ and every pair of morphisms $g, h\colon A \rightrightarrows C$, the following property holds:*

$$(g \circ f = h \circ f) \Rightarrow (g = h).$$

We shall generally use the notation $f\colon B \longrightarrow\mkern-14mu\rightarrow A$ to emphasize the fact that f is an epimorphism.

Proposition 1.8.2 *In a category $\mathscr{C}$,*

(1) every identity morphism is an epimorphism,

(2) the composite of two epimorphisms is an epimorphism,

(3) if the composite $f \circ k$ of two morphisms is an epimorphism, then f is an epimorphism.

Proof We use the notation of 1.8.1 and consider another morphism $k\colon D \longrightarrow B$.

(1) is obvious.

(2) If f and k are epimorphisms,

$$h \circ f \circ k = g \circ f \circ k \Rightarrow h \circ f = g \circ f \Rightarrow h = g.$$

(3) If $f \circ k$ is an epimorphism,

$$g \circ f = h \circ f \Rightarrow g \circ f \circ k = h \circ f \circ k \Rightarrow g = h. \qquad \square$$

Proposition 1.8.3 *In a category, every retraction is an epimorphism.*

Proof By 1.8.2.(1,3). $\qquad \square$

Transposing definition 1.7.5 to the case of epimorphisms, we obtain

Proposition 1.8.4 *A faithful functor reflects epimorphisms.*

Proof Consider a faithful functor $F: \mathscr{A} \longrightarrow \mathscr{B}$, a morphism $f: A' \longrightarrow A$ and suppose Ff is an epimorphism in $\mathscr{B}$. Choose another object $A'' \in \mathscr{A}$ and two morphisms $g, h: A \rightrightarrows A''$ in $\mathscr{A}$. Then

$$g \circ f = h \circ g \Rightarrow Fg \circ Ff = Fh \circ Ff$$
$$\Rightarrow Fg = Fh$$
$$\Rightarrow g = h,$$

where the second implication holds since Ff is an epimorphism and the last one follows from the faithfulness of F. $\qquad \square$

The similarity of the previous proofs with those of section 1.7 is striking: this is a special instance of the "duality principle" described in section 1.10.

Examples 1.8.5

1.8.5.a In the category Set of sets and mappings, the epimorphisms are exactly the surjective mappings. Choose $f: A \longrightarrow B$ a surjective mapping and $g, h: B \rightrightarrows C$ two mappings such that $g \circ f = h \circ f$. For every element $b \in B$, we can find an element $a \in A$ such that $f(a) = b$; therefore

$$g(b) = g\big(f(a)\big) = h\big(f(a)\big) = h(b),$$

which proves the equality $g = h$.

Conversely, if $f: A \longrightarrow B$ is an epimorphism, consider the two-element set $\{0, 1\}$ and the following mappings $g, h: B \rightrightarrows \{0, 1\}$:

$$g(b) = 1 \text{ if } b \in f(A),$$
$$g(b) = 0 \text{ if } b \notin f(A),$$
$$h(b) = 1 \text{ for every } b \in B.$$

Clearly $g \circ f = h \circ f$ is the constant mapping on 1; therefore $g = h$ and $f(A) = B$.

1.8.5.b In the category **Top** of topological spaces and continuous mappings, the epimorphisms are exactly the surjective continuous mappings. The previous proof applies when $\{0, 1\}$ is provided with the indiscrete topology.

1.8.5.c In the category **Haus** of Hausdorff topological spaces and continuous mappings between them, the epimorphisms are exactly the continuous mappings with a dense image. We recall that a continuous mapping $f : A \longrightarrow B$ has a dense image precisely when every element $b \in B$ is the limit of a net of elements of $f(A)$, i.e. a set of elements indexed by a filtered poset (see 2.13.1); when B is a Hausdorff space, the limit of a converging net is unique. Suppose $f : A \longrightarrow B$ has a dense image and choose $g, h : B \rightrightarrows C$ such that $g \circ f = h \circ f$. Given an element $b \in B$, choose a net $(a_i)_{i \in I}$ of elements in A such that $b = \lim f(a_i)$. By continuity of g, h we have

$$g(b) = \lim(g \circ f)(a_i), \quad h(b) = \lim(h \circ f)(a_i).$$

Since $g \circ f = h \circ f$ and the limit is unique, we conclude that $g(b) = h(b)$ and thus $g = h$.

Conversely if $f : A \longrightarrow B$ is an epimorphism, and B is not empty, A cannot be empty. Indeed if $B \amalg B$ is the space constituted by two disjoint copies of B, $B \amalg B$ is a Hausdorff space and the two canonical inclusions $i_1, i_2 : B \rightrightarrows B \amalg B$ are continuous and distinct. A empty would yield $i_1 \circ f = i_2 \circ f$ and thus $i_1 = i_2$, since f is an epimorphism. Now consider the quotient of B which identifies with a single point the closure $\overline{f(A)}$ of the image of A; this is a Hausdorff space as a quotient of a Hausdorff space by a closed subspace; write $p : B \longrightarrow B/\overline{f(A)}$ for the corresponding continuous projection. Since $f(A)$ is not empty, we can consider as well the constant mapping $q : B \longrightarrow B/\overline{f(A)}$ on the equivalence class of the elements of $f(A)$. Clearly $p \circ f = q \circ f$ and therefore $p = q$, which proves the equality $\overline{f(A)} = B$.

1.8.5.d In the category **Gr** of groups and their homomorphisms, the epimorphisms are exactly the surjective homomorphisms. Indeed, a surjective homomorphism is clearly an epimorphism. Conversely suppose $f : A \longrightarrow B$ is an epimorphism. We can factor f through its image

$$A \longrightarrow f(A) \longrightarrow B,$$

thus through a surjection followed by an injection. By 1.8.2.(3), the injective part is an epimorphism and so the problem reduces to proving that an epimorphic inclusion is an identity.

Given two groups G, H with a common subgroup K, it is possible to construct the amalgamation of G and H over K: this is the group $G \star_K H$ of words constructed with the "letters" of G and H, the two copies of a "letter" of K being identified in $G \star_K H$. The amalgamation property for groups tells us that the two canonical morphisms

$$G \longrightarrow G \star_K H, \quad H \longrightarrow G \star_K H,$$

are injective and that two "letters" of G and H are identified in $G \star_K H$ just when they are the two copies of a "letter" in K (see **Kuroš**). If we apply that amalgamation property choosing the inclusion $f(A) \hookrightarrow B$ twice, we first deduce the equality of the two canonical inclusions

$$i_1 \colon B \longrightarrow B \star_{f(A)} B, \quad i_2 \colon B \longrightarrow B \star_{f(A)} B$$

since they coincide on $f(A)$ and $f(A) \longrightarrow B$ is an epimorphism. But then each element of B is already in $f(A)$ by the amalgamation property.

1.8.5.e Consider a ring R with unit. In the category Mod_R of right R-modules, the epimorphisms are exactly the surjective linear mappings. In particular, choosing $R = \mathbb{Z}$, the epimorphisms of the category of abelian groups are exactly the surjective homomorphisms. Again a surjective linear mapping is clearly an epimorphism. Conversely if $f \colon A \longrightarrow B$ is an epimorphism, consider both the quotient mapping and the zero mapping

$$p \colon B \longrightarrow B/f(A), \quad 0 \colon B \longrightarrow B/f(A).$$

From the equality

$$p \circ f = 0 = 0 \circ f$$

we deduce $p = 0$ and thus $B = f(A)$.

1.8.5.f The form of epimorphisms in the category of commutative rings with unit is known (see exercise 1.11.13); let us just emphasize the fact that epimorphisms of rings are not necessarily surjective. Consider the inclusion of the ring $\mathbb{Z}$ of integers in the ring $\mathbb{Q}$ of rational numbers, $i \colon \mathbb{Z} \longrightarrow \mathbb{Q}$. This is clearly not a surjection but it is an epimorphism of rings. Indeed given another ring A and two ring homomorphisms $f, g \colon \mathbb{Q} \rightrightarrows A$ which agree on the integers, we deduce first that for every integer $0 \neq z \in \mathbb{Z}$, z is invertible in $\mathbb{Q}$ and therefore $f(z)$ and $g(z)$ are invertible in A; clearly

$$\frac{1}{f(z)} = f\left(\frac{1}{z}\right) \quad \text{and} \quad \frac{1}{g(z)} = g\left(\frac{1}{z}\right).$$

Since f and g agree on the integers, $f\left(\frac{1}{z}\right) = g\left(\frac{1}{z}\right)$ and finally,

$$f\left(\frac{z'}{z}\right) = f\left(z' \cdot \frac{1}{z}\right) = f(z') \cdot f\left(\frac{1}{z}\right)$$

$$= g(z') \cdot g\left(\frac{1}{z}\right) = g\left(z' \cdot \frac{1}{z}\right) = g\left(\frac{z'}{z}\right).$$

1.8.5.g In the category Ban_1 of Banach spaces and linear contractions, the epimorphisms are the linear contractions with dense image. Choose $f\colon A \longrightarrow B$ with a dense image and $g, h\colon B \rightrightarrows C$ such that $g \circ f = h \circ f$. Since g and h agree on $f(A)$, by continuity g, h agree on on $\overline{f(A)} = B$ as well; therefore $g = h$. Conversely if $f\colon A \longrightarrow B$ is an epimorphism, the quotient of B by the closed subspace $\overline{f(A)}$ is a Banach space and both the quotient mapping p and the zero mapping are linear contractions:

$$p\colon B \longrightarrow B/\overline{f(A)}, \quad 0\colon B \longrightarrow B/\overline{f(A)}.$$

From the equalities $p \circ f = 0 = 0 \circ f$, we deduce $p = 0$ and thus $B = \overline{f(A)}$.

1.9 Isomorphisms

We consider finally the case of those morphisms of a category which are invertible.

Definition 1.9.1 *A morphism $f\colon A \longrightarrow B$ in a category $\mathscr{C}$ is called an isomorphism when there exists a morphism $g\colon B \longrightarrow A$ of $\mathscr{C}$ which satisfies the relations*

$$f \circ g = 1_B, \quad g \circ f = 1_A.$$

Clearly such a morphism g is necessarily unique; indeed if $h\colon B \longrightarrow A$ is another morphism with the same properties

$$f \circ h = 1_B, \quad h \circ f = 1_A,$$

we conclude that

$$g = g \circ 1_B = g \circ f \circ h = 1_A \circ h = h.$$

Therefore we shall call such a morphism g "the" inverse of f and we shall denote it by f^{-1}.

Proposition 1.9.2 *In a category,*

(1) every identity is an isomorphism,

(2) the composite of two isomorphisms is an isomorphism,

(3) an isomorphism is both a monomorphism and an epimorphism.

Proof

(1) is obvious.
(2) If $f: A \longrightarrow B$ and $g: B \longrightarrow C$ are isomorphisms, so is $g \circ f$ and $(g \circ f)^{-1} = f^{-1} \circ g^{-1}$.
(3) is just the conjunction of 1.7.4 and 1.8.3 ☐

Proposition 1.9.3 *In a category, if a section is an epimorphism, it is an isomorphism.*

Proof If $g \circ f = 1_A$ and $f: A \longrightarrow B$ is an epimorphism, from $f \circ g \circ f = f$ we deduce $f \circ g = 1_B$. ☐

Proposition 1.9.4 *Every functor preserves isomorphisms.*

Proof Obvious. ☐

Transposing definition 1.7.5 to the case of isomorphisms, we obtain

Proposition 1.9.5 *A full and faithful functor reflects isomorphisms.*

Proof Obvious. ☐

Examples 1.9.6

1.9.6.a In the category Set of sets, the isomorphisms are exactly the bijections.

1.9.6.b In the category Top of topological spaces, the isomorphisms are exactly the homeomorphisms. Since a continuous bijection is in general not a homeomorphism, this provides an example where the converse of statement 1.9.2.(3) does not hold (see 1.7.7.b and 1.8.5.b).

1.9.6.c In the categories Gr of groups, Ab of abelian groups and Rng of commutative rings with unit, the isomorphisms are the bijective homomorphisms.

1.9.6.d In the category Mod_R of right modules over a ring R, the isomorphisms are the bijective R-linear mappings.

1.9.6.e In the category Ban_∞ of real Banach spaces and bounded linear mappings, the isomorphisms are the bounded linear bijections. An isomorphism is obviously bijective. Conversely if $f: A \longrightarrow B$ is a bounded linear bijection, the inverse mapping $f^{-1}: B \longrightarrow A$ is certainly linear. By the open mapping theorem, f is open because it is surjective; but "f open" means precisely "f^{-1} continuous" and thus f^{-1} is bounded.

1.9.6.f In the category Ban₁ of real Banach spaces and linear contractions, the isomorphisms are exactly the isometric bijections. An isometric bijection is obviously an isomorphism. Conversely if the linear contraction $f: A \longrightarrow B$ has an inverse mapping $f^{-1}: B \longrightarrow A$ which is also a linear contraction, then for every element $a \in A$

$$\|a\| = \|f^{-1}f(a)\| \leq \|f(a)\|$$

and thus $\|a\| = \|f(a)\|$ since f is contracting.

1.9.6.g In the category Cat of small categories and functors, the isomorphisms are those defined in 1.5.1.

1.9.6.h Going back to example 1.2.6.d, a group can be seen as a category with a single object all of whose morphisms are isomorphisms.

1.10 The duality principle

At this point the reader will have noticed that every result proved for covariant functors has its counterpart for contravariant functors and every result proved for monomorphisms has its counterpart for epimorphisms. These facts are just special instances of a very general principle.

Definition 1.10.1 *Given a category $\mathscr{A}$, the dual category $\mathscr{A}^*$ is defined in the following way:*

(1) $|\mathscr{A}^*| = |\mathscr{A}|$

 (both categories have the same objects);

(2) for all objects A, B of $\mathscr{A}^*$, $\mathscr{A}^*(A, B) = \mathscr{A}(B, A)$

 (the morphisms of $\mathscr{A}^*$ are those of $\mathscr{A}$ "written in the reverse direction"; to avoid confusion, we shall write $f^*: A \longrightarrow B$ for the morphism of $\mathscr{A}^*$ corresponding to the morphism $f: B \longrightarrow A$ of $\mathscr{A}$);

(3) the composition law of $\mathscr{A}^*$ is given by

$$f^* \circ g^* = (g \circ f)^*.$$

Metatheorem 1.10.2 *(Duality principle) Suppose the validity, in every category, of a statement expressing the existence of some objects or morphisms or the equality of some composites. Then the "dual statement" is also valid in every category; this dual statement is obtained by reversing the direction of every arrow and replacing every composite $f \circ g$ by the composite $g \circ f$.*

Proof If S denotes the given statement and S^* denotes its dual statement, proving the statement S^* in a category $\mathscr{A}$ is equivalent to proving the statement S in the category $\mathscr{A}^*$, and this is supposed to be valid.

□

For example, the notion of $f: A \longrightarrow B$ being a monomorphism in $\mathscr{A}$ means

$$\forall C \in \mathscr{A} \ \ \forall g, h \in \mathscr{A}(C, A) \ \ f \circ g = f \circ h \Rightarrow g = h.$$

The dual notion is thus that of a morphism $f: B \longrightarrow A$ which satisfies

$$\forall C \in \mathscr{A} \ \ \forall g, h \in \mathscr{A}(A, C) \ \ g \circ f = h \circ f \Rightarrow g = h$$

...which is exactly the notion of an epimorphism. With that remark in mind, it is obvious that all the results of section 1.8 are just the dual statements of the results of section 1.7: so, formally, the validity of the latter follows at once from the validity of the former via the duality principle.

The case of contravariant functors can also be reduced to the case of covariant functors via the consideration of the dual category: a contravariant functor from $\mathscr{A}$ to $\mathscr{B}$ is just a covariant functor from $\mathscr{A}^*$ to $\mathscr{B}$ (or, equivalently, a covariant functor from $\mathscr{A}$ to $\mathscr{B}^*$).

It is interesting to notice that, in category theory, some notions are their own dual. For example $f: A \longrightarrow B$ is an isomorphism when

$$\exists g: B \longrightarrow A \ \ g \circ f = 1_A, \ \ f \circ g = 1_B.$$

The dual notion is that of a morphism $f: B \longrightarrow A$ with the property

$$\exists g: A \longrightarrow B \ \ f \circ g = 1_A, \ \ g \circ f = 1_B$$

...but this is again the definition of f being an isomorphism.

Examples 1.10.3

1.10.3.a With every category $\mathscr{A}$ we can associate a bifunctor, still written $\mathscr{A}$,

$$\mathscr{A}: \mathscr{A}^* \times \mathscr{A} \longrightarrow \mathsf{Set},$$

defined by the following formulas:

- $\mathscr{A}(A, B)$ is the set of morphisms from A to B;
- if $f: A' \longrightarrow A$ and $g: B \longrightarrow B'$ are morphisms of $\mathscr{A}$,

$$\mathscr{A}(f, g): \mathscr{A}(A, B) \longrightarrow \mathscr{A}(A', B'), \ \ \mathscr{A}(f, g)(h) = g \circ h \circ f.$$

Fixing the first variable A we obtain the covariant functor defined in 1.2.8.d and fixing the second variable B we obtain the contravariant functor defined in 1.4.3.b. The bifunctor $\mathscr{A}$ is called the "Hom-functor" of the category $\mathscr{A}$ (from "homomorphism"); it is "contravariant in the first variable and covariant in the second variable".

1.10.3.b The dual of the category of sets and mappings is equivalent to the category of complete atomic boolean algebras and $(\vee - \wedge)$-preserving homomorphisms. Indeed, writing **CBA** for the second category, the contravariant power set functor can be seen as a contravariant functor $\mathcal{P}^*\colon \mathsf{Set} \longrightarrow \mathsf{CBA}$. It is well-known that every complete atomic boolean algebra B is isomorphic to the power-set $\mathcal{P}X$ of its set X of atoms. Let us prove now that $\mathcal{P}^*$ is a full and faithful functor. Given two sets X and Y, the mapping

$$\mathsf{Set}(X, Y) \longrightarrow \mathsf{CBA}(\mathcal{P}^*Y, \mathsf{P}^*X), \quad f \mapsto f^{-1}$$

is obviously injective. To prove it is surjective, let us consider a morphism $g\colon \mathcal{P}^*Y \longrightarrow \mathcal{P}^*X$ in **CBA** and an element $x \in X = g(Y)$ (g preserves the top element). Now Y is the union of its singletons and g preserves unions, so there exists some $y \in Y$ such that $x \in g(\{y\})$. Such an element y is necessarily unique since $x \in g(\{y'\})$ with $y' \neq y$ would imply

$$x \in g(\{y\} \cap \{y'\}) = g(\emptyset) = \emptyset,$$

because g preserves intersections and the bottom element. Writing $f(x)$ for that element y, it follows easily that g is just f^{-1}.

1.10.3.c The dual of the category of abelian groups and their homomorphisms is equivalent to the category of compact abelian groups and continuous homomorphisms. This is just the Pontryagin duality theorem: with every abelian group A is associated its group of characters $\hat{A} = \mathrm{Hom}\,(A, U)$ where U is the circle group and the topology of $\hat{A}$ is that induced by the product topology U^A; with every homomorphism $f\colon A \longrightarrow B$ is associated the morphism $\hat{f}\colon \hat{B} \longrightarrow \hat{A}$ of composition with f.

1.10.3.d The category of finite abelian groups and their homomorphisms is equivalent to its own dual category. Indeed, it suffices to particularize the Pontryagin duality to the case of finite groups: when A is finite, $\hat{A}$ is isomorphic to A as a group and therefore is finite. But the finite compact groups are just the finite discrete groups, thus finally just the finite groups.

1.11 Exercises

1.11.1 If two ordered sets A, B are viewed as categories (see 1.2.6.b), prove that a functor from A to B is just an order preserving mapping. If $f, g: A \rightrightarrows B$ are two such functors, prove that there exists a (single) natural transformation from f to g if and only if for every element $a \in A$, $f(a) \le g(a)$.

1.11.2 If two monoids M and N are viewed as categories (see 1.2.6.d), prove that a functor from M to N is just an homomorphism of monoids. What is a natural transformation between two such functors?

1.11.3 In exercise 1.11.2, if M and N are groups, show the existence of a natural transformation between two functors $f, g: M \rightrightarrows N$ if and only if f and g are conjugate:

$$\exists n \in N \ \forall m \in M \ \ f(m) = n^{-1} \cdot g(m) \circ n.$$

1.11.4 If G is a group considered as a category (see 1.9.6.h), prove that a natural transformation on the identity functor of G is just an element of the centre of G.

1.11.5 Prove that a covariant representable functor preserves monomorphisms.

1.11.6 Prove that a contravariant representable functor maps an epimorphism to a monomorphism.

1.11.7 Prove that the forgetful functor $\mathsf{Rng} \longrightarrow \mathsf{Set}$ which maps a ring to its underlying set is faithful and representable by the ring $\mathbb{Z}[X]$, but does not preserve epimorphisms. [Hint: see 1.8.5.f.]

1.11.8 If $\mathscr{A}$, $\mathscr{B}$, $\mathscr{C}$ are small categories, prove the isomorphism of categories

$$\mathsf{Fun}(\mathscr{A} \times \mathscr{B}, \mathscr{C}) \cong \mathsf{Fun}\big(\mathscr{A}, \mathsf{Fun}(\mathscr{B}, \mathscr{C})\big),$$

where Fun denotes the category of functors and natural transformations.

1.11.9 Prove that a retraction which is a monomorphism is necessarily an isomorphism.

1.11.10 Determine the nature of the monomorphisms, epimorphisms and isomorphisms in examples 1.2.7.

1.11.11 Consider a small category $\mathscr{A}$ and the corresponding functor category $\mathsf{Fun}(\mathscr{A}, \mathsf{Set})$. Prove that a morphism α of $\mathsf{Fun}(\mathscr{A}, \mathsf{Set})$ (a natural transformation) is a monomorphism if and only if each component α_A, $A \in \mathscr{A}$, is a monomorphism in Set. [Hint: use the Yoneda lemma].

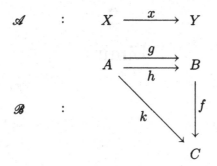

Diagram 1.14

1.11.12 The statement in 1.11.11 is no longer valid when Set is replaced by an arbitrary category $\mathscr{B}$. Consider the categories of diagram 1.14 (as a convention, identity arrows are not shown) where, in $\mathscr{B}$, the two composites $f \circ g$ and $f \circ h$ are equal to k. The category $\mathsf{Fun}(\mathscr{A}, \mathscr{B})$ is the category of arrows of $\mathscr{B}$ (see 1.2.7.c). The pair $(1_B, f) \colon (B, 1_B, B) \longrightarrow (B, f, C)$ is a monomorphism in $\mathsf{Fun}(\mathscr{A}, \mathscr{B})$ while f is not a monomorphism in $\mathscr{B}$.

1.11.13 Consider the category Rng of commutative rings with unit. A morphism $f \colon A \longrightarrow B$ is an epimorphism precisely when given any element $b \in B$, the equality $1 \otimes b = b \otimes 1$ holds in $B \otimes_A B$. This is also equivalent to saying that the morphism $B \longrightarrow B \otimes_A B$ is surjective, or again equivalently is an epimorphism.

2

Limits

We have seen in chapter 1 that the models of a mathematical theory and the corresponding homomorphisms very often constitute an interesting example of category: the category of sets and mappings, the category of vector spaces and linear mappings, the category of topological spaces and continuous mappings, and so on.

With a given mathematical structure are very often associated "operations on models or homomorphisms": cartesian product, quotient, kernel, union, intersection, and so on... It is the aim of this chapter to develop a general theory containing most of those constructions as particular cases.

2.1 Products

Everybody knows how to construct the cartesian product of two sets A and B; this is just

$$A \times B = \big\{(a,b) \,\big|\, a \in A \,;\, b \in B \big\}\,.$$

This "cartesian product" is provided with two "canonical" projections

$$p_A \colon A \times B \longrightarrow A, \quad p_A(a,b) = a,$$
$$p_B \colon A \times B \longrightarrow B, \quad p_B(a,b) = b.$$

Moreover, if C is a set and $f \colon C \longrightarrow A$, $g \colon C \longrightarrow B$ are arbitrary mappings, there exists a unique mapping $h \colon C \longrightarrow A \times B$ which makes diagram 2.1 commutative. Indeed, $h(c) = \big(f(c), g(c)\big)$.

Replacing "set" by "category" and "mapping" by "functor", the situation of diagram 2.1 recaptures precisely the fundamental property of the "product of two categories" as studied in 1.6.5 and 1.6.6. This fact is much more general and makes sense in every category.

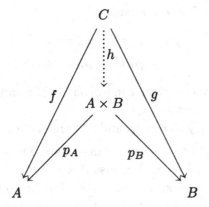

Diagram 2.1

Definition 2.1.1 *Let $\mathscr{C}$ be a category and $A, B \in \mathscr{C}$ two objects of $\mathscr{C}$. A (cartesian) product of A and B is, by definition, a triple (P, p_A, p_B) where*

(1) $P \in \mathscr{C}$ is an object,

(2) $p_A : P \longrightarrow A$ and $p_B : P \longrightarrow B$ are morphisms,

and this triple is such that for every other triple (Q, q_A, q_B) where

(1) $Q \in \mathscr{C}$ is an object,

(2) $q_A : Q \longrightarrow A$ and $q_B : Q \longrightarrow B$ are morphisms,

there exists a unique morphism $r : Q \longrightarrow P$ such that $q_A = p_A \circ r$ and $q_B = p_B \circ r$.

It is a fundamental observation that:

Proposition 2.1.2 *In a category, the cartesian product of two objects (when it exists) is unique up to isomorphism.*

Proof In the category $\mathscr{C}$, consider two products

$$(P, p_A, p_B) \text{ and } (Q, q_A, q_B)$$

of the same objects A, B. Since (P, p_A, p_B) is a product, there exists a morphism $r : Q \longrightarrow P$ such that $q_A = p_A \circ r$ and $q_B = p_B \circ r$. Since (Q, q_A, q_B) is a product, there exists a morphism $s : P \longrightarrow Q$ such that $p_A = q_A \circ s$ and $p_B = q_B \circ s$.

Applying definition 2.1.1 to the triple (P, p_A, p_B) seen both as a product and as "another triple", we deduce the existence of a unique morphism $t : P \longrightarrow P$ such that $p_A = p_A \circ t$ and $p_B = p_B \circ t$. Clearly $t = 1_P$

is such a morphism. But the relations

$$p_A = q_A \circ s = p_A \circ r \circ s,$$
$$p_B = q_B \circ s = p_B \circ r \circ s$$

indicate that $t = r \circ s$ is another morphism of this kind. The uniqueness of t therefore implies $r \circ s = 1_P$.

In an analogous way $s \circ r = 1_Q$ and r, s are inverse isomorphisms. □

In view of the previous theorem, it makes sense to write "the" product of A and B as $(A \times B, p_A, p_B)$ or just $A \times B$.

It is a common mistake to think that the projections p_A, p_B of a product are epimorphisms. This is not true, not even in the category of sets and mappings! The projection $p_A: A \times B \longrightarrow A$ in the category of sets is not surjective when B is empty and A is non-emtpy; thus it is not an epimorphism (see example 1.8.5.a).

Another common mistake is to think that once the object $A \times B$ in a product has been fixed, the corresponding projections p_A, p_B are necessarily unique. This is not true at all (see example 2.1.7.i): in the definition of a product both the object $A \times B$ and the projections p_A, p_B are defined up to isomorphism.

Proposition 2.1.3 *In a category, when the corresponding products exist, the following isomorphisms hold:*

$$A \times B \cong B \times A;$$
$$A \times (B \times C) \cong (A \times B) \times C.$$

Proof If $(A \times B, p_A, p_B)$ is a product of A and B, it suffices to notice that $(A \times B, p_B, p_A)$ is a product of B and A and apply proposition 2.1.2. An analogous argument holds in the second case. □

Writing $(A \times B, p_A, p_B)$ and $(B \times A, p'_A, p'_B)$ for the products of A, B and B, A, the unique morphism $\tau: A \times B \longrightarrow B \times A$ such that $p_A = p'_A \circ \tau$ and $p_B = p'_B \circ \tau$ is generally called the "twisting isomorphism". It is indeed an isomorphism as proved in 2.1.3. In the case of the category Set of sets and mappings, it is the usual twisting bijection $\tau(a, b) = (b, a)$.

Proposition 2.1.3 indicates in particular that the existence of all binary products allows the definition (up to isomorphism) of the product of n objects of the category ($n \in \mathbb{N}, n \geq 2$). In fact, it makes sense to define the product of an arbitrary family of objects in a given category.

Definition 2.1.4 *Let I be a set and $(C_i)_{i \in I}$ a family of objects in a given category $\mathscr{C}$. A product of that family is a pair $\left(P, (p_i)_{i \in I}\right)$ where*

(1) P is an object of $\mathscr{C}$,

(2) for every $i \in I$, $p_i\colon P \longrightarrow C_i$ is a morphism of $\mathscr{C}$,

and this pair is such that for every other pair $\left(Q, (q_i)_{i\in I}\right)$ where

(1) Q is an object of $\mathscr{C}$,

(2) for every $i \in I$, $q_i\colon Q \longrightarrow C_i$ is a morphism of $\mathscr{C}$,

there exists a unique morphism $r\colon Q \longrightarrow P$ such that for every index i, $q_i = p_i \circ r$.

The arguments of proposition 2.1.2 generalize immediately to show that:

Proposition 2.1.5 *When the product of a family of objects exists in a category, it is unique up to an isomorphism.* □

We shall generally write $\prod_{i\in I} C_i$ for "the" product of a family $(C_i)_{i\in I}$ of objects. The definition of a product is by nature independent of any ordering on the set of indices; it obviously generalizes the situation of definition 2.1.1. The following generalized associative law holds for products:

Proposition 2.1.6 *Consider a set I and a partition $I = \bigcup_{k\in K} J_k$ of this set into disjoint subsets $(J_k)_{k\in K}$. Consider a family $(C_i)_{i\in I}$ of objects in a category $\mathscr{C}$. When all the products involved exist, the following isomorphism holds:*

$$\prod_{i\in I} C_i \cong \prod_{k\in K} \left(\prod_{j\in J_k} C_j \right).$$

Proof   Just show that the right-hand side satisfies the definition of $\prod_{i\in I} C_i$ and apply 2.1.5. □

It should be noticed that the existence of the product of a family of objects does not imply the existence of the product of a subfamily of those objects. For example, choose a natural number $n \in N$ and consider the full subcategory $\mathscr{C}_n$ of Set whose objects are the sets with fewer than n elements. It is easy to prove that products in $\mathscr{C}_n$, when they exist, are just cartesian products of sets. Therefore the product of two sets with k and l elements exists in $\mathscr{C}_n$ precisely when $kl \leq n$. But the product of an arbitrary family containing the empty set always exists: it is just the empty set.

Examples 2.1.7

2.1.7.a   In the category Set of sets and mappings, the cartesian product of a family $(C_i)_{i \in I}$ is just the cartesian product

$$\prod_{i \in I} C_i = \{(x_i)_{i \in I} \mid x_i \in C_i\}$$

with projections $p_{i_0}\big((x_i)_{i \in I}\big) = x_{i_0}$.

2.1.7.b In the category **Cat** of small categories and functors, the product of two categories is that described in 1.6.5, 1.6.6. This construction admits an obvious generalization to the case of an arbitrary family of small categories.

2.1.7.c In the categories of groups, abelian groups, rings, modules, algebras, boolean algebras, and so on... the product of a family of objects is just their cartesian product provided with the pointwise operations. For example in the case of groups

$$\prod_{i \in I} C_i = \{(x_i)_{i \in I} \mid x_i \in C_i\}$$

$$(x_i)_{i \in I} + (y_i)_{i \in I} = (x_i + y_i)_{i \in I}.$$

2.1.7.d In the category **Ban$_1$** of real Banach spaces and linear contractions, the product of a family $(C_i)_{i \in I}$ is given by

$$\prod_{i \in I} C_i = \left\{ (x_i)_{i \in I} \mid x_i \in C_i \ ; \ \sup_{i \in I} \|x_i\| < \infty \right\}$$

$$\|(x_i)_{i \in I}\| = \sup_{i \in I} \|x_i\|.$$

Indeed each projection $p_i \colon \prod_{i \in I} C_i \longrightarrow C_i$ is continuous and has norm 1. Moreover if $(f_i \colon B \longrightarrow C_i)_{i \in I}$ is a family of linear contractions and $b \in B$, $\|b\| \le 1$, then for every $i \in I$, $\|f_i(b)\| \le 1$ and thus $\|\sup_{i \in I} f_i(b)\| \le 1$. Therefore the obvious factorization $f \colon B \longrightarrow \prod_{i \in I} C_i$, $f(b) = \big(f_i(b)\big)_{i \in I}$, is a linear contraction.

2.1.7.e In the category **Ban$_\infty$** of real Banach spaces and bounded linear mappings, the product of a finite family of objects exists and is computed as in **Ban$_1$**; but the product of an infinite family of objects does not exist in general. For example the "power" $\mathbb{R} \times \mathbb{R} \times \mathbb{R} \times \ldots$ of the constant family $(\mathbb{R})_{n \in \mathbb{N}}$ does not exist. Indeed, assume this product does exist and is given by $\big(P, (p_n \colon P \longrightarrow \mathbb{R})_{n \in \mathbb{N}}\big)$. The constant family $(\mathrm{id} \colon \mathbb{R} \longrightarrow \mathbb{R})_{n \in \mathbb{N}}$ factors through this product via a morphism $\Delta \colon \mathbb{R} \longrightarrow P$; from $\mathrm{id} = p_n \circ \Delta$ we deduce that no p_n can be the zero mapping. Next consider the family

$$f_n \colon \mathbb{R} \longrightarrow \mathbb{R}, \quad f_n(r) = n \cdot \|p_n\| \cdot r,$$

which factors through the product via a morphism $f\colon \mathbb{R} \longrightarrow P$. From $f_n = p_n \circ f$ we deduce $n \cdot \|p_n\| = \|f_n\| \le \|p_n\| \cdot \|f\|$ and thus $n \le \|f\|$ for every n, which is impossible.

2.1.7.f In the category **Top** of topological spaces and continuous mappings, the product of a family $(X_i, \mathcal{T}_i)_{i \in I}$ of topological spaces is given by $(X, \mathcal{T})$ where the set X is just

$$X = \prod_{i \in I} X_i = \{(x_i)_{i \in I} \,|\, x_i \in X_i\}\,.$$

We choose as basic open subsets those of the form

$$\prod_{i \in I} U_i = \{(x_i)_{i \in I} \,|\, x_i \in U_i\}$$

where $U_i \in \mathcal{T}_i$ and

$$\{i \in I \,|\, U_i \ne X_i\} \quad \text{is finite.}$$

This family of basic open subsets is closed under finite intersections; the topology $\mathcal{T}$ consists of all the unions of basic open subsets. The obvious projections

$$p_{i_0}\colon X \longrightarrow X_{i_0}, \quad p_{i_0}\big((x_i)_{i \in I}\big) = x_{i_0}$$

are continuous since, for $U \in \mathcal{T}_{i_0}$,

$$p_{i_0}^{-1}(U) = \prod_{i \in I} U_i$$

where $U_{i_0} = U$ and $U_i = X_i$ for $i \ne i_0$. Next, given a family

$$f_i\colon (Y, \mathcal{S}) \longrightarrow (X_i, \mathcal{T}_i)$$

of continuous mappings, the unique factorization

$$f\colon Y \longrightarrow X, \quad f(y) = \big(f_i(y)\big)_{i \in I}$$

is continuous since, for a fundamental open subset $\prod_{i \in I} U_i \in \mathcal{T}$,

$$f^{-1}\left(\prod_{i \in I} U_i\right) = \{y \in Y \,|\, \forall i \in I \ \ f_i(y) \in U_i\}$$

$$= \bigcap_{i \in I} f_i^{-1}(U_i).$$

Each $f_i^{-1}(U_i)$ is open since f_i is continuous and U_i is open. Moreover, when $U_i = X_i$, we get $f_i^{-1}(U_i) = Y$ and this term does not play any role in the intersection. Therefore we can write

$$f^{-1}\left(\prod_{i \in I} U_i\right) = \bigcap\{f_i^{-1}(U_i) \mid i \in I \; ; \; U_i \neq X_i\} \, ,$$

which is now a finite intersection of open subsets, thus an open subset.

2.1.7.g　In the category Comp of compact Hausdorff spaces and continuous mappings, the product of a family of objects can be computed just as in Top (Tychonoff theorem).

2.1.7.h　Consider a poset $(X, \leq)$ viewed as a category (see example 1.2.6.b). Given a family of elements $x_i \in X$, one checks immediately that defining the product of this family is just defining its infimum.

2.1.7.i　Consider, in the category of sets and mappings, the two sets $\mathbb{Z}$ of integers and $\mathbb{R}$ of real numbers. Consider the usual cartesian product

$$\mathbb{Z} \times \mathbb{R} = \{(z, r) \mid z \in \mathbb{Z} \, , \, r \in \mathbb{R}\}$$

and the two mappings

$$p_{z_0} \colon \mathbb{Z} \times \mathbb{R} \longrightarrow \mathbb{Z} \;, \quad p_{z_0}(z, r) = z + z_0,$$
$$p_{r_0} \colon \mathbb{Z} \times \mathbb{R} \longrightarrow \mathbb{R} \;, \quad p_{r_0}(z, r) = r + r_0,$$

where $z_0 \in \mathbb{Z}$, $r_0 \in \mathbb{R}$ are fixed numbers.

For any choice of z_0, r_0, $(\mathbb{Z} \times \mathbb{R}, p_{z_0}, p_{r_0})$ is a product of $\mathbb{Z}$ and $\mathbb{R}$ in the category of sets. This product is indeed just the usual cartesian product $(\mathbb{Z} \times \mathbb{R}, p_{\mathbb{Z}}, p_{\mathbb{R}})$ defined in 2.1.7.a composed with the isomorphism

$$\mathbb{Z} \times \mathbb{R} \longrightarrow \mathbb{Z} \times \mathbb{R} \;, \quad (z, r) \mapsto (z + z_0, r + r_0).$$

2.2 Coproducts

The dual notion of "product" is that of "coproduct". Thus:

Definition 2.2.1 *Let I be a set and $(C_i)_{i \in I}$ a family of objects in a given category $\mathscr{C}$. A coproduct of that family is a pair $\big(P, (s_i)_{i \in I}\big)$ where*
(1) P is an object of $\mathscr{C}$,
(2) for every $i \in I$, $s_i \colon C_i \longrightarrow P$ is a morphism of $\mathscr{C}$,
and this pair is such that for every other pair $\big(Q, (t_i)_{i \in I}\big)$ where
(1) Q is an object of $\mathscr{C}$,
(2) for every $i \in I$, $t_i \colon C_i \longrightarrow Q$ is a morphism of $\mathscr{C}$,
there exists a unique morphism $r = P \longrightarrow Q$ such that for every index i, $t_i = r \circ s_i$.

Applying the results of sections 1.10 and 2.1, we get

Proposition 2.2.2 *When the coproduct of a family of objects exists in a category, it is unique up to an isomorphism.* □

We shall often write $\coprod_{i \in I} C_i$ for the coproduct of the family $(C_i)_{i \in I}$. The following generalized associativity law is obtained from 2.1.6 by duality.

Proposition 2.2.3 *Consider a set I and a partition $I = \bigcup_{k \in K} J_k$ of this set into disjoint subsets $(J_k)_{k \in K}$. Consider a family $(C_i)_{i \in I}$ of objects in a category $\mathscr{C}$. When all the coproducts involved exist, the following isomorphism holds:*

$$\coprod_{i \in I} C_i \cong \coprod_{k \in K} \left(\coprod_{j \in J_k} C_j \right).$$ □

Examples 2.2.4

2.2.4.a In the category Set of sets and mappings, the coproduct of a family $(C_i)_{i \in I}$ is just its "disjoint union", i.e. the union of the sets C_i considered as disjoint sets. When the various C_i's are not disjoint, we replace them first by isomorphic disjoint sets

$$C_i' = \{(x, i) \mid x \in C_i\}$$
$$= C_i \times \{i\}$$

and we perform the usual union of these sets C_i'. Thus in short

$$\coprod_{i \in I} C_i = \{(x, i) \mid i \in I ;\ x \in C_i\}.$$

The canonical mappings $s_i : C_i \longrightarrow \coprod_{i \in I} C_i$ are just the obvious inclusions: $s_i(x) = (x, i)$.

2.2.4.b In the category Top of topological spaces and continuous mappings, the coproduct of a family $(X_i, \mathcal{T}_i)_{i \in I}$ is just $(X, \mathcal{T})$ where X is the disjoint union of the X_i and $\mathcal{T}$ is the topology generated by the disjoint union of the $\mathcal{T}_i$.

2.2.4.c In the category Comp of compact Hausdorff spaces and continuous mappings, the coproduct of a finite family of objects is computed as in Top. The existence of arbitrary coproducts holds but proving this requires more sophisticated arguments (see chapter 3).

2.2.4.d In the category Cat of small categories and functors, the coproduct of a family of categories is just their disjoint union.

2.2.4.e   In the category Gr of groups and group homomorphisms, the coproduct of a family $(G_i)_{i \in I}$ of groups is obtained as follows. Consider first the disjoint union V of the sets G_i and then the set W of "words" of V; thus W is the set of all finite sequences of elements in V. On W, introduce the equivalence relation generated by the following data:

- the unit element of each group G_i is equivalent to the empty sequence
- if a sequence contains two consecutive elements belonging to the same component G_i, the original sequence is equivalent to the sequence obtained by replacing the two elements by their composite in G_i.

Write $\coprod_{i \in I} G_i$ for the quotient of W by this equivalence relation. Concatenation on W induces an associative composition law on $\coprod_{i \in I} G_i$, with the empty sequence as a unit. This is a group structure: the inverse of a sequence is the sequence of inverses of its elements, in reversed order. Each group G_i is mapped into $\coprod_{i \in I} G_i$, the element $x \in G_i$ going to the equivalence class $[x]$ of the sequence consisting of that single element. $\coprod_{i \in I} G_i$ is easily seen to be the coproduct of the G_i's. In group theory, $\coprod_{i \in I} G_i$ is generally called the "free product" of the G_i's.

2.2.4.f   In the category Ab of abelian groups and group homomorphisms, the coproduct of a family $(G_i)_{i \in I}$ is just their direct sum

$$\coprod_{i \in I} G_i = \left\{ (x_i)_{i \in I} \,\middle|\, x_i \in G_i, \ \{i \mid x_i \neq 0\} \text{ is finite } \right\}.$$

The composition law is defined componentwise and the canonical morphisms $s_{i_0} \colon G_{i_0} \longrightarrow \coprod_{i \in I} G_i$ are defined by $s_{i_0}(x) = (x_i)_{i \in I}$ where $x_{i_0} = x$ and the other components are just 0. If H is an abelian group and $f_i \colon G_i \longrightarrow H$ is a family of group homomorphisms, the unique factorization $f \colon \coprod_{i \in I} G_i \longrightarrow H$ is given by $f\big((x_i)_{i \in I}\big) = \sum_{i \in I} f_i(x_i)$; this sum makes sense since it contains just finitely many non-zero terms.

2.2.4.g   In the category Rng of commutative rings with unit and corresponding homomorphisms, the coproduct of a family $(R_i)_{i \in I}$ is just their generalized tensor product. We describe the construction in the case of the coproduct $R \amalg S$ of two rings and leave the infinite case to the reader. So we define $R \amalg S$ to be the tensor product $R \otimes_{\mathbb{Z}} S$ of the underlying abelian groups. We provide it with the multiplication generated by

$$(r \otimes s) \cdot (r' \otimes s') = (rr') \otimes (ss').$$

This is easily seen to be a ring multiplication and we define

$$s_R: R \longrightarrow R \amalg S, \quad s_R(r) = r \otimes 1,$$
$$s_S: S \longrightarrow R \amalg S, \quad s_S(s) = 1 \otimes s,$$

to complete the definition of the coproduct. If $f: R \longrightarrow T$, $g: S \longrightarrow T$ are ring homomorphisms, we get the unique factorization $h: R \amalg S \longrightarrow T$ as $h\left(\sum_{i=1}^{n} r_i \otimes s_i\right) = \sum_{i=1}^{n} f(r_i) g(s_i)$.

2.2.4.h In the category $\mathbf{Ban}_1$ of real Banach spaces and linear contractions, the coproduct of a family $(C_i)_{i \in I}$ is given by

$$\coprod_{i \in I} C_i = \left\{ (x_i)_{i \in I} \,\Big|\, x_i \in C_i \,;\, \sum_{i \in I} \|x_i\| < \infty \right\},$$
$$\|(x_i)_{i \in I}\| = \sum_{i \in I} \|x_i\|.$$

The canonical inclusions $s_i: C_i \longrightarrow \coprod_{i \in I} C_i$ are defined as in the case of abelian groups and are continuous with norm 1. If $f_i: C_i \longrightarrow B$ is a family of linear contractions, we define

$$f: \coprod_{i \in I} C_i \longrightarrow B, \quad f\big((x_i)_{i \in I}\big) = \sum_{i \in I} f_i(x_i).$$

From $\sum_{i \in I} \|f_i(x_i)\| \leq \sum_{i \in I} \|x_i\| < \infty$ we conclude that the expression $\sum_{i \in I} f_i(x_i)$ defining f makes sense and from

$$\left\| \sum_{i \in I} f_i(x_i) \right\| \leq \sum_{i \in I} \|f_i(x_i)\| \leq \sum_{i \in I} \|x_i\|,$$

we deduce that f is in fact a linear contraction.

2.2.4.i If $(X, \leq)$ is a poset viewed as a category (see example 1.2.6.b), defining the coproduct of a family is just defining its supremum.

2.3 Initial and terminal objects

Consider an empty family of objects in a category $\mathscr{C}$. What does it mean for the product of this family to exist? Well, it must be a pair $\big(\mathbf{1}, (\;)_{i \in \emptyset}\big)$ such that for each other pair $\big(C, (\;)_{i \in \emptyset}\big)$ there is a unique morphism $C \longrightarrow \mathbf{1} \ldots$ satisfying an empty condition! In short, the product of an empty family is an object $\mathbf{1}$ with the property that every object C is provided with exactly one morphism $C \longrightarrow \mathbf{1}$. Dually for an empty coproduct.

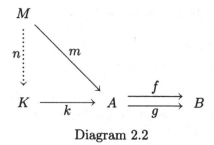

Diagram 2.2

Definition 2.3.1 *An object **1** of a category is terminal (or final) when every object C is provided with exactly one arrow from C to **1**.*
*An object **0** of a category is initial when every object C is provided with exactly one arrow from **0** to C.*

Examples 2.3.2

2.3.2.a In the category Set of sets and mappings, the empty set is the initial object and a singleton is a terminal object. The same holds in the category Top of topological spaces and continuous mappings.

2.3.2.b In the categories of groups, abelian groups, vector spaces, Banach spaces, and so on, (0) is both the initial and the terminal object.

2.3.2.c In the category Rng of commutative rings with a unit, (0) is the terminal object and $\mathbb{Z}$ is the initial object.

2.4 Equalizers, coequalizers

The notion of "product" defines a "limit object" (the product), starting with a given family of objects. We want now to define "limit objects" starting with data containing both objects and arrows.

Definition 2.4.1 *Consider two arrows $f, g: A \rightrightarrows B$ in a category $\mathscr{C}$. An equalizer of f, g is a pair (K, k) where*
(1) K is an object of $\mathscr{C}$,
(2) $k: K \longrightarrow A$ is an arrow of $\mathscr{C}$ such that $f \circ k = g \circ k$,
and such that for every pair (M, m) where
(1) M is an object of $\mathscr{C}$,
(2) $m: M \longrightarrow A$ is an arrow of $\mathscr{C}$ such that $f \circ m = g \circ m$,
there exists a unique morphism $n: M \longrightarrow K$ such that $m = k \circ n$ (see diagram 2.2).

Proposition 2.4.2 *When it exists, the equalizer of two morphisms is unique up to isomorphism.*

Proof With the notation of 2.4.1, suppose (K, k) and (M, m) are both equalizers of the pair f, g. Since (M, m) is an equalizer, there is a morphism $l \colon K \longrightarrow M$ such that $k = m \circ l$. Therefore we have $k = k \circ 1_K$ and $k = k \circ (n \circ l)$; since k is an equalizer, $n \circ l = 1_K$. In the same way $l \circ n = 1_M$. □

We shall write $\mathsf{Ker}\,(f, g)$ for "the" equalizer of f, g.

Proposition 2.4.3 *In a category $\mathscr{C}$, when two arrows $f, g \colon A \rightrightarrows B$ have an equalizer (K, k), the morphism $k \colon K \longrightarrow A$ is a monomorphism.*

Proof Consider $u, v \colon C \rightrightarrows K$ such that $k \circ u = k \circ v$. The morphism $k \circ u$ is such that $f \circ k \circ u = g \circ k \circ u$ and factors in two ways (u and v) through k; therefore $u = v$. □

By duality, one defines the "coequalizer" of two morphisms; when it exists, it is unique up to isomorphism and is an epimorphism. We write $\mathsf{Coker}\,(f, g)$ for "the" coequalizer of f, g.

Proposition 2.4.4 *Let $f \colon A \longrightarrow B$ be an arrow in a category $\mathscr{C}$. The equalizer of the pair (f, f) always exists and is just the identity on A.* □

Proposition 2.4.5 *In a category $\mathscr{C}$, suppose the arrow $f \colon A \longrightarrow B$ is both an epimorphism and an equalizer. Then f is an isomorphism.*

Proof Assume f is the equalizer of u, v. Then $u \circ f = v \circ f$ and since f is an epimorphism, $u = v$. But the equalizer of $u = v \colon B \rightrightarrows C$ is 1_B (see 2.4.4). Therefore f is isomorphic to the identity on B and so is an isomorphism. □

Examples 2.4.6

2.4.6.a In most "concrete" categories (Set, Top, Gr, Ab, Ban_1, Ban_∞, Rng, ...) the equalizer of two morphisms $f, g \colon A \rightrightarrows B$ is just given by

$$\mathsf{Ker}\,(f, g) = \{a \in A \,|\, f(a) = g(a)\}$$

provided with the structure induced by that of A.

2.4.6.b In the category Set of sets and mappings, the coequalizer of

$$f, g \colon A \rightrightarrows B$$

is the quotient of B by the equivalence relation generated by the pairs $\big((f(a), g(a)\big)$ for every element $a \in A$.

2.4.6.c In the category of abelian groups, consider first a group homomorphism $f\colon A \longrightarrow B$. The coequalizer of f and the zero homomorphism is just the quotient of B by the subgroup $f(A)$. More generally, the coequalizer of two morphisms $f, g\colon A \xrightarrow{\quad} B$ is the coequalizer of $f - g\colon A \longrightarrow B$ and the zero morphism. An analogous description of the coequalizer holds for categories of vector spaces or modules.

2.4.6.d In many "algebraic-like" structures, the situation is less simple (groups, rings, ...). The general procedure in those cases for computing the coequalizer of $f, g\colon A \xrightarrow{\quad} B$, is to construct the quotient of B by the congruence generated by all the pairs $\big(f(a), g(a)\big)$ for $a \in A$. This congruence is thus the smallest equivalence relation containing all those pairs and closed under all the algebraic operations (see chapter 3, volume 2).

2.4.6.e In the category Top of topological spaces and continuous mappings, the coequalizer is constructed as in Set and provided with the quotient topology.

2.4.6.f In the category Haus of Hausdorff spaces or the category Comp of compact Hausdorff spaces, the coequalizer of two continuous mappings $f, g\colon A \xrightarrow{\quad} B$ is the quotient of B by the closure $\overline{R} \subseteq B \times B$ of the equivalence relation R generated by the pairs $\big(f(a), g(a)\big)$ for $a \in A$. The quotient of a Hausdorff space B by a closed equivalence relation $\overline{R}$ is indeed another Hausdorff space and the quotient is compact as long as B is compact (continuous image of a compact set). Now choose C Hausdorff and $h\colon B \longrightarrow C$ such that $h \circ f = h \circ g$. The diagonal $\Delta_C \subseteq C \times C$ is closed and therefore $(h \times h)^{-1}(\Delta_C)$ is a closed equivalence relation on B containing all the pairs $\big(f(a), g(a)\big)$ for $a \in A$; thus it contains $\overline{R}$ and h factors through the quotient.

2.4.6.g In the category Ban$_1$ of Banach spaces and linear contractions, the coequalizer of a linear contraction $f\colon A \longrightarrow B$ and the zero mapping is just the quotient of B by the closure of $f(A)$. The quotient by a closed subspace indeed produces a Banach space. Moreover, if $h\colon B \longrightarrow C$ is a linear contraction such that $h \circ f = 0$, then $h^{-1}(0)$ is a closed subspace containing $f(A)$, thus also $\overline{f(A)}$; so h factors through the quotient. More generally, the coequalizer of two linear contractions $f, g\colon A \xrightarrow{\quad} B$ is the same as the coequalizer of the linear contraction $\frac{1}{2}(f - g)$ and the zero mapping.

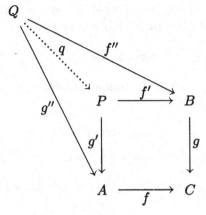

Diagram 2.3

2.5 Pullbacks, pushouts

This is another example of a "limit object" constructed from objects and arrows.

Definition 2.5.1 *Consider two morphisms* $f: A \longrightarrow C$, $g: B \longrightarrow C$ *in a category* $\mathscr{C}$. *A pullback of* (f, g) *is a triple* (P, f', g') *where*

(1) P *is an object of* $\mathscr{C}$,

(2) $f': P \longrightarrow B$, $g': P \longrightarrow A$ *are morphisms of* $\mathscr{C}$ *such that* $f \circ g' = g \circ f'$,

and for every other triple (Q, f'', g'') *where*

(1) Q *is an object of* $\mathscr{C}$,

(2) $f'': Q \longrightarrow B$, $g'': Q \longrightarrow A$ *are morphisms of* $\mathscr{C}$ *such that* $f \circ g'' = g \circ f''$,

there exists a unique morphism $q: Q \longrightarrow P$ *such that* $f'' = f' \circ q$ *and* $g'' = g' \circ q$ *(see diagram 2.3).*

Proposition 2.5.2 *When it exists, the pullback of two arrows is unique up to isomorphism.*

Proof Analoguous to that of 2.1.2. $\qquad \Box$

Proposition 2.5.3 *With the conditions of 2.5.1, let us consider the pullback* (P, f', g') *of* (f, g).

(1) If g *is a monomorphism,* g' *is a monomorphism as well;*

(2) If g *is an isomorphism,* g' *is an isomorphism as well.*

Proof If g is a monomorphism, consider two morphisms $u, v \colon Q \rightrightarrows P$ such that $g' \circ u = g' \circ v$. Put $g'' = g' \circ u$ and $f'' = f' \circ u$. One has

$$f \circ g'' = f \circ g' \circ u = g \circ f' \circ u = g \circ f''$$

and $u \colon Q \longrightarrow P$ is therefore the unique possible factorization of (g'', f'') through (g', f'). But

$$g' \circ v = g' \circ u = g'',$$
$$g \circ f' \circ v = f \circ g' \circ v = f \circ g' \circ u = f \circ g'' = g \circ f'',$$

and since g is a monomorphism, $f' \circ v = f''$. Thus v is another factorization and $u = v$.

Now if g is an isomorphism, again with the notation of 2.5.1 let us put $Q = A$, $f'' = g^{-1} \circ f$, $g'' = 1_A$. We get a unique morphism q such that $g' \circ q = 1_A$ and $f' \circ q = g^{-1} \circ f$. Now one computes immediately from this that

$$g' \circ q \circ g' = 1_A \circ g' = g' = g' \circ 1_P,$$
$$f' \circ q \circ g' = g^{-1} \circ f \circ g' = g^{-1} \circ g \circ f' = f' = f' \circ 1_P,$$

from which $q \circ g' = 1_P$ by the uniqueness condition in the definition of a pullback. So $g' \circ q = 1_A$ and $q \circ g' = 1_P$, thus $q = (g')^{-1}$. $\square$

Statement 2.5.3.(1) is very often referred to as the fact that "the pullback of a monomorphism is a monomorphism". The dual notion of a "pullback" is that of a "pushout". In particular, "the pushout of an epimorphism is an epimorphism".

Observe that the pullback of an epimorphism is in general not an epimorphism. For example in the category **Haus** of Hausdorff spaces and continuous mappings, $f \colon X \longrightarrow Y$ is an epimorphism when $f(X)$ is dense in Y (see 1.8.5.c). In particular f need not be surjective; assume it is not. Choose $y \in Y \setminus f(X)$; the pullback of y and f is just the empty set (see diagram 2.4). The empty set is by no means dense in the singleton so that the left vertical arrow is no longer an epimorphism.

Definition 2.5.4 *In a category $\mathscr{A}$, the kernel pair of an arrow*
$$f \colon A \longrightarrow B$$
is (when it exists) the pullback (P, α, β) of f with itself.

Proposition 2.5.5 *In a category $\mathscr{A}$, if the kernel pair (P, α, β) of an arrow f exists, α and β are epimorphisms.*

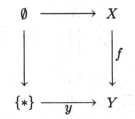

Diagram 2.4

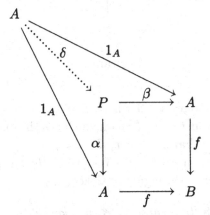

Diagram 2.5

Proof From $f \circ 1_A = f \circ 1_A$ one gets a unique factorization δ such that $\alpha \circ \delta = 1_A$, $\beta \circ \delta = 1_A$. Thus δ is a monomorphism and α, β are epimorphisms (see 1.7.2, 1.8.2 and diagram 2.5).   □

The following result is obvious, but will prove to have interesting consequences.

Proposition 2.5.6 *Consider a morphism $f: A \longrightarrow B$ in a category $\mathscr{C}$. The following conditions are equivalent:*

(1) f is a monomorphism;
(2) the kernel pair of f exists and is given by $(A, 1_A, 1_A)$;
(3) the kernel pair (P, α, β) of f exists and is such that $\alpha = \beta$.   □

Let us now indicate two interesting properties relating kernel pairs and coequalizers.

Proposition 2.5.7 *In a category $\mathscr{C}$, if a coequalizer has a kernel pair, it is the coequalizer of its kernel pair.*

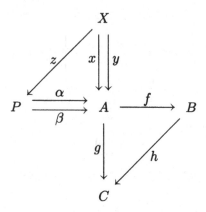

Diagram 2.6

Proof   Consider diagram 2.6 where $f = \mathsf{Coker}\,(x,y)$ and α, β is the kernel pair of f. From $f \circ x = f \circ y$ one deduces the existence of a unique factorization z such that $\alpha \circ z = x$, $\beta \circ z = y$.

If $g \circ \alpha = g \circ \beta$, then $g \circ x = g \circ \alpha \circ z = g \circ \beta \circ z = g \circ y$ and we get a unique factorization h through $f = \mathsf{Coker}\,(x,y)$.   □

Proposition 2.5.8 *In a category $\mathscr{C}$, if a kernel pair has a coequalizer, it is the kernel pair of its coequalizer.*

Proof   We again consider diagram 2.6, supposing now that (α, β) is the kernel pair of g and $f = \mathsf{Coker}\,(\alpha, \beta)$. Since $g \circ \alpha = g \circ \beta$, one gets a unique factorization h through $f = \mathsf{Coker}\,(\alpha, \beta)$.

Now take x, y such that $f \circ x = f \circ y$. This implies $g \circ x = h \circ f \circ x = h \circ f \circ y = g \circ y$, from which there is a unique factorization z such that $\alpha \circ z = x$, $\beta \circ z = y$.   □

Let us conclude the body of this section with the so-called "associativity property" of pullbacks.

Proposition 2.5.9 *In a category $\mathscr{C}$ consider diagram 2.7, which is commutative.*

(1) *If the squares (I) and (II) are pullbacks, the outer rectangle is a pullback*

(2) *If $\mathscr{C}$ has pullbacks, if the square (II) is a pullback and the outer rectangle is a pullback, then the square (I) is also a pullback.*

Proof   If (I) and (II) are pullbacks and $g \circ f \circ x = e \circ y$, there is a unique z such that $b \circ z = y$, $d \circ z = f \circ x$. From $d \circ z = f \circ x$ we find a unique

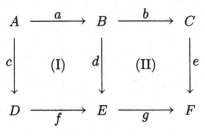

Diagram 2.7

w such that $a \circ w = z$, $c \circ w = x$. In particular $b \circ a \circ w = b \circ z = y$. If w' is another morphism such that $b \circ a \circ w' = y$ and $c \circ w' = x$, we have $b \circ (a \circ w') = b \circ (a \circ w)$ and $d \circ (a \circ w') = f \circ c \circ w' = f \circ x = f \circ c \circ w = d \circ (a \circ w)$. Since (II) is a pullback, $a \circ w' = a \circ w$. On the other hand $c \circ w' = x = c \circ w$, thus $w = w'$ since (I) is a pullback.

Under the assumptions of (2), compute the pullback (A', c', a') of f and d. Since $d \circ a = f \circ c$, one gets a unique factorization $h \colon A \longrightarrow A'$ such that $a' \circ h = a$, $c' \circ h = c$. The triple $(A, c, b \circ a)$ is, by assumption, the pullback of $(g \circ f, e)$. Applying (1), we know that $(A', c', b \circ a')$ is another such pullback. h is a factorization beween those pullbacks because $c' \circ h = c$ and $b \circ a' \circ h = b \circ a$. By uniqueness of the pullback (see 1.5.2), h is an isomorphism and thus (I) is a pullback. $\square$

Examples 2.5.10

2.5.10.a With the notation of 2.5.1, the pullback of the pair (f, g) in the category **Set** of sets and mappings is given by

$$P = \big\{ (a, b) \,\big|\, a \in A \,,\, b \in B \,,\, f(a) = g(b) \big\} \,,$$
$$g'(a, b) = a, \quad f'(a, b) = b.$$

2.5.10.b Under the conditions of 2.5.10.a, when B is a subset of C and g is the canonical inclusion, P is isomorphic to $f^{-1}(B)$, the inverse image of B along f.

2.5.10.c Under the conditions of 2.5.10.a, if both A and B are subsets of C with f, g the canonical inclusions, P is isomorphic to the intersection $A \cap B$.

2.5.10.d In the category **Set** of sets and mappings, the kernel pair (P, α, β) of a morphism $f \colon A \longrightarrow B$ is given by

$$P = \big\{ (a_1, a_2) \,\big|\, a_1 \in A, \ a_2 \in A, \ f(a_1) = f(a_2) \big\} \,,$$
$$\alpha(a_1, a_2) = a_1,; \quad \beta(a_1, a_2) = a_2.$$

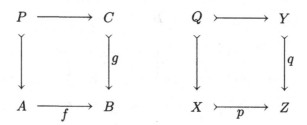

Diagram 2.8

It is the equivalence relation on A determined by f. In other words, the image $f(A)$ of f is the quotient of A by that equivalence relation.

Due to 2.5.3 and 2.5.10.b,c, when the squares of diagram 2.8 are pullbacks with g, p, q monomorphisms, we shall write P as $f^{-1}(C)$ and call it "the inverse image of C along f" (to be precise, we should refer to g instead of C); in the same way we shall write Q as $X \cap Y$ and call it the intersection of X and Y (in fact, of p and q).

2.6 Limits and colimits

In this section, we introduce the general definition of "limit of a functor" which will turn out to contain as special cases the various constructions of the previous sections of this chapter.

Definition 2.6.1 *Given a functor* $F: \mathscr{D} \longrightarrow \mathscr{C}$, *a cone on* F *consists of*
(1) an object $C \in \mathscr{C}$,
(2) for every object $D \in \mathscr{D}$, *a morphism* $p_D: C \longrightarrow FD$ *in* $\mathscr{C}$,
in such a way that for every morphism $d: D \longrightarrow D'$ *in* $\mathscr{D}$, $p_{D'} = Fd \circ p_D$.

Definition 2.6.2 *Given a functor* $F: \mathscr{D} \longrightarrow \mathscr{C}$, *a limit of* F *is a cone* $\left(L, (p_D)_{D \in \mathscr{D}}\right)$ *on* F *such that, for every cone* $\left(M, (q_D)_{D \in \mathscr{D}}\right)$ *on* F, *there exists a unique morphism* $m: M \longrightarrow L$ *such that for every object* $D \in \mathscr{D}$, $q_D = p_D \circ m$.

Again, a proof analoguous to that of 2.1.2 yields:

Proposition 2.6.3 *When a functor* $F: \mathscr{D} \longrightarrow \mathscr{C}$ *admits a limit, that limit is unique up to isomorphism.* $\square$

In the same way one has:

Proposition 2.6.4 *If* $\left(L, (p_D)_{D \in \mathscr{D}}\right)$ *is a limit of the functor* $F: \mathscr{D} \longrightarrow \mathscr{C}$, *two morphisms* $f, g: M \rightrightarrows L$ *in* $\mathscr{C}$ *are equal as long as for every object* $D \in \mathscr{D}$, $p_D \circ f = p_D \circ g$.

$$\mathcal{K}: \quad A \xrightarrow[\beta]{\alpha} B$$

Diagram 2.9

Proof f and g are two factorizations of the cone $\left(M, (p_D \circ f)\right)_{D \in \mathscr{D}}$ through the limit. $\qquad \square$

Due to the importance of the notion of limit, it is probably worth while to write out explicitly the dual notion of "colimit".

Definition 2.6.5 *Given a functor* $F: \mathscr{D} \longrightarrow \mathscr{C}$, *a cocone on F consists in*

(1) *an object $C \in \mathscr{C}$,*
(2) *for every object $D \in \mathscr{D}$, a morphism $s_D: FD \longrightarrow C$ in $\mathscr{C}$,*

in such a way that for every morphism $d: D' \longrightarrow D$ in $\mathscr{D}$, $s_{D'} = s_D \circ Fd$.

Definition 2.6.6 *Given a functor* $F: \mathscr{D} \longrightarrow \mathscr{C}$, *a colimit of F is a cocone* $\left(L, (s_D)_{D \in \mathscr{D}}\right)$ *on F such that, for every cocone* $\left(M, (t_D)_{D \in \mathscr{D}}\right)$ *on F, there exists a unique morphism $m: L \longrightarrow M$ such that for every object $D \in \mathscr{D}$, $t_D = m \circ s_D$.*

Let us now observe that the constructions of the previous sections are special cases of the notion of limit.

Examples 2.6.7

2.6.7.a Given a set I, let us view it as a discrete category $\mathscr{I}$ (see 1.2.6.c). Giving a functor $F: \mathscr{I} \longrightarrow \mathscr{C}$ to a category $\mathscr{C}$ is just giving a family $Fi \in \mathscr{C}$, $i \in I$, of objects and defining the limit of F is just defining the product $\prod_{i \in I} Fi$.

2.6.7.b Consider the category $\mathscr{K}$ defined by

$$|\mathscr{K}| = \{A, B\},$$
$$\mathscr{K}(A, A) = \{1_A\}, \quad \mathscr{K}(B, B) = \{1_B\},$$
$$\mathscr{K}(A, B) = \{\alpha, \beta\}, \quad \mathscr{K}(B, A) = \emptyset,$$

and sketched in diagram 2.9. Giving a functor F from $\mathscr{K}$ to a category $\mathscr{C}$ is just giving two arrows $F\alpha, F\beta: FA \rightrightarrows FB$ in $\mathscr{C}$ and defining the limit of F is just defining the equalizer of $F\alpha$, $F\beta$.

2.6.7.c Consider the category $\mathscr{P}$ defined by

$$\mathscr{P} = \{A, B, C\},$$
$$\mathscr{P}(A, A) = \{1_A\}, \quad \mathscr{P}(B, B) = \{1_B\}, \quad \mathscr{P}(C, C) = \{1_C\},$$

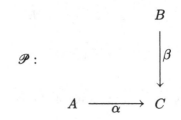

Diagram 2.10

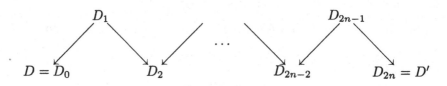

Diagram 2.11

$$\mathscr{P}(A,C) = \{\alpha\}, \quad \mathscr{P}(B,C) = \{\beta\},$$
$$\mathscr{P}(C,A) = \mathscr{P}(C,B) = \mathscr{P}(A,B) = \mathscr{P}(B,A) = \emptyset,$$

and sketched in diagram 2.10. Giving a functor F from $\mathscr{P}$ to a category $\mathscr{C}$ is just giving a pair

$$F\alpha \colon FA \longrightarrow FC, \quad F\beta \colon FB \longrightarrow FC$$

of arrows in $\mathscr{C}$ and defining the limit of F is just defining the pullback of $F\alpha$, $F\beta$.

2.6.7.d The previous examples can be dualized to present the notions of coproduct, coequalizer and pushout as special cases of the general notion of colimit.

2.6.7.e A category $\mathscr{D}$ is connected when it is non-empty and, given two objects $D, D' \in \mathscr{D}$, there exists a finite "zigzag" in $\mathscr{D}$ as in diagram 2.11. Consider an object A of a category $\mathscr{A}$ and the corresponding constant functor $\Delta_A \colon \mathscr{D} \longrightarrow \mathscr{A}$ on $\mathscr{A}$ (see 1.2.8.e). If $\left(f_D \colon \Delta_A(D) \longrightarrow M\right)_{D \in \mathscr{D}}$ is a cocone on F and D, D' are connected by the above zigzag, one immediately gets

$$f_D = f_{D_0} = f_{D_1} = f_{D_2} = \cdots = f_{D_{2n-2}} = f_{D_{2n-1}} = f_{D_{2n}} = f_{D'};$$

thus the cocone is a constant one. Therefore $\left(A, (1_A)_{D \in \mathscr{D}}\right)$ is the colimit of Δ_A.

2.6.7.f Observe that the colimit of a constant functor $\Delta_A \colon \mathscr{D} \longrightarrow \mathscr{A}$ is generally not given by the object A. Indeed, in general $A \amalg A \ncong A$.

2.7 Complete categories

It could sound reasonable to define a complete category as one such that every functor into it admits a limit. Unfortunately this doesn't produce any relevant notion, due to set theoretical reasons. For example if $\mathscr{D}$ is a discrete category and $F\colon \mathscr{D} \longrightarrow \mathsf{Set}$ is a functor to the category of sets, the limit of F should be the cartesian product of all sets $FD\dots$ which does not exist (in general) if the indices (the objects of $\mathscr{D}$) do not themselves constitute a set. More precisely:

Proposition 2.7.1 *Consider a category $\mathscr{C}$ such that, for every category $\mathscr{D}$ and every functor $F\colon \mathscr{D} \longrightarrow \mathscr{C}$, the limit of F exists. In that case, $\mathscr{C}$ is just a preordered class. The same conclusion holds when "category" is replaced everywhere by "small category" or "finite category".*

Proof Let us use the axiom system of universes (see section 1.1) so that the objects of $\mathscr{C}$ constitute a set in some universe. We must prove that for any two objects C_1, C_2 of $\mathscr{C}$, $\mathscr{C}(C_1, C_2)$ has at most one element. If this were not the case, consider two distinct morphisms $f, g\colon C_1 \rightrightarrows C_2$ in $\mathscr{C}$ between some fixed objects C_1, C_2. By assumption, the "power object" $(C_2)^{\#\mathscr{C}}$, product of $\#\mathscr{C}$ copies of C_2, exists, where $\#\mathscr{C}$ denotes the cardinal of the set of arrows of $\mathscr{C}$. Using just the arrows f and g, we can already construct $2^{\#\mathscr{C}}$ distinct cones $(C_1 \longrightarrow C_2)_{\#\mathscr{C}}$ and therefore $2^{\#\mathscr{C}}$ distinct factorizations $C_1 \longrightarrow (C_2)^{\#\mathscr{C}}$. But this set of factorizations is a subset of the set of all arrows, thus $2^{\#\mathscr{C}} \leq \#\mathscr{C}$. This contradicts the Cantor Theorem. The same proof applies to the cases of small or finite categories. $\qquad\square$

The pertinent definition is in fact

Definition 2.7.2 *A category $\mathscr{C}$ is complete when every functor*
$$F\colon \mathscr{D} \longrightarrow \mathscr{C},$$
with $\mathscr{D}$ a small category, has a limit.
The category $\mathscr{C}$ is finitely complete when every functor
$$F\colon \mathscr{D} \longrightarrow \mathscr{C},$$
with $\mathscr{D}$ a finite category, has a limit.

By duality, we get the notion of a cocomplete category. See 1.2.3 for the notion of a small category.

When the limit of a functor $F\colon \mathscr{D} \longrightarrow \mathscr{C}$ exists with $\mathscr{D}$ a "large" (= non-small) category, we shall sometimes call it a "large limit". For example the product of all sets exists ... and is just the empty set!

2.8 Existence theorem for limits

Consider a functor $F: \mathscr{D} \longrightarrow$ Set where $\mathscr{D}$ is a small category and Set is the category of sets. It is rather straightforward to verify that the set

$$L = \{(x_D)_{D \in \mathscr{D}} \mid x_D \in FD \; ; \; \forall f: D \longrightarrow D' \text{ in } \mathscr{D} \; Ff(x_D) = x_{D'}\}$$

provided with the obvious projections $p_D: L \longrightarrow FD$ is the limit of F. It should be observed that L is in fact a subobject of the product $\prod_{D \in \mathscr{D}} FD$ and, more precisely, L is the equalizer of the two mappings α, β:

$$L \rightarrowtail \prod_{D \in \mathscr{D}} FD \overset{\alpha}{\underset{\beta}{\rightrightarrows}} \prod_{f \in \mathscr{D}} F(\text{target of } f),$$

where
$$\alpha\big((x_D)_{D \in \mathscr{D}}\big) = \big(x_{\text{target of } f}\big)_{f \in \mathscr{D}},$$
$$\beta\big((x_D)_{D \in \mathscr{D}}\big) = \big(Ff(x_{\text{source of } f})\big)_{f \in \mathscr{D}}.$$

We shall now prove that this construction generalizes to an arbitrary category, from which the existence of limits will follow from that of products and equalizers. For brevity, we abbreviate "target of f" and "source of f" just as $t(f)$, $s(f)$.

Theorem 2.8.1 *A category $\mathscr{C}$ is complete precisely when each family of objects has a product and each pair of parallel arrows has an equalizer.*

Proof Let us first make clear that by "family", we always mean a set indexed family. We know already that completeness implies the existence of products and equalizers (see examples 2.6.7).

Conversely, consider a small category $\mathscr{D}$ and a functor $F: \mathscr{D} \longrightarrow \mathscr{C}$. We construct the products

$$\left(\prod_{D \in \mathscr{D}} FD, (p'_D)_{D \in \mathscr{D}}\right) \text{ and } \left(\prod_{f \in \mathscr{D}} F(t(f)), (p''_{t(f)})_{f \in \mathscr{D}}\right).$$

α is the unique factorization such that $p''_f \circ \alpha = p'_{t(f)}$ for every $f \in \mathscr{D}$; β is the unique factorization such that $p''_f \circ \beta = Ff \circ p'_{s(f)}$ for every $f \in \mathscr{D}$ and (L, l) is the equalizer of the pair (α, β) (see diagram 2.12). We define $p_D = p'_D \circ l$ and we shall prove that $\big(L, (p_D)_{D \in \mathscr{D}}\big)$ is the limit of the functor F.

First of all, for a morphism $f: D \longrightarrow D'$ in $\mathscr{D}$ we have

$$Ff \circ p_D = Ff \circ p'_D \circ l = p''_f \circ \beta \circ l = p''_f \circ \alpha \circ l = p'_{D'} \circ l = p_{D'},$$

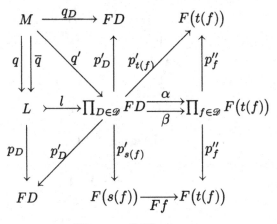

Diagram 2.12

so that $(L, (p_D)_{D \in \mathscr{D}})$ is indeed a cone on F. Moreover, if $(M, (q_D)_{D \in \mathscr{D}})$ is another cone on F, there exists a unique factorization q' such that $p'_D \circ q' = q_D$ for every $D \in \mathscr{D}$. But for every $f \colon D \longrightarrow D'$ in $\mathscr{D}$ one has

$$
\begin{aligned}
p''_f \circ \alpha \circ q' &= p'_{D'} \circ q' \\
&= q_{D'} \\
&= Ff \circ q_D \\
&= Ff \circ p'_D \circ q' \\
&= p''_f \circ \beta \circ q',
\end{aligned}
$$

from which $\alpha \circ q' = \beta \circ q'$. This implies the existence of a unique factorization $q \colon M \longrightarrow L$ such that $l \circ q = q'$. Putting these relations together yields

$$
p_D \circ q = p'_D \circ l \circ q = p'_D \circ q' = q_D,
$$

so that q is indeed the required factorization. To prove its uniqueness, consider another morphism $\bar{q}$ such that $p_D \circ \bar{q} = q_D$ for every $D \in \mathscr{D}$. Since l is a monomorphism (see 2.4.3), it remains to prove that $l \circ q = l \circ \bar{q}$; this is equivalent to proving $p'_D \circ l \circ q = p'_D \circ l \circ \bar{q}$ for each $D \in \mathscr{D}$ (see 2.6.4). This last equality holds since

$$
p'_D \circ l \circ q = p'_D \circ q' = q_D = p_D \circ \bar{q} = p'_D \circ l \circ \bar{q}. \qquad \square
$$

It should be noticed that while the existence of all limits implies that

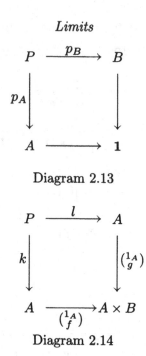

Diagram 2.13

Diagram 2.14

of all products, the existence of the single limit $(L, (p_D)_{D \in \mathscr{D}})$ in diagram 2.12 does not imply the existence of the two corresponding products used in the proof (see exercise 2.17.4). But when those two products and the limit exist, the limit is again the equalizer of α, β (see exercise 2.17.10).

Proposition 2.8.2 *For a category $\mathscr{C}$, the following conditions are equivalent:*

(1) $\mathscr{C}$ is finitely complete;

(2) $\mathscr{C}$ has a terminal object, binary products and equalizers;

(3) $\mathscr{C}$ has a terminal object and pullbacks.

Proof   (1) $\Rightarrow$ (2) and (1) $\Rightarrow$ (3) are obvious. Let us now assume (2). By associativity of products (see 2.1.3) the product of a finite (and nonempty) family of objects exists as long as binary products do. Observing the proof of 2.8.1, it suffices now to notice that when $\mathscr{D}$ is finite, so are the products involved in the proof.

Assuming (3), observe that in the pullback of diagram 2.13 (P, p_A, p_B) is the product of A, B, since $\mathbf{1}$ is terminal. It remains to prove the existence of equalizers to establish conditions (2). Given $f, g \colon A \rightrightarrows B$, compute the pullback of diagram 2.14. Composing with the two projections of the product $A \times B$, we get

$$k = p_A \circ \binom{1_A}{f} \circ k = p_A \circ \binom{1_A}{g} \circ l = l,$$
$$f \circ k = p_B \circ \binom{1_A}{f} \circ k = p_B \circ \binom{1_A}{g} \circ l = g \circ l = g \circ k.$$

Moreover, if $x \colon X \longrightarrow A$ is such that $f \circ x = g \circ x$, one has $\binom{1_A}{f} \circ x = \binom{1_A}{g} \circ x$ from which there is a unique $y \colon X \longrightarrow P$ such that $x = k \circ y = l \circ y$. Thus $k = \mathsf{Ker}\,(f, g)$. $\qquad\qquad\square$

We have introduced the notion of limit for a functor $F \colon \mathscr{D} \longrightarrow \mathscr{A}$ defined on a category $\mathscr{D}$. An apparently more general approach would have been to replace the category $\mathscr{D}$ by "a graph with commutativity conditions" (see section 5.1); in fact the notion of limit on such a graph is equivalent to that of limit on the category generated by the graph. Our choice generally makes life easier and is good enough for the applications, if one observes the following property whose proof is obvious.

Proposition 2.8.3 *Consider a functor $F \colon \mathscr{D} \longrightarrow \mathscr{A}$ and a family $(f_i)_{i \in I}$ of morphisms of $\mathscr{D}$ such that each morphism of the category $\mathscr{D}$ is a composite $f_{i_1} \circ \cdots \circ f_{i_n}$. A cone on F is just a pair $\bigl(A, (p_D \colon A \longrightarrow FD)_{D \in \mathscr{D}}\bigr)$ where, for each $f_i \colon D \longrightarrow D'$ of the given family, $F f_i \circ p_D = p_{D'}$.* $\qquad\square$

Definition 2.8.4 *A category $\mathscr{D}$ is finitely generated when*

(1) $\mathscr{D}$ has finitely many objects,
(2) there are finitely many arrows $f_1, \ldots, f_n$ such that each arrow of $\mathscr{D}$ is the composite of finitely many of these f_i.

Proposition 2.8.5 *Let $F \colon \mathscr{D} \longrightarrow \mathscr{A}$ be a functor, with $\mathscr{A}$ finitely complete and $\mathscr{D}$ finitely generated. Then the limit of F exists.*

Proof Because of 2.8.3 and with the notation of 2.8.4, it suffices in the proof of 2.8.1 to take as second product $\prod_{i=1}^{n} F\bigl(t(f_i)\bigr)$. $\qquad\square$

Example 2.8.6

It follows immediately from 2.1.7, 2.2.4, 2.4.6 and 2.8.1 that the following categories are complete and cocomplete: Set, Ab, Rng, Ban$_1$, ... A poset is complete as a category precisely when it is complete as a poset (see 2.1.7.h, 2.4.4 and 2.8.1).

2.9 Limit preserving functors

This section is devoted to some observations on functors which commute with the construction of limits.

Definition 2.9.1 *A functor* $F: \mathcal{A} \longrightarrow \mathcal{B}$ *preserves limits when, for every small category* $\mathcal{D}$ *and every functor* $G: \mathcal{D} \longrightarrow \mathcal{A}$, *if the limit* $(L, (p_D)_{D \in \mathcal{D}})$ *of* G *exists, then* $(FL, (Fp_D)_{D \in \mathcal{D}})$ *is the limit of* $F \circ G$.

As an immediate consequence of 2.8.1, we get

Proposition 2.9.2 *Let* $\mathcal{A}$ *be a (finitely) complete category and* $\mathcal{B}$ *an arbitrary category. A functor* $F: \mathcal{A} \longrightarrow \mathcal{B}$ *preserves (finite) limits precisely when it preserves (finite) products and equalizers.* $\square$

In fact this result can be improved, requiring just the existence of products in $\mathcal{A}$ (see exercise 2.17.6).

Proposition 2.9.3 *A functor which preserves pullbacks also preserves monomorphisms.*

Proof By 2.5.6. $\square$

There are interesting situations in which a functor preserves limits, even when not all limits exist. Here is a first basic example.

Proposition 2.9.4 *Consider a category* $\mathcal{C}$ *and an object* $C \in \mathcal{C}$. *The representable functor* $\mathcal{C}(C, -): \mathcal{C} \longrightarrow$ Set *preserves all existing limits, including large ones. In particular, it preserves monomorphisms.*

Proof Consider a functor $F: \mathcal{D} \longrightarrow \mathcal{C}$ with limit $(L, (p_D)_{D \in \mathcal{D}})$ and a cone $(q_D: M \longrightarrow \mathcal{C}(C, FD))_{D \in \mathcal{D}}$ over $\mathcal{C}(C, F-)$ in the category of sets. For each element $m \in M$, the family $(q_D(m): C \longrightarrow FD)_{D \in \mathcal{D}}$ is a cone on F and therefore there exists a unique morphism $q(m): C \longrightarrow L$ in $\mathcal{C}$ such that for each $D \in \mathcal{D}$, $p_D \circ q(m) = q_D(m)$. This defines a mapping $q: M \longrightarrow \mathcal{C}(C, L)$ with the property $\mathcal{C}(C, p_D) \circ q = q_D$ for each $D \in \mathcal{D}$. The uniqueness of q results immediately from that of the $q(m)$'s. The last assertion follows from 2.9.3. $\square$

It is worth dualizing the previous result. Applying it to the dual category $\mathcal{C}^*$ yields

The functors $\mathcal{C}^*(C, -): \mathcal{C}^* \longrightarrow$ Set preserve limits.

Therefore, in terms of $\mathcal{C}$, we obtain

Proposition 2.9.5 *Consider a category* $\mathcal{C}$ *and an object* $C \in \mathcal{C}$. *The representable functor* $\mathcal{C}(-, C): \mathcal{C} \longrightarrow$ Set *transforms existing colimits into limits and in particular epimorphisms into monomorphisms.* $\square$

Let us recall that $\mathcal{C}(-, C)$ is a contravariant functor, thus reverses the direction of morphisms!

Definition 2.9.6 *Let $F: \mathscr{A} \longrightarrow \mathscr{B}$ be a functor. F reflects limits when, for every functor $G: \mathscr{D} \longrightarrow \mathscr{A}$ with $\mathscr{D}$ a small category and every cone $\big(L, (p_D)_{D \in \mathscr{D}}\big)$ on G, if $\big(FL, (Fp_D)_{D \in \mathscr{D}}\big)$ is the limit of $F \circ G$ in $\mathscr{B}$, then $\big(L, (p_D)_{D \in \mathscr{D}}\big)$ is the limit of G in $\mathscr{A}$.*

Proposition 2.9.7 *Let $F: \mathscr{A} \longrightarrow \mathscr{B}$ be a limit preserving functor. If $\mathscr{A}$ is complete and F reflects isomorphisms, F also reflects limits.*

Proof Consider a functor $G: \mathscr{D} \longrightarrow \mathscr{A}$ with $\mathscr{D}$ a small category. Consider the limit $\big(L, (p_D)_{D \in \mathscr{D}}\big)$ of G; $\big(FL, (Fp_D)_{D \in \mathscr{D}}\big)$ is thus the limit of $F \circ G$. Consider now another cone $\big(M, (q_D)_{D \in \mathscr{D}}\big)$ such that $\big(FM, (Fq_D)_{D \in \mathscr{D}}\big)$ is also the limit of $F \circ G$. In $\mathscr{A}$ we have a unique factorization $f: M \longrightarrow L$ of the second cone through the limit. In $\mathscr{B}$, Ff is just a factorization between two limits of $F \circ G$, thus Ff is an isomorphism. Therefore f itself is an isomorphism and $\big(M, (q_D)_{D \in \mathscr{D}}\big)$ is a limit of G. $\square$

Let us also consider the case of finitely generated limits (see 2.8.4).

Proposition 2.9.8 *Let $\mathscr{A}, \mathscr{B}$ be finitely complete categories and*
$$F: \mathscr{A} \longrightarrow \mathscr{B}$$
a functor which preserves (or reflects) finite limits. Then F preserves (or reflects) finitely generated limits.

Proof A finitely generated limit can be expressed via equalizers and finite products (see 2.8.5), from which the result follows. $\square$

Finally let us observe that without any further assumption:

Proposition 2.9.9 *A full and faithful functor $F: \mathscr{A} \longrightarrow \mathscr{B}$ reflects limits.*

Proof Let $G: \mathscr{D} \longrightarrow \mathscr{A}$ be a functor and $(p_D: L \longrightarrow GD)_{D \in \mathscr{D}}$ a cone on G such that $(Fp_D: FL \longrightarrow FGD)_{D \in \mathscr{D}}$ is a limit cone. Given another cone $(q_D: M \longrightarrow GD)_{D \in \mathscr{D}}$, we get a unique factorization $l: FM \longrightarrow FL$ such that $Fp_D \circ l = Fq_D$. Since F is full and faithful, there exists a unique $m: M \longrightarrow L$ such that $F(m) = l$ and therefore $p_D \circ m = q_D$. $\square$

Applying proposition 2.9.2 and examples 2.1.7, 2.2.4 and 2.4.6, we get the following examples.

Examples 2.9.10

2.9.10.a The forgetful functor $U: \mathsf{Top} \longrightarrow \mathsf{Set}$ mapping a topological space to its underlying set preserves limits and colimits.

2.9.10.b The forgetful functor $U: \mathsf{Ab} \longrightarrow \mathsf{Set}$ mapping an abelian group to its underlying set preserves limits.

2.9.10.c   Consider Ban_1, the category of Banach spaces and linear contractions, and the obvious forgetful functor $U: Ban_1 \longrightarrow Set$ which maps a Banach space to its underlying set and a linear contraction to the corresponding mapping. Example 2.1.7.d shows that U does not preserve arbitrary products, but just finite products. Now consider the functor $B: Ban_1 \longrightarrow Set$ mapping a Banach space to its closed unit ball and a linear contraction to its restriction at the level of unit balls. Example 2.1.7.d shows immediately that for a family $(C_i)_{i \in I}$ of Banach spaces, the unit ball of the Banach space $\prod_{i \in I} C_i$ is just the usual cartesian product of the unit balls of the various C_i's; indeed

$$\sup_{i \in I} \|x_i\| \leq 1 \Leftrightarrow \forall i \in I \ \|x_i\| \leq 1.$$

Therefore the functor B preserves arbitrary products. It also preserves equalizers (example 2.4.6.a), so it preserves limits (see 2.9.2). Another proof consists in observing that the "unit ball functor" is just the functor represented by $\mathbb{R}$ (see 2.9.4).

2.9.10.d   The category Ab of abelian groups is complete and the forgetful functor $U: Ab \longrightarrow Set$ reflects isomorphisms, so it reflects limits (see 2.9.7).

2.9.10.e   The category Ban_1 is complete and the "unit ball functor" $B: Ban_1 \longrightarrow Set$ reflects isomorphisms (a linear mapping between Banach spaces is an isometry precisely when it induces a bijection between the unit balls). Therefore B reflects limits (see 2.9.7).

2.10 Absolute colimits

In the previous section we were concerned with a functor preserving all limits. Now we shall have a look at those limits preserved by all functors. In fact we shall develop the theory in the case of colimits since this is the case most commonly referred to in the examples.

Definition 2.10.1 *Consider a functor* $G: \mathscr{D} \longrightarrow \mathscr{A}$ *with colimit*
$$\left(L, (p_D)_{D \in \mathscr{D}}\right).$$
That colimit is absolute when for every functor $F: \mathscr{A} \longrightarrow \mathscr{B}$,
$$\left(FL, (Fp_D)_{D \in \mathscr{D}}\right)$$
is the colimit of $F \circ G$.

Here is the most famous example of an absolute colimit.

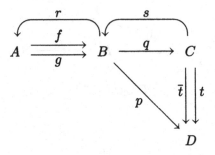

Diagram 2.15

Proposition 2.10.2 *In a category $\mathscr{C}$, consider arrows f, g, q, r, s as in diagram 2.15. When the relations*

$$q \circ f = q \circ g, \quad g \circ r = 1_B, \quad s \circ q = f \circ r, \quad q \circ s = 1_C,$$

hold, (C, q) is the coequalizer of the pair (f, g) and this coequalizer is absolute.

Proof By assumption, $q \circ f = q \circ g$. If $p \colon B \longrightarrow D$ is such that $p \circ f = p \circ g$, define $t = p \circ s$. One has

$$t \circ q = p \circ s \circ q$$
$$= p \circ f \circ r$$
$$= p \circ g \circ r$$
$$= p,$$

and if $\bar{t}$ is such that $\bar{t} \circ q = p$,

$$\bar{t} = \bar{t} \circ q \circ s$$
$$= p \circ s$$
$$= t.$$

So $(C, q) = \operatorname{Coker}(f, g)$ and since the equalities of the statement are preserved by any functor, the same conclusion applies to the image of diagram 2.15 under any functor. $\square$

The conditions in proposition 2.10.2 are not necessary for having an absolute coequalizer (see exercise 2.17.7).

Examples 2.10.3

2.10.3.a In the category Set of sets, consider an equivalence relation $R \subseteq B \times B$ on the set B (see diagram 2.16). Write p_1, p_2 for the two projections and (Q, q) for their coequalizer, thus for the quotient of B by

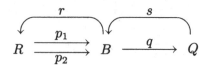

Diagram 2.16

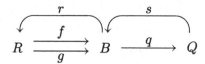

Diagram 2.17

R (see example 2.4.4.b). Using the axiom of choice, for every equivalence class $[b] \in Q$ choose an element $s(b) \in [b]$; this defines s such that $q \circ s = 1_Q$. Now define $r(b) = ((s \circ q)(b), b)$; since $(q \circ s \circ q)(b) = q(b)$, r takes values in R. Since moreover $p_1 \circ r = s \circ q$ and $p_2 \circ r = 1_B$, the coequalizer is absolute.

2.10.3.b In the category Vec_K of vector spaces over a field K, an argument analogous to that in (a) can be developed, using for R a subspace of $B \times B$ which is an equivalence relation on B (what is called a "congruence" on B). The existence of s is again a consequence of the axiom of choice: given a basis $(e_i)_{i \in I}$ of Q, choose $s(e_i)$ a representative of the class e_i.

2.10.3.c More generally suppose that in a category you have diagram 2.17, where q is the coequalizer of (f, g), (f, g) is the kernel pair of q and $q \circ s = 1_Q$. In this case the pair $(s \circ q, 1_B) \colon B \rightrightarrows B$ factors through the pullback via a morphism $r \colon B \longrightarrow R$ and we get an absolute coequalizer.

2.10.3.d Let M be a left module on the ring R with unit. In the category of abelian groups, the scalar multiplication on M yields a morphism $\mu \colon R \otimes M \longrightarrow M$, while the unit and the multiplication of R yield morphisms $e \colon \mathbb{Z} \longrightarrow R$, $m \colon R \otimes R \longrightarrow R$ (all tensor products are over $\mathbb{Z}$). In the category of abelian groups, diagram 2.18 satisfies the conditions of proposition 2.10.2 and thus (M, μ) is the absolute coequalizer of the pair $(1 \otimes \mu, m \otimes 1)$.

$$\overbrace{\xrightarrow{e \otimes 1_R \otimes 1_M}}\qquad \overbrace{\xrightarrow{e \otimes 1_M}}$$

$$R \otimes R \otimes M \underset{m \otimes 1_M}{\overset{1_R \otimes \mu}{\rightrightarrows}} R \otimes M \xrightarrow{\;\mu\;} M$$

Diagram 2.18

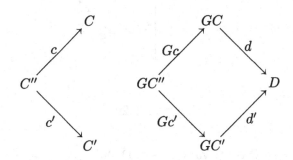

Diagram 2.19

2.11 Final functors

The main result of this section should be compared with the classical fact, in real or complex analysis, that when a sequence converges, any sub-sequence of it converges to the same limit.

Definition 2.11.1 *A functor $G\colon \mathscr{C} \longrightarrow \mathscr{D}$ is final when the following conditions are satisfied for every category $\mathscr{A}$ and every functor $F\colon \mathscr{D} \longrightarrow \mathscr{A}$:*

(1) *if the limit $\big(L,(p_D)_{D\in\mathscr{D}}\big)$ of F exists, then $\big(L,(p_{GC})_{C\in\mathscr{C}}\big)$ is the limit of $F \circ G$;*

(2) *if the limit $\big(L,(q_C)_{C\in\mathscr{C}}\big)$ of $F \circ G$ exists, then the limit of F exists as well.*

Observe that in condition (2), applying condition (1) implies immediately that the limit of F has the form $\big(L,(p_D)_{D\in\mathscr{D}}\big)$ with $p_{GC} = q_C$. Very often, one abbreviates this definition by just saying that *the limit of F exists if and only if the limit of $F \circ G$ exists and those limits are equal.*

The next proposition gives a sufficient condition for being a final functor; this condition is not necessary (see exercise 2.17.8).

Proposition 2.11.2 *A functor $G\colon \mathscr{C} \longrightarrow \mathscr{D}$ is final as long as it satisfies the following two conditions (see diagram 2.19):*

(1) *$\forall D \in \mathscr{D}\ \exists C \in \mathscr{C}\ \exists d\colon GC \longrightarrow D$;*

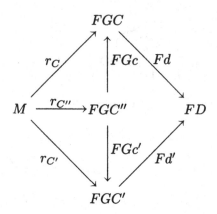

Diagram 2.20

(2) $\forall C, C' \in \mathscr{C}$ $\forall D \in \mathscr{D}$ $\forall d\colon GC \longrightarrow D$ $\forall d'\colon GC' \longrightarrow D$
    $\exists C'' \in \mathscr{C}$ $\exists c\colon C'' \longrightarrow C$ $\exists c'\colon C'' \longrightarrow C'$ such that $d{\circ}Gc = d'{\circ}Gc'$.

Proof   Let $F\colon \mathscr{D} \longrightarrow \mathscr{A}$ be a functor. Every cone $\big(M, (q_D)_{D\in\mathscr{D}}\big)$ on
F immediately induces a cone $\big(M, (q_{GC})_{C\in\mathscr{C}}\big)$ on $F \circ G$. Conversely,
consider a cone $\big(M, (r_C)_{C\in\mathscr{C}}\big)$ on $F \circ G$; we shall prove that it induces
a unique cone $\big(M, (q_D)_{D\in\mathscr{D}}\big)$ on F such that $q_{GC} = r_C$. Given $D \in \mathscr{D}$,
choose $C \in \mathscr{C}$ and $d\colon GC \longrightarrow D$. Define $q_D = Fd \circ r_C$. First of all, this
definition is independent of the choices of C and d. Indeed, choosing
$C' \in \mathscr{C}$ and $d'\colon GC' \longrightarrow D$, assumption (2) ensures the existence of c, c'
in $\mathscr{C}$ such that $d{\circ}Gc = d'{\circ}Gc'$. The three pieces of diagram 2.20 are thus
commutative, from which $Fd \circ r_C = Fd' \circ r_{C'}$. Notice that in particular
$q_{GC} = r_C$ (choose $d = 1_{GC}$). On the other hand this requirement $q_{GC} =
r_C$ ensures that, in the previous situation, $q_D = Fd \circ q_{GC} = Fd \circ r_C$
from which the required uniqueness of the cone $(q_D)_{D\in\mathscr{D}}$ follows.

   The rest of the proof is straightforward computation. If $\big(L, (p_D)_{D\in\mathscr{D}}\big)$
is the limit of F, $\big(L, (p_{GC})_{C\in\mathscr{C}}\big)$ is a cone on $F \circ G$; it is a limit cone
because every other cone $\big(M, (r_C)_{C\in\mathscr{C}}\big)$ on $F \circ G$ can be "extended" to
a cone $\big(M, (q_D)_{D\in\mathscr{C}}\big)$ which factors uniquely through the limit of F. An
analogous argument proves the converse implication. $\square$

   Here is a special case of interest.

Proposition 2.11.3 *Consider a category $\mathscr{D}$ with pullbacks and a full
subcategory $\mathscr{C} \subseteq \mathscr{D}$ which satisfies the condition*
$$\forall D \in \mathscr{D} \; \exists C \in \mathscr{C} \; \exists d\colon C \longrightarrow D.$$
Then the inclusion $\mathscr{C} \hookrightarrow \mathscr{D}$ is a final functor.

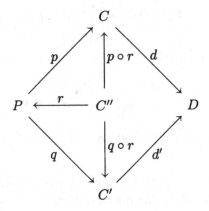

Diagram 2.21

Proof Consider objects $C, C' \in \mathscr{C}$ and morphisms d, d' as in diagram 2.21. Construct (P, p, q), the pullback of d and d' in $\mathscr{D}$, and choose a morphism $r: C'' \longrightarrow P$ with $C'' \in \mathscr{C}$. Since $\mathscr{C}$ is full in $\mathscr{D}$, $p \circ r$ and $q \circ r$ are in $\mathscr{C}$ and $d \circ p \circ r = d' \circ q \circ r$. Deduce the conclusion by 2.11.2. $\square$

Another useful case is

Proposition 2.11.4 *Consider a category $\mathscr{D}$ with an initial object **0**. The inclusion of the subcategory $\{0\}$ in $\mathscr{D}$ is a final functor.*

Proof Obvious by 2.11.2 and the definition of an initial object (see 2.3.1). $\square$

Corollary 2.11.5 *Consider a category $\mathscr{D}$ with an initial object **0** and an arbitrary functor $F: \mathscr{D} \longrightarrow \mathscr{A}$. Let us write $0_D: \mathbf{0} \longrightarrow D$ for the unique arrow from **0** to D in $\mathscr{D}$. In these conditions, the limit of F exists and is given by $\big(F\mathbf{0}, (F0_D)_{D \in \mathscr{D}}\big)$.* $\square$

Corollary 2.11.6 *Consider a category $\mathscr{D}$ with an initial object **0**. With the notation of the previous corollary, $\big(\mathbf{0}, (0_D)_{D \in \mathscr{D}}\big)$ is the limit of the identity functor $\mathscr{D} = \!\!= \mathscr{D}$.* $\square$

Let us comment on this last result. Roughly speaking, it says that the colimit of the empty functor is also the limit of the identity functor. This fact admits an interesting generalization (see exercise 2.17.2) showing that the colimit of a functor $F: \mathscr{D} \longrightarrow \mathscr{A}$ can always be described canonically via the limit of another functor $G: \mathscr{C} \longrightarrow \mathscr{A}$. The price to

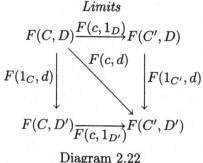

Diagram 2.22

pay is that when $\mathscr{D}$ is a small category, $\mathscr{C}$ has no longer any reason to be small. In the example of the initial object, $\mathscr{D}$ is empty but $\mathscr{C}$ is equal to $\mathscr{A}$.

2.12 Interchange of limits

Given a functor $F\colon \mathscr{D} \longrightarrow \mathscr{A}$ and its limit $\big(L, (p_D)_{D\in\mathscr{D}}\big)$, one often writes $\lim FD$ for the object L. This notation is somewhat ambiguous since it does not mention the arrows of $\mathscr{D}$, but used with care, it will turn out to be useful when computing limits and colimits.

We are now interested in the study of the limit of a functor

$$F\colon \mathscr{C} \times \mathscr{D} \longrightarrow \mathscr{A}$$

and we would like to prove the interchange property

$$\lim_{C\in\mathscr{C}} \big(\lim_{D\in\mathscr{D}} F(C,D)\big) \cong \lim_{D\in\mathscr{D}} \big(\lim_{C\in\mathscr{C}} F(C,D)\big),$$

as long as all the involved limits exist. We had better give a precise meaning to this statement.

For every fixed object $C \in \mathscr{C}$, there is a functor

$$F(C,-)\colon \mathscr{D} \longrightarrow \mathscr{A}$$

defined by

$$F(C,-)(D) = F(C,D), \quad F(C,-)(d) = F(1_C, d),$$

for an object D and an arrow d of $\mathscr{D}$. By $\lim_{D\in\mathscr{D}} F(C,D)$ we mean the limit of that functor $F(C,-)$. Now every morphism $c\colon C \longrightarrow C'$ in $\mathscr{C}$ induces an arrow $F(c,1_D)$ of $\mathscr{A}$ for every object $D \in \mathscr{D}$; moreover diagram 2.22 commutes for every arrow $d\colon D \longrightarrow D'$ in $\mathscr{D}$. In other words, the arrow $c\colon C \longrightarrow C'$ induces a natural transformation

$$F(c,-)\colon F(C,-) \Rightarrow F(C',-).$$

Now suppose $\lim_{D \in \mathscr{D}} F(C, D)$ and $\lim_{D \in \mathscr{D}} F(C', D)$ exist. The composites

$$\lim_{D \in \mathscr{D}} F(C, D) \xrightarrow{\; p_D \;} F(C, D) \xrightarrow{\; F(c, 1_D) \;} F(C', D)$$

obviously constitute a cone on the functor $F(C', -)$, from which it follows that there is a unique factorization written

$$\lim_{D \in \mathscr{D}} F(c, 1_D) \colon \lim_{D \in \mathscr{D}} F(C, D) \longrightarrow \lim_{D \in \mathscr{D}} F(C', D)$$

and such that $p'_D \circ \lim_{D \in \mathscr{D}} F(c, 1_D) = F(c, 1_D) \circ p_D$. (We have written p_D and p'_D for the projections of the limits of $F(C, -)$ and $F(C', -)$.) When all the functors $F(C, -)$ have a limit, we can define a new functor

$$L \colon \mathscr{C} \longrightarrow \mathscr{A},$$

$$L(C) = \lim_{D \in \mathscr{D}} F(C, D), \quad L(c) = \lim_{D \in \mathscr{D}} F(c, 1_D),$$

where $C \in |\mathscr{C}|$ and c is an arrow of $\mathscr{C}$. We must prove that L is indeed a functor. For example given two arrows $C \xrightarrow{\; c \;} C' \xrightarrow{\; c' \;} C''$ in $\mathscr{C}$, we deduce immediately that

$$\begin{aligned}
p''_D \circ Lc' \circ Lc &= F(c', 1_D) \circ p'_D \circ Lc \\
&= F(c', 1_D) \circ F(c, 1_D) \circ p_D \\
&= F(c' \circ c, 1_D) \circ p_D,
\end{aligned}$$

so that $Lc' \circ Lc = L(c' \circ c)$. An analogous argument holds for the identity axiom.

The limit of the previous functor L, when it exists, is exactly what we denote by $\lim_{C \in \mathscr{C}} \left(\lim_{D \in \mathscr{D}} F(C, D) \right)$. An analogous description holds for $\lim_{D \in \mathscr{D}} \left(\lim_{C \in \mathscr{C}} F(C, D) \right)$. The interchange property

$$\lim_{C \in \mathscr{C}} \left(\lim_{D \in \mathscr{D}} F(C, D) \right) \cong \lim_{D \in \mathscr{D}} \left(\lim_{C \in \mathscr{C}} F(C, D) \right)$$

means that the "canonical morphisms" connecting these two limits are in fact isomorphisms (this is indeed stronger than the existence of "some isomorphism"). Let us just describe these "canonical morphisms" between the two limits.

Starting with the limit of L described above, we have the corresponding projections

$$\lim L \xrightarrow{\; p_C \;} \lim_{D \in \mathscr{D}} F(C, D) \xrightarrow{\; p_D \;} F(C, D).$$

For a fixed object $D \in \mathscr{D}$ and a morphism $c: C \longrightarrow C'$ in $\mathscr{C}$ we have, using the previous notation

$$F(c, 1_D) \circ p_D \circ p_C = p'_D \circ \lim_{D \in \mathscr{D}} F(c, 1_D) \circ p_C$$
$$= p'_D \circ p_{C'}.$$

This shows that the composites $p_D \circ p_C$ constitute a cone on the functor $F(-, D)$ from which it follows that there is a unique factorization $\lambda_D: \lim L \longrightarrow \lim_{C \in \mathscr{C}} F(C, D)$ such that $\bar{p}_C \circ \lambda_D = p_D \circ p_C$, where the $\bar{p}_C$ denote the canonical projections of the limit $\lim_{C \in \mathscr{C}} F(C, D)$. Given an arrow $d: D \longrightarrow D'$ in $\mathscr{D}$ we have also (writing $\bar{p}'_C$ for the projections of the limit $\lim_{C \in \mathscr{C}} F(C, D')$)

$$\bar{p}'_C \circ \lim_{C \in \mathscr{C}} F(1_C, d) \circ \lambda_D = F(1_C, d) \circ \bar{p}_C \circ \lambda_D$$
$$= F(1_C, d) \circ p_D \circ p_C$$
$$= p_{D'} \circ p_C$$
$$= \bar{p}'_C \circ \lambda_{D'},$$

from which $\lim_{C \in \mathscr{C}} F(1_C, d) \circ \lambda_D = \lambda_{D'}$ and we get the fact that those arrows $(\lambda_D)_{D \in \mathscr{D}}$ constitute a cone. This implies the existence of a unique factorization $\lambda: \lim L \longrightarrow \lim_{D \in \mathscr{D}} \big(\lim_{C \in \mathscr{C}} F(C, D) \big)$. Analogously we can define a canonical morphism in the other direction

$$\mu: \lim_{D \in \mathscr{D}} \big(\lim_{C \in \mathscr{C}} F(C, D) \big) \longrightarrow \lim_{C \in \mathscr{C}} \big(\lim_{D \in \mathscr{D}} F(C, D) \big).$$

The precise meaning of the interchange property for limits is the fact that λ, μ are inverse isomorphisms.

Proposition 2.12.1 *Consider a complete category $\mathscr{A}$ and two small categories $\mathscr{C}, \mathscr{D}$. Given a functor $F: \mathscr{C} \times \mathscr{D} \longrightarrow \mathscr{A}$ and using the previous notations, the following interchange property holds:*

$$\lim_{C \in \mathscr{C}} \big(\lim_{D \in \mathscr{D}} F(C, D) \big) \cong \lim_{D \in \mathscr{D}} \big(\lim_{C \in \mathscr{C}} F(C, D) \big).$$

Proof We want to prove that the two composites λ, μ in the previous discussion are inverse isomorphisms. By 2.6.4, $\mu \circ \lambda = 1$ reduces to the equalities

$$p_D \circ p_C \circ \mu \circ \lambda = p_D \circ p_C$$

which are straightforward from the definitions of λ and μ; an analogous argument holds for $\lambda \circ \mu$. $\square$

Examples 2.12.2

2.12.2.a   Consider two sets I and J viewed as discrete categories $\mathscr{I}$ and $\mathscr{J}$; a functor $F: \mathscr{I} \times \mathscr{J} \longrightarrow \mathscr{A}$ is just a family F_{ij} of objects of $\mathscr{A}$

(see example 2.6.7.a). When $\mathscr{A}$ has products, the interchange property reduces to the formula

$$\prod_{i \in I}\left(\prod_{j \in J} F_{ij}\right) \cong \prod_{j \in J}\left(\prod_{i \in I} F_{ij}\right)$$

which can be seen as an instance of the general associativity law for products (see 2.1.6); that law asserts in fact that both expressions are isomorphic to $\prod_{i,j} F_{ij}$.

2.12.2.b Consider a set I, viewed as a discrete category $\mathscr{I}$, and the category $\mathscr{K} = \{\bullet \rightrightarrows \bullet\}$ of example 2.6.7.b. A functor $F: \mathscr{I} \times \mathscr{K} \longrightarrow \mathscr{A}$ is a family $(f_i, g_i: A_i \rightrightarrows B_i)_{i \in I}$ of pairs of arrows in $\mathscr{A}$. When $\mathscr{A}$ is complete, the interchange property reduces to

$$\mathsf{Ker}\left(\prod_{i \in I} f_i, \prod_{i \in I} g_i\right) \cong \prod_{i \in I} \mathsf{Ker}\,(f_i, g_i).$$

2.13 Filtered colimits

Let us consider again a functor $F: \mathscr{C} \times \mathscr{D} \longrightarrow \mathscr{A}$ as in the previous section, and let us look this time at the mixed interchange property

$$\mathrm{colim}_{C \in \mathscr{C}}\left(\lim_{D \in \mathscr{D}} F(C, D)\right) \cong \lim_{D \in \mathscr{D}}\left(\mathrm{colim}_{C \in \mathscr{C}} F(C, D)\right)$$

when all the limits and colimits involved in this formula exist. The precise meaning of each side of the formula is obvious from the considerations of the previous section. Let us also note the existence of a canonical morphism

$$\lambda: \mathrm{colim}_{C \in \mathscr{C}}\left(\lim_{D \in \mathscr{D}} F(C, D)\right) \longrightarrow \lim_{D \in \mathscr{D}}\left(\mathrm{colim}_{C \in \mathscr{C}} F(C, D)\right).$$

The existence of λ is equivalent to the existence of a cone

$$\lambda_D: \mathrm{colim}_{C \in \mathscr{C}}\left(\lim_{D \in \mathscr{D}} F(C, D)\right) \longrightarrow \mathrm{colim}_{C \in \mathscr{C}} F(C, D)$$

and the existence of each λ_D reduces to the existence of a cocone

$$(\lambda_D)_C: \lim_{D \in \mathscr{D}} F(C, D) \longrightarrow \mathrm{colim}_{C \in \mathscr{C}} F(C, D).$$

This last arrow is just the composite

$$\lim_{D \in \mathscr{D}} F(C, D) \xrightarrow{\;p_D\;} F(C, D) \xrightarrow{\;s_C\;} \mathrm{colim}_{C \in \mathscr{C}} F(C, D)$$

where p_D is the canonical projection of the limit and s_C is the canonical injection of the colimit. Straightforward computations, perfectly analogous to that of section 2.12, prove that the $(\lambda_D)_C$ constitute a cocone in C and the λ_D constitute a cone in $\mathscr{D}$. Thus λ is well-defined and the mixed interchange property refers to the fact of λ being an isomorphism.

But the bad point about the mixed interchange property is that it does not hold in general! For example if I and J are sets viewed as discrete categories and $\mathscr{A}$ is the category of sets, the mixed interchange property reduces to

$$\coprod_{i \in I} \left(\prod_{j \in J} F_{ij} \right) \cong \prod_{j \in J} \left(\coprod_{i \in I} F_{ij} \right)$$

for a family $(F_{ij})_{(i,j) \in I \times J}$ of sets. A special instance of this formula would be

$$(A_{11} \times A_{12}) \amalg (A_{21} \times A_{22}) \cong (A_{11} \amalg A_{21}) \times (A_{12} \amalg A_{22}),$$

which is easily seen to be false just by a cardinality argument.

Now there is a very important case in which the mixed interchange property holds in the category of sets (and in many "algebraic-like" categories as we shall see later): this is the case where $\mathscr{C}$ is "filtered" and $\mathscr{D}$ is "finite". By "$\mathscr{D}$ finite" we mean clearly that $\mathscr{D}$ has just a finite number of objects and arrows.

Definition 2.13.1 *A category $\mathscr{C}$ is filtered when*

(1) $\exists C \in \mathscr{C}$ ("$\mathscr{C}$ is not empty"),

(2) $\forall C_1, C_2 \in \mathscr{C}$ $\exists C_3 \in \mathscr{C}$ $\exists f \colon C_1 \longrightarrow C_3$ $\exists g \colon C_2 \longrightarrow C_3$,

(3) $\forall C_1, C_2 \in \mathscr{C}$ $\forall f, g \colon C_1 \rightrightarrows C_2$ $\exists C_3 \in \mathscr{C}$ $\exists h \colon C_2 \longrightarrow C_3$ $h \circ f = h \circ g$.

By a "filtered colimit" we mean the colimit of a functor defined on a filtered category. We say that a category $\mathscr{A}$ "has filtered colimits" when for every small filtered category $\mathscr{C}$ and every functor $F \colon \mathscr{C} \longrightarrow \mathscr{A}$, the colimit of F exists.

First of all, let us prove a useful lemma.

Lemma 2.13.2 *Let $\mathscr{C}$ be a filtered category. For every finite category $\mathscr{D}$ and every functor $F \colon \mathscr{D} \longrightarrow \mathscr{C}$, there exists a cocone on F.*

Proof First observation: given a finite family $(C_i)_{i \in I}$ of objects of $\mathscr{C}$, it is possible to find $C \in \mathscr{C}$ and morphisms $C_i \longrightarrow C$. We prove this by induction on the cardinal of I. When I is empty, this is just condition (1) in definition 2.13.1. If the result is valid in the case of $n-1$ indices, while $I = \{i_1, \ldots, i_n\}$, choose an object C' and morphisms $C_{i_k} \longrightarrow C'$ for $k = 1, \ldots, n-1$. Applying condition 2.13.1.(2) we choose $C \in \mathscr{C}$ provided with morphisms $C' \longrightarrow C, C_{i_n} \longrightarrow C$; this answers the question.

Next, let us consider a finite family $(f_i: C \longrightarrow C')_{i \in I}$ of parallel arrows in $\mathscr{C}$. It is possible to find $C'' \in \mathscr{C}$ and $f: C' \longrightarrow C''$ such that for every pair (i, j) of indices, $f \circ f_i = f \circ f_j$. Again we prove this by induction on the number of elements of I. When I is empty, we again find condition (1) in definition 2.13.1. If the result holds for $n - 1$ indices and $I = \{i_1, \ldots, i_n\}$, choose C'' and f such that $f \circ f_i = f \circ f_j$ for all indices $i, j \leq n - 1$. We get in this way a pair of parallel arrows $(f \circ f_1, f \circ f_n): C \rightrightarrows C''$ and using condition (3) in definition 2.13.1, we choose $C''' \in \mathscr{C}$ and $f': C'' \longrightarrow C'''$ such that $f' \circ f \circ f_1 = f' \circ f \circ f_n$. The composite $f' \circ f$ is then the required morphism.

Now apply the first part of the proof to the family $(FD)_{D \in \mathscr{D}}$, getting an object $C \in \mathscr{C}$ and for every $D \in \mathscr{D}$, a morphism $f_D: FD \longrightarrow C$. For every arrow $d: D \longrightarrow D'$ in $\mathscr{D}$, we obtain a pair $(f_D, f_{D'} \circ Fd): FD \rightrightarrows C$ and using condition (3) in definition 2.13.1, we choose $C_d \in \mathscr{C}$ and $g_d: C \longrightarrow C_d$ such that $g_d \circ f_D = g_d \circ f_{D'} \circ Fd$. Using again the first part of the proof we choose an object $C' \in \mathscr{C}$ and arrows $h_d: C_d \longrightarrow C'$. We have now finitely many arrows $(h_d \circ g_d: C \longrightarrow C')_{d \in \mathscr{D}}$ and using the second part of the proof, we choose an object $C'' \in \mathscr{C}$ and a morphism $k: C' \longrightarrow C''$ such that $k \circ h_d \circ g_d = k \circ h_{d'} \circ g_{d'}$ for every pair d, d' of arrows of $\mathscr{D}$. Let us write l for this single composite from C to C''. The family $(l \circ f_D)_{D \in \mathscr{D}}$ is the required cocone on F. □

Most often, we shall apply this lemma to the inclusion of an arbitrary finite subcategory $\mathscr{D} \subseteq \mathscr{C}$.

The construction of a colimit reduces to that of two coproducts and a coequalizer (see 2.8.1), but in the category Set of sets the explicit description of a coequalizer is generally very technical since it involves the description of the equivalence relation generated by a family of pairs (see example 2.4.6.b). But in the case of filtered colimits, the corresponding equivalence relation admits a very easy description.

Proposition 2.13.3 *Consider a small filtered category $\mathscr{C}$ and a functor $F: \mathscr{C} \longrightarrow$ Set to the category of sets and mappings. The colimit $(L, (s_C)_{C \in \mathscr{C}})$ of F is given by*

$$L = \coprod_{C \in \mathscr{C}} FC / \approx, \quad s_C: FC \longrightarrow L, \quad s_C(x) = [x],$$

where $[x]$ denotes the equivalence class of x and $\approx$ is the equivalence relation defined as follows:

$$(x \in FC) \approx (x' \in FC') \text{ precisely when}$$
$$\exists C'' \in \mathscr{C} \; \exists f: C \longrightarrow C'' \; \exists g: C' \longrightarrow C'' \; Ff(x) = Fg(x').$$

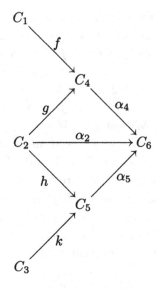

Diagram 2.23

Proof First of all let us observe that $\approx$ is indeed an equivalence relation. It is obviously reflexive and symmetric. To prove the transitivity, choose $(x \in FC_1) \approx (x' \in FC_2)$ and $(x' \in FC_2) \approx (x'' \in FC_3)$ (see diagram 2.23). This means the existence of morphisms f, g, h, k such that $Ff(x) = Fg(x')$ and $Fh(x') = Fk(x'')$. Applying lemma 2.13.2 we find a cocone $(\alpha_i \colon C_i \longrightarrow C_6)$ on the diagram constituted of the objects C_1 to C_5 and the morphisms f, g, h, k. In particular

$$F\alpha_4 \circ Ff(x) = F\alpha_4 \circ Fg(x') = F\alpha_2(x')$$
$$= F\alpha_5 \circ Fh(x') = F\alpha_5 \circ Fk(x'')$$

which proves that $x \approx x''$.

The mappings $s_C \colon FC \longrightarrow L$ do constitute a cone since for every morphism $f \colon C \longrightarrow C'$ in $\mathscr{C}$ and every element $x \in FC$, the equality $Ff(x) = F(1_{C'})(Ff(x))$ indicates that $[x] = [Ff(x)]$. Now given another cone $(t_C \colon FC \longrightarrow M)_{C \in \mathscr{C}}$ on F, we define $t \colon L \longrightarrow M$ by $t([x]) = t_C(x)$ for an element $x \in FC$. This definition is unambiguous since given another element $x' \in FC', [x'] = [x]$, we can find morphisms

$$f \colon C \longrightarrow C'', \quad g \colon C' \longrightarrow C''$$

such that $Ff(x) = Fg(x')$. In particular since t is a cone on F

$$t_C(x) = t_{C''} \circ Ff(x) = t_{C''} \circ Fg(x') = t_{C'}(x').$$

By definition, $t \circ s_C = t_C$ and this relation forces the previous definition of t. $\square$

Here is the key property of filtered colimits.

Theorem 2.13.4 *Consider a small filtered category $\mathscr{C}$ and a finite category $\mathscr{D}$. Given a functor $F: \mathscr{C} \times \mathscr{D} \longrightarrow$ Set to the category of sets and mappings, the following mixed interchange property holds:*

$$\text{colim}_{C \in \mathscr{C}} \left(\lim_{D \in \mathscr{D}} F(C,D) \right) \cong \lim_{D \in \mathscr{D}} \left(\text{colim}_{C \in \mathscr{C}} F(C,D) \right).$$

Proof By a finite category, we clearly mean a category with finitely many objects and finitely many arrows. In section 2.8, we gave a description of limits in Set and in 2.13.3, a description of filtered colimits. Using them, the morphism λ defined at the beginning of this section is given by

$$\lambda \Big(\big[(x_D)_{D \in \mathscr{D}} \big] \Big) = \big([x_D] \big)_{D \in \mathscr{D}}$$

where $x_D \in F(C,D)$. We must prove that λ is bijective.

Let us prove first that λ is injective. Consider $(x_D)_{D \in \mathscr{D}} \in F(C,D)$ and $(y_D)_{D \in \mathscr{D}} \in F(C',D)$ with the property $[x_D] = [y_D]$ for every index D. This means the existence of morphisms $f_D: C \longrightarrow C_D$ and $g_D: C' \longrightarrow C_D$ such that $F(f_D, 1_D)(x_D) = F(g_D, 1_D)(y_D)$. Applying lemma 2.13.2 to the diagram constituted of all the morphisms f_D, g_D, we get in particular two composite morphisms $f: C \longrightarrow C'''$, $g: C' \longrightarrow C'''$ such that for each index D, $F(f, 1_D)(x_D) = F(g, 1_D)(y_D)$. But this means precisely that

$$\lim_{D \in \mathscr{D}} F(f, 1_D)\big((x_D)_{D \in \mathscr{D}} \big) = \lim_{D \in \mathscr{D}} F(g, 1_D)\big((y_D)_{D \in \mathscr{D}} \big);$$

thus the equality $\big[(x_D)_{D \in \mathscr{D}} \big] = \big[(y_D)_{D \in \mathscr{D}} \big]$ holds.

Let us now prove that λ is surjective. Given a family $\big([x_D] \big)_{D \in \mathscr{D}}$ in the right-hand side, we choose a representative element $x_D \in F(C_D, D)$, for each index D. Applying lemma 2.13.2, we choose also a morphism $f_D: C_D \longrightarrow C$, for each $D \in \mathscr{D}$, and we know already that $[x_D] = \big[F(f_D, 1_D)(x_D) \big]$. Thus the elements $F(f_D, 1_D)(x_D) \in F(C,D)$ are also representatives of the original family $\big([x_D] \big)_{D \in \mathscr{D}}$. Now given an arrow $d: D \longrightarrow D'$, the elements $F(1_{C_D}, d)(x_D)$ and $x_{D'}$ are identified in the colimit, thus the equivalent elements

$$F(f_D, d)(x_D), \quad F(f_{D'}, 1_D)(x_{D'})$$

are identified as well. The existence of morphisms $g_d, h_d \colon C \rightrightarrows C_d$ such that

$$F(g_d \circ f_D, d)(x_D) = F(h_d \circ f_{D'}, 1_D)(x_{D'})$$

follows at once. Applying lemma 2.13.2 to the diagram constituted of all the morphisms g_d, h_d, we finally find a single morphism $k \colon C \longrightarrow C'$ such that

$$F(k \circ f_D, d)(x_D) = F(k \circ f_{D'}, 1_{D'})(x_{D'})$$

for all arrows d. Therefore the family $\big(F(d \circ f_D, 1_D)(x_D)\big)_{D \in \mathscr{D}}$ is an element of $\lim_{D \in \mathscr{D}} F(C, D)$ and its equivalence class is still mapped by λ to $\big([x_D]\big)_{D \in \mathscr{D}}$. $\square$

To give a flavour of why the previous result generalizes to algebraic contexts, let us handle the case of abelian groups.

Proposition 2.13.5 *The forgetful functor* $U \colon \mathsf{Ab} \longrightarrow \mathsf{Set}$ *from the category of abelian groups to the category of sets preserves and reflects filtered colimits.*

Proof Consider a small filtered category $\mathscr{C}$ and a functor $F \colon \mathscr{C} \longrightarrow \mathsf{Ab}$. Define $\big(L, (s_C)_{C \in \mathscr{C}}\big)$ to be the colimit of the composite $U \circ F$ in the category of sets. Given two elements $[x \in FC]$ and $[y \in FC']$ in L, we choose morphisms $f \colon C \longrightarrow C''$, $g \colon C' \longrightarrow C''$ and get $[x] = \big[Ff(x)\big]$, $[y] = \big[Fg(y)\big]$. In order for s_C to become a group homomorphism, we must define

$$[x] + [y] = \big[Ff(x) + Fg(y)\big].$$

Applying 2.13.2, it is now a straightforward computation to verify that this definition is independent of the various choices we have made and that L has eventually been provided with a group structure making all the s_C's group homomorphisms. The s_C's constitute a cone since the underlying mappings do.

Now given another cocone $\big(M, (t_C)_{C \in \mathscr{C}}\big)$ on F, the unique factorization $t \colon L \longrightarrow M$ which exists at the level of underlying mappings is a group homomorphism since, with the previous notation

$$
\begin{aligned}
t\big([x] + [y]\big) &= t\big([Ff(x) + Fg(y)]\big) \\
&= t_{C''}\big(Ff(x) + Fg(y)\big) \\
&= t_{C''}\big(Ff(x)\big) + t_{C''}\big(Fg(y)\big) \\
&= t\big([Ff(x)]\big) + t\big([Fg(y)]\big) \\
&= t\big([x]\big) + t\big([y]\big).
\end{aligned}
$$

Thus we have indeed obtained the colimit of F in **Ab**.

By construction of the filtered colimit in **Ab**, it is preserved by U. On the other hand the uniqueness of the group structure on L implies that filtered colimits are also reflected by U (equivalently, use the fact that U reflects isomorphisms; see 2.9.7). $\qquad\square$

Corollary 2.13.6 *In the category* **Ab** *of abelian groups, finite limits commute with filtered colimits.* $\qquad\square$

The expression "finite limits commute with filtered colimits" is just another expression for stating the mixed interchange property between finite limits and filtered colimits.

Let us observe now a result valid in every category. Roughly speaking, *an arbitrary colimit is the filtered colimit of its finitely generated partial colimits.*

Proposition 2.13.7 *Consider a functor* $F\colon \mathscr{D} \longrightarrow \mathscr{C}$, *with* $\mathscr{C}$ *finitely complete. Write* $\mathscr{F}$ *for the poset of finitely generated subcategories of* $\mathscr{D}$; $\mathscr{F}$ *is filtered. Given* $\mathscr{X} \in \mathscr{F}$, *consider the colimit* $\lambda\mathscr{X}$ *of* $F\colon \mathscr{X} \longrightarrow \mathscr{C}$; *this extends to a functor* $\lambda\colon \mathscr{F} \longrightarrow \mathscr{C}$. *This functor* λ *has a colimit if and only if* F *has a colimit and the two colimit objects coincide.*

Proof Each colimit $\big(\lambda\mathscr{X}, (\sigma_X^{\mathscr{X}})_{X\in\mathscr{X}}\big)$ exists by 2.8.5. If $\mathscr{X} \subseteq \mathscr{Y}$ are finitely generated subcategories of $\mathscr{D}$, $(\sigma_X^{\mathscr{Y}})_{X\in\mathscr{X}}$ is a cocone on $F\mathscr{X}$, from which there is a factorization $\lambda\mathscr{X} \longrightarrow \lambda\mathscr{Y}$, making $\lambda\colon \mathscr{F} \longrightarrow \mathscr{C}$ a functor. The subcategory generated by the union of two finitely generated categories is obviously finitely generated, thus $\mathscr{F}$ is filtered.

If the colimit $\big(L, (\Sigma_{\mathscr{X}})_{\mathscr{X}\in\mathscr{F}}\big)$ of λ exists, for each $D \in \mathscr{D}$ consider the one-point category $\langle D \rangle$; the colimit of $F\colon \langle D \rangle \longrightarrow \mathscr{C}$ is just $(FD, 1_{F_D})$. The morphisms $\Sigma_{\langle D \rangle}\colon FD \longrightarrow L$ constitute a cocone on F. Indeed every arrow $d\colon D \longrightarrow D'$ can be identified with a finite subcategory $\langle d \rangle \subseteq \mathscr{D}$ and the colimit of $F\colon \langle d \rangle \longrightarrow \mathscr{C}$ is just $\big(FD', (Fd, 1_{FD'})\big)$ (see 2.11.4). Since $\langle D \rangle$ and $\langle D' \rangle$ are contained in $\langle d \rangle$, with corresponding factorizations $Fd\colon \lambda\langle D \rangle \longrightarrow \lambda\langle d \rangle$, $1_{FD'}\colon \lambda\langle D' \rangle \longrightarrow \lambda\langle d \rangle$, one has

$$\Sigma_{\langle D' \rangle} \circ Fd = \Sigma_{\langle d \rangle} \circ 1_{FD'} \circ Fd = \Sigma_{\langle d \rangle} \circ Fd = \Sigma_{\langle D \rangle}.$$

Thus the $\Sigma_{\langle D \rangle}$ constitute a cocone on F. If $\gamma_D\colon FD \longrightarrow M$ is another cocone on F, for every $\mathscr{X} \in \mathscr{F}$ the $(\gamma_D)_{D\in\mathscr{X}}$ constitute a cocone on $F\mathscr{X}$, from which a factorization $\Gamma_{\mathscr{X}}\colon \lambda\mathscr{X} \longrightarrow M$. If $\mathscr{X} \subseteq \mathscr{Y}$, the relations

$$\Gamma_{\mathscr{Y}} \circ \lambda(\mathscr{X} \subseteq \mathscr{Y}) \circ \sigma_{\mathscr{X}}^X = \Gamma_{\mathscr{Y}} \circ \sigma_{\mathscr{Y}}^X = \gamma_X = \Gamma_{\mathscr{X}} \circ \sigma_{\mathscr{X}}^X$$

for each $X \in \mathscr{X}$ imply $\Gamma_{\mathscr{Y}} \circ \lambda(\mathscr{X} \subseteq \mathscr{Y}) = \Gamma_{\mathscr{X}}$. So the various $\Gamma_{\mathscr{X}}$ constitute a cocone on λ and we get the required factorization $L \longrightarrow M$.

Conversely suppose $(L, (\theta_D)_{D \in \mathscr{D}})$ is the colimit of F. Given $\mathscr{X} \in \mathscr{F}$, $(\theta_X)_{X \in \mathscr{X}}$ is a cocone on $F\mathscr{X}$ from which we get a unique factorization $\Sigma_{\mathscr{X}} \colon \lambda\mathscr{X} \longrightarrow L$ such that $\Sigma_{\mathscr{X}} \circ \sigma_{\mathscr{X}}^X = \theta_X$. Given $\mathscr{X} \subseteq \mathscr{Y}$, the relations

$$\Sigma_{\mathscr{Y}} \circ \lambda(\mathscr{X} \subseteq \mathscr{Y}) \circ \sigma_{\mathscr{X}}^X = \Sigma_{\mathscr{Y}} \circ \sigma_{\mathscr{Y}}^X = \theta_X = \Sigma_{\mathscr{X}} \circ \sigma_{\mathscr{X}}^X$$

imply $\Sigma_{\mathscr{Y}} \circ \lambda(\mathscr{X} \subseteq \mathscr{Y}) = \Sigma_{\mathscr{X}}$, from which it follows that the $(\Sigma_{\mathscr{X}})_{\mathscr{X} \in \mathscr{F}}$ constitute a cocone on λ. If $\pi_{\mathscr{X}} \colon \lambda\mathscr{X} \longrightarrow M$ is another cocone on λ, $\left(\pi_{\langle D \rangle}\right)_{D \in \mathscr{D}}$ is a cocone on F. Indeed given a morphism $d \colon D \longrightarrow D'$ in $\mathscr{D}$, one has

$$\pi_{\langle D' \rangle} \circ Fd = \pi_{\langle d \rangle} \circ 1_{FD'} \circ Fd = \pi_{\langle d \rangle} \circ Fd = \pi_{\langle D \rangle}.$$

Therefore we get the expected factorization $L \longrightarrow M$. □

Examples 2.13.8

2.13.8.a   In the category Set of sets, consider a set X and the diagram $\mathscr{D}$ constituted of the finite subsets of X and the canonical inclusions between them. This diagram is filtered since $\emptyset$ is a finite subset of X and the union of two finite subsets is finite. Notice the diagram never contains two different parallel arrows. The filtered colimit of this diagram is obviously X.

2.13.8.b   In the category Ab of abelian groups, the finitely generated subgroups of a group A and the canonical inclusions between them again constitute a filtered diagram, whose colimit is obviously the group A itself.

2.13.8.c   Consider the poset $(\mathbb{N}, \leq)$ viewed as a category (see example 1.2.6.b); it is obviously a filtered category. On the other hand consider a finite set I viewed as a discrete category (see example 1.2.6.c). A functor

$$F \colon \mathbb{N} \times I \longrightarrow \mathsf{Set}$$

is just a family of sequences

$$\left(A_{0,i} \xrightarrow{f_{0,i}} A_{1,i} \xrightarrow{f_{1,i}} A_{2,i} \xrightarrow{f_{2,i}} \ldots\right)_{i \in I}.$$

The mixed interchange property applies, showing that

$$\prod_{i \in I} \operatorname{colim}_{n \in \mathbb{N}} A_{n,i} \cong \operatorname{colim}_{n \in \mathbb{N}} \prod_{i \in I} A_{n,i}.$$

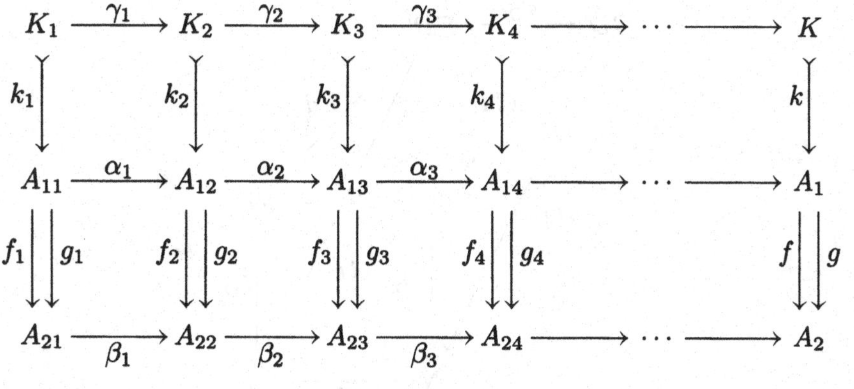

$$K_1 \xrightarrow{\gamma_1} K_2 \xrightarrow{\gamma_2} K_3 \xrightarrow{\gamma_3} K_4 \longrightarrow \cdots \longrightarrow K$$

$$k_1 \downarrow \qquad k_2 \downarrow \qquad k_3 \downarrow \qquad k_4 \downarrow \qquad \qquad k \downarrow$$

$$A_{11} \xrightarrow{\alpha_1} A_{12} \xrightarrow{\alpha_2} A_{13} \xrightarrow{\alpha_3} A_{14} \longrightarrow \cdots \longrightarrow A_1$$

$$f_1 \Big\Downarrow g_1 \qquad f_2 \Big\Downarrow g_2 \qquad f_3 \Big\Downarrow g_3 \qquad f_4 \Big\Downarrow g_4 \qquad f \Big\Downarrow g$$

$$A_{21} \xrightarrow[\beta_1]{} A_{22} \xrightarrow[\beta_2]{} A_{23} \xrightarrow[\beta_3]{} A_{24} \longrightarrow \cdots \longrightarrow A_2$$

Diagram 2.24

2.13.8.d A category $\mathscr{C}$ with a terminal object is obviously filtered, but computing the colimit of a functor defined on $\mathscr{C}$ is not very relevant (see corollary 2.11.5).

Counterexample 2.13.9

The mixed interchange property between finite limits and filtered colimits is not a general fact. In many categories, in particular when topologies are involved, it does not hold. Consider for example diagram 2.24 in the category Top of topological spaces and continuous mappings. The objects A_{nm} are defined by

$$A_{1n} = \left[0, \frac{1}{n}\right],$$

$$A_{2n} = \left(\left[0, \frac{1}{n}\right] \amalg \left[0, \frac{1}{n}\right]\right)\Big/ \approx,$$

where the equivalence relation identifies the two copies of 0 as well as the two copies of $\frac{1}{n}$. The morphisms α_n, β_n are

$$\alpha_n(x) = \begin{cases} x & \text{if } x \leq \frac{1}{n+1}, \\ \frac{1}{n+1} & \text{if } x \geq \frac{1}{n+1}, \end{cases}$$

$$\beta_n = (\alpha_n \amalg \alpha_n)\big/ \approx.$$

The morphisms f_n, g_n are the two canonical injections of A_{1n} in A_{2n}.

Let us consider the poset $(\mathbb{N}^*, \leq)$ as a category $\mathscr{N}^*$ as well as the category $\mathscr{K}$ defined in example 2.6.7.b. We have just defined a functor

$$F \colon \mathscr{N}^* \times \mathscr{K} \longrightarrow \text{Set}.$$

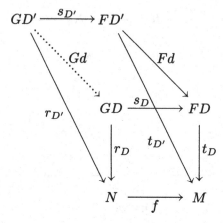

Diagram 2.25

From this we can compute $\lim_{K\in\mathscr{K}} F(n,K)$ for each n and the corresponding factorizations between those objects. This produces the family of equalizers $K_n = \mathsf{Ker}\,(f_n, g_n)$ to which the morphisms α_n restrict. It is immediate that each K_n is just the discrete two-point space, so that

$$K = \mathrm{colim}_{\,n\in\mathscr{N}^*} K_n = \mathrm{colim}_{\,n\in\mathscr{N}^*}\bigl(\lim_{K\in\mathscr{K}} F(n,K)\bigr)$$

is just the discrete two-point space. On the other hand, an easy calculation shows that $A_1 = \mathrm{colim}_{\,n\in\mathscr{N}^*} A_{1n}$ and $A_2 = \mathrm{colim}_{\,n\in\mathscr{N}^*} A_{2n}$ are both the two point space $\{0,\varepsilon\}$ provided with the topology for which ε is an open point. The families $(f_n)_{n\in\mathbb{N}}$ and $(g_n)_{n\in\mathbb{N}}$ both induce the identity as factorizations between those colimits, so that the corresponding equalizer of those factorizations is again the space $\{0,\varepsilon\}$ where ε is open (the "Sierpinski space"). But this is precisely $\lim_{K\in\mathscr{K}}\bigl(\mathrm{colim}_{\,n\in\mathscr{N}^*} F(n,K)\bigr)$, which proves that the interchange property does not hold.

2.14 Universality of colimits

This section points out another important compatibility condition between pullbacks and arbitrary colimits in the category Set of sets and mappings.

Let us consider a category $\mathscr{C}$ with pullbacks and an arbitrary functor $F\colon\mathscr{D}\longrightarrow\mathscr{C}$. Given a cocone $(t_D\colon FD\longrightarrow M)_{D\in\mathscr{D}}$ on F and a morphism $f\colon N\longrightarrow M$ in $\mathscr{C}$, we can compute the various pullbacks (GD, r_D, s_D) of t_D along f (see diagram 2.25). Moreover, given a morphism $d\colon D'\longrightarrow D$

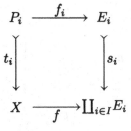

Diagram 2.26

in $\mathcal{D}$, the equalities

$$t_D \circ Fd \circ s_{D'} = t_{D'} \circ s_{D'} = f \circ r_{D'}$$

imply the existence of a unique factorization Gd making diagram 2.25 commutative. In particular, we have defined a functor $G: \mathcal{D} \longrightarrow \mathcal{C}$, a cocone $(r_D: GD \longrightarrow N)_{D \in \mathcal{D}}$ on this functor G and a natural transformation $s: G \Rightarrow F$.

Definition 2.14.1 *We use the previous notation and consider a category $\mathcal{C}$ with pullbacks and an arbitrary category $\mathcal{D}$. Given a functor $F: \mathcal{D} \longrightarrow \mathcal{C}$ with colimit $\big(M, (t_D)_{D \in \mathcal{D}}\big)$, this colimit is universal when for every morphism $f: N \longrightarrow M$ in $\mathcal{C}$, the cocone $\big(N, (r_D)_{D \in \mathcal{D}}\big)$ constructed above is a colimit of the corresponding functor G.*

Theorem 2.14.2 *In the category Set of sets, small colimits are universal.*

Proof It is an immediate consequence of the dual of theorem 2.8.1 that it suffices to prove the result separately for coproducts and coequalizers.

Let us thus consider a coproduct $\coprod_{i \in I} E_i$ of sets and a mapping

$$f: X \longrightarrow \coprod_{i \in I} E.$$

We have to compute the pullbacks of diagram 2.26. Since s_i is injective (see 2.2.4.a), so is t_i (see 2.5.3) and in fact

$$P_i = \{x \in X \,|\, f(x) \in E_i\}\,.$$

The subsets P_i of X are disjoint since the subsets E_i of $\coprod_{i \in I} E_i$ are; moreover they cover X since the E_i's cover $\coprod_{i \in I} E_i$.

For the case of a coequalizer, we refer to diagram 2.27 where $(Q, q) = \text{Coker}\,(f, g)$ and α is an arbitrary morphism. (D, p, β) is the pullback of (α, q) and (C, l, γ) is the pullback of $(\alpha, q \circ f = q \circ g)$. From $q \circ f \circ \gamma = \alpha \circ l$

Limits

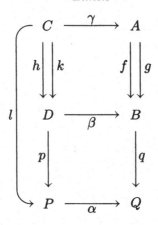

Diagram 2.27

we get the factorization h such that $\beta \circ h = f \circ \gamma$ and $p \circ h = l$ and from $q \circ g \circ \gamma = \alpha \circ l$ the factorization k such that $\beta \circ k = g \circ \gamma$ and $p \circ k = l$. By 2.5.9, (h, γ) is the pullback of β, f and (k, γ) is the pullback of β, g. We must prove that $(P, p) = \mathsf{Coker}\,(h, k)$.

First of all if $x \in P$, $\alpha(x) = [b]$ for some $b \in B$ and thus $(x, b) \in D$ with $p(x, b) = x$. Therefore p is surjective and P is the quotient of D by the equivalence relation

$$u R v \ \text{ iff } \ p(u) = p(v) \ \ (u, v \in D).$$

Since $p \circ h = p \circ k$, R contains the equivalence relation S defining the coequalizer of the pair (h, k) (see 2.4.6.b). It remains to prove that conversely, R is contained in S. So let us consider two elements $u = (x, b)$, $v = (x', b')$ of D such that $p(u) = p(v)$; this means $x = x'$. One has

$$q(b) = (q \circ \beta)(x, b) = (\alpha \circ p)(x, b) = \alpha(x),$$
$$q(b') = (q \circ \beta)(x, b') = (\alpha \circ p)(x, b') = \alpha(x),$$

so that $q(b) = q(b')$ and thus the pair (b, b') is in the equivalence relation T on B generated by the pairs $\big(f(a), g(a)\big)$, $a \in A$. Notice that given an element $a \in A$ and an element $x \in P$ such that $\alpha(x) = (q \circ f)(a) = (q \circ g)(a)$, one has $(x, a) \in C$ so that $\big(x, f(a)\big) = h(x, a)$ and $\big(x, g(a)\big) = k(x, a)$ are S-equivalent. But since the pairs $\big((x, f(a)), (x, g(a))\big)$ are in S, so is the equivalence relation generated by those pairs and in particular the pairs $\big((x, b), (x, b')\big)$ with $(b, b') \in T$. And we have just seen that this contains R. $\qquad\square$

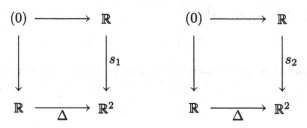

Diagram 2.28

The universality of colimits is a very peculiar property which is much less common than the mixed interchange property of section 2.13. For example, colimits are not universal in the category Ab of abelian groups. Consider indeed the pullbacks of diagram 2.28 where $\mathbb{R}$ stands for the additive group of real numbers, $s_1(r) = (r, 0)$, $s_2(r) = (0, r)$ and $\Delta(r) = (r, r)$. Both pullbacks are just the zero group. Now the coproduct $\mathbb{R} \amalg \mathbb{R}$ is just $(\mathbb{R}^2, s_1, s_2)$ (see 2.2.4.f) but $\mathbb{R}$ is not the coproduct $(0) \amalg (0)$.

2.15 Limits in categories of functors

We investigate the existence of limits in categories of functors and natural transformations.

Proposition 2.15.1 *Consider categories* $\mathscr{A}, \mathscr{C}, \mathscr{D}$, *with* $\mathscr{C}$ *and* $\mathscr{D}$ *small. Let* $F: \mathscr{D} \longrightarrow \mathsf{Fun}(\mathscr{C}, \mathscr{A})$ *be a functor, where* $\mathsf{Fun}(\mathscr{C}, \mathscr{A})$ *is the category of functors and natural transformations from* $\mathscr{C}$ *to* $\mathscr{A}$. *If for every object* $C \in \mathscr{C}$ *the functor* $F(-)(C): \mathscr{D} \longrightarrow \mathscr{A}$ *has a limit, then* F *has a limit as well and this limit is computed pointwise.*

Proof   The precise meaning of "being computed pointwise" will be explained in the proof.

Consider the small category $\mathscr{D}$ and the functor $F: \mathscr{D} \longrightarrow \mathsf{Fun}(\mathscr{C}, \mathscr{A})$. For each fixed object $C \in \mathscr{C}$ we get a functor

$$F(-)(C): \mathscr{D} \longrightarrow \mathscr{A},$$

and for each morphism $f: C \longrightarrow C'$, a natural transformation

$$F(-)(f): F(-)(C) \Rightarrow F(-)(C').$$

For each $C \in \mathscr{C}$ let us consider the limit $\left(L(C), (p_D^C)_{D \in \mathscr{D}}\right)$ of the functor $F(-)(C)$. For each morphism $f: C \longrightarrow C'$ in $\mathscr{C}$, the natural transformation $F(-)(f)$ induces a factorization $L(f): L(C) \longrightarrow L(C')$ between the limits, with the property that

$$p_D^{C'} \circ L(f) = F(D)(f) \circ p_D^{C}.$$

Straightforward computations analogous to those of section 2.12 show that

$$L: \mathscr{C} \longrightarrow \mathscr{A}$$

is a functor and $\left(p_D^C: L(C) \longrightarrow F(D)(C)\right)_{C \in \mathscr{C}}$ is a natural transformation $p_D: L \Rightarrow F(D)$. The theorem asserts that $\left(L, (p_D)_{D \in \mathscr{D}}\right)$ is the limit of F. The pointwise character of this construction is precisely expressed by the formula

$$\left(\lim_{D \in \mathscr{D}} F(D)\right)(C) = \lim_{D \in \mathscr{D}}\left(F(D)(C)\right).$$

In other words, the value of the limit $\lim_{D \in \mathscr{D}} F(D)$ at an object C is the limit of the values of $F(D)$ at C.

First of all let us observe that the p_D's constitute a cone on F. If $d: D \longrightarrow D'$ is a mapping of $\mathscr{D}$, we have for every $C \in \mathscr{C}$

$$F(d)(C) \circ p_D^C = p_{D'}^C$$

by definition of the morphism p_D^C, thus indeed $F(d) \circ p_D = p_{D'}$. Now if $\left(M, (q_D)_{D \in \mathscr{D}}\right)$ is another cone on F, each $\left(M(C), (q_{D,C})_{D \in \mathscr{D}}\right)$ is a cone on $F(-)(C)$, which yields a unique factorization $r_C: M(C) \longrightarrow L(C)$ such that $p_D^C \circ r_C = q_{D,C}$. These r_C's constitute a natural transformation $r: M \Rightarrow L$ since, given a morphism $f: C \longrightarrow C'$,

$$\begin{aligned}
p_D^{C'} \circ L(f) \circ r_C &= F(D)(f) \circ p_D^C \circ r_C \\
&= F(D)(f) \circ q_{D,C} \\
&= q_{D,C'} \circ M(f) \\
&= p_D^{C'} \circ r_{C'} \circ M(f),
\end{aligned}$$

and therefore $L(f) \circ r_C = r_{C'} \circ M(f)$ (see 2.6.4). By definition, the natural transformation $r: M \Rightarrow L$ satisfies the relation $p_D \circ r = q_D$; it is the only one with this property, due to the uniqueness condition satisfied by the arrows r_C.   □

As an immediate corollary we get

Theorem 2.15.2 *Consider a complete category $\mathscr{A}$ and a small category $\mathscr{C}$. Under these conditions, the category $\mathsf{Fun}(\mathscr{C}, \mathscr{A})$ of functors from $\mathscr{C}$ to $\mathscr{A}$ and natural transformations between them is complete and limits in it are computed pointwise.*   □

Corollary 2.15.3 *Consider a category $\mathscr{A}$ with pullbacks and a small category $\mathscr{C}$. Given two functors $F, G: \mathscr{C} \rightrightarrows \mathscr{A}$ and a natural transformation $\alpha: F \Rightarrow G$, α is a monomorphism in $\mathsf{Fun}(\mathscr{C}, \mathscr{A})$ if and only if for each object $C \in \mathscr{C}$, $\alpha_C: FC \longrightarrow GC$ is a monomorphism in $\mathscr{A}$.*

Proof   In 2.15.1, choose for $\mathscr{D}$ the category $\mathscr{P} = \{\bullet \longrightarrow \bullet \longleftarrow \bullet\}$ defining pullbacks (see 2.6.7.c). Pullbacks are thus computed pointwise in $\mathsf{Fun}(\mathscr{C}, \mathscr{A})$ and the result follows immediately from 2.5.6.   □

Corollary 2.15.4 *Consider a small category $\mathscr{C}$ and the corresponding category $\mathsf{Fun}(\mathscr{C}, \mathsf{Set})$ of Set-valued functors.*

(1) $\mathsf{Fun}(\mathscr{C}, \mathsf{Set})$ is complete and cocomplete.
(2) In $\mathsf{Fun}(\mathscr{C}, \mathsf{Set})$, finite limits commute with filtered colimits.
(3) In $\mathsf{Fun}(\mathscr{C}, \mathsf{Set})$, colimits are universal.

Proof By 2.15.1, 2.8.2, 2.13.4 and 2.14.2.   □

Proposition 2.15.5 *Consider a small category $\mathscr{C}$ and the covariant Yoneda embedding (see 1.4.3.d)*

$$Y: \mathscr{C} \longrightarrow \mathsf{Fun}(\mathscr{C}^*, \mathsf{Set}), \quad Y(C) = \mathscr{C}(-, C).$$

This functor Y preserves limits.

Proof   Consider a functor $F: \mathscr{D} \longrightarrow \mathscr{C}$ with limit $\left(L, (p_D)_{D \in \mathscr{D}}\right)$. We must prove that $\left(\mathscr{C}(-, L), (\mathscr{C}(-, p_D))_{D \in \mathscr{D}}\right)$ is the limit of $Y \circ F$. Applying proposition 2.15.1, we must prove that $\left(\mathscr{C}(C, L), (\mathscr{C}(C, p_D))_{D \in \mathscr{D}}\right)$ is the limit of the functor $\mathscr{C}(C, F-): \mathscr{C} \longrightarrow \mathsf{Set}$. This holds by proposition 2.9.4.   □

Theorem 2.15.6 *Consider a small category $\mathscr{C}$ and a functor F from $\mathscr{C}$ to the category Set of sets. In the category $\mathsf{Fun}(\mathscr{C}, \mathsf{Set})$ of functors and natural transformations, F can be presented as the colimit of a diagram just constituted of representable functors and representable natural transformations.*

Proof Let us consider the composite functor

$$\mathsf{Elts}(F) \xrightarrow{\ \phi_F\ } \mathscr{C} \xrightarrow{\ Y^*\ } \mathsf{Fun}(\mathscr{C}, \mathsf{Set}),$$

where $\mathsf{Elts}(F)$ is the category of elements of F defined in 1.6.4, ϕ_F is the corresponding forgetful functor and Y^* is the Yoneda embedding referred to in example 1.4.3.a. We shall prove that F is exactly the (object part of the) colimit of $Y^* \circ \phi_F$. This composite is contravariant, so that the reader could prefer replacing $\mathscr{C}$ and $\mathsf{Elts}(F)$ by the dual categories; but this just makes notations heavier.

An object in $\mathsf{Elts}(F)$ is a pair (A, a) where $A \in \mathscr{C}$ and $a \in FA$; by the Yoneda lemma (see 1.3.3) this corresponds to a natural transformation $s_{(A,a)} \colon \mathscr{C}(A, -) \Rightarrow F$. If $f \colon (A, a) \longrightarrow (B, b)$ is a morphism of $\mathsf{Elts}(F)$, we have $F(f)(a) = b$ and by the naturality in A of the Yoneda isomorphisms (see 1.3.3), this is equivalent to the relation $s_{(A,a)} \circ \mathscr{C}(f, -) = s_{(B,b)}$. Therefore the family $\big(s_{(A,a)}\big)_{(A,a)\in\mathrm{ELTS}(F)}$ is a cocone on the functor $Y^* \circ \phi_F$.

Choose another cocone $\big(G, t_{(A,a)}\big)_{(A,a)\in\mathrm{ELTS}(F)}$ on the same functor $Y^* \circ \phi_F$. We want to produce first a natural transformation $\alpha \colon F \Rightarrow G$. For each object $C \in \mathscr{C}$ we must define a mapping $\alpha_C \colon FC \longrightarrow GC$. Given an element $x \in FC$, we consider the corresponding object (C, x) in $\mathsf{Elts}(F)$; the natural transformation $t_{(C,x)} \colon \mathscr{C}(C, -) \Rightarrow G$ corresponds by the Yoneda lemma to a unique element of GC, which we define to be $\alpha_C(x)$. To prove the naturality of α, we choose a morphism $g \colon C \longrightarrow D$ in $\mathscr{C}$, which yields a morphism $g \colon (C, x) \longrightarrow (D, Fg(x))$ in $\mathsf{Elts}(F)$. Since the arrows $t_{(A,a)}$ constitute a cocone on $Y^* \circ \phi_F$, we have $t_{(C,x)} \circ \mathscr{C}(g, -) = t_{(D,Fg(x))}$ which, again by naturality of the Yoneda isomorphisms, implies $G(g)\big(\alpha_C(x)\big) = \alpha_D\big(Fg(x)\big)$. This expresses precisely the naturality of α.

Given (C, x) in $\mathsf{Elts}(F)$, we must prove that $\alpha \circ s_{(C,x)} = t_{(C,x)}$. Via the Yoneda isomorphisms, both sides indeed correspond to $\alpha_C(x)$. Moreover, if $\beta \colon F \Rightarrow G$ is another natural transformation satisfying $\beta \circ s_{(C,x)} = t_{(C,x)}$, applying the Yoneda isomorphisms to both sides yields $\beta_C(x) = \alpha_C(x)$, from which follows the uniqueness of α. $\qquad\Box$

Examples 2.15.7

2.15.7.a   Consider a group G (written multiplicatively) and the corresponding category of G-sets. A G-set is a pair $(E, \bullet)$ where E is a set and $\bullet$ is an action

$$\bullet \colon E \times G \longrightarrow E \ , \quad (e, g) \mapsto eg,$$

satisfying the axioms

$$e1 = e, \quad (eg)g' = e(gg'), \quad \text{for all } e \in E, \ g, g' \in G.$$

A morphism $f \colon (E, \bullet) \longrightarrow (F, \bullet)$ of G-sets is a mapping $f \colon E \longrightarrow F$ satisfying the axiom

$$f(eg) = f(e)g, \quad \text{for all } e \in E, \ g \in G.$$

We can view G as a category $\mathscr{G}$ with a single object $*$ and arrows $\mathscr{G}(*,*) = G$, the composition being given by the multiplication of G. The category of G-sets and corresponding homomorphisms is exactly the category $\mathsf{Fun}(G^*, \mathsf{Set})$ of contravariant functors and natural transformations from G to Set. Indeed giving an action $E \times G \longrightarrow E$ is just the same as giving the various multiplications

$$E \longrightarrow E, \quad e \mapsto eg,$$

for each individual element $g \in G$.

From 2.15.3 we deduce that the category of G-sets and their homomorphisms is complete and cocomplete; finite limits of G-sets commute with filtered colimits and colimits of G-sets are universal.

Let us observe that the unique representable functor $G \longrightarrow \mathsf{Set}$ corresponds exactly to the G-set $(G, \bullet)$ where the scalar multiplication is just the multiplication of the group. Each G-set can thus be presented as the colimit of a diagram involving simply the basic G-set $(G, \bullet)$ (see 2.15.4).

2.15.7.b A category $\mathsf{Fun}(\mathscr{C}, \mathscr{A})$ of functors can be complete even when $\mathscr{A}$ is not. An obvious example is obtained by taking $\mathscr{A}$ and $\mathscr{C}$ to be empty: $\mathscr{A}$ is not complete or cocomplete, since it does not have a terminal or an initial object. But $\mathsf{Fun}(\mathscr{C}, \mathscr{A})$ is the category with just one single object (the empty functor) and the identity on it; that category is obviously both complete and cocomplete. And since $\mathscr{C}$ doesn't have any object ... limits in $\mathsf{Fun}(\mathscr{C}, \mathscr{A})$ are still pointwise! See exercise 2.17.10 for a non-pointwise limit.

2.16 Limits in comma categories

Comma categories were introduced in section 1.6.

Proposition 2.16.1 *Consider two complete categories $\mathscr{A}, \mathscr{B}$ and two limit preserving functors $F: \mathscr{A} \longrightarrow \mathscr{C}$, $G: \mathscr{B} \longrightarrow \mathscr{C}$. The comma category (F, G) is then complete and the projection functors $U: (F, G) \longrightarrow \mathscr{A}$ and $V: (F, G) \longrightarrow \mathscr{B}$ are limit preserving.*

Proof Given a small category $\mathscr{D}$ and a functor $H: \mathscr{D} \longrightarrow (F, G)$, consider the limit $\big(L, (p_D)_{D \in \mathscr{D}}\big)$ of $U \circ H$ and the limit $\big(M, (q_D)_{D \in \mathscr{D}}\big)$ of $V \circ H$. Our assumptions imply that

$$\lim FUH = \big(FL, (Fp_D)_{D \in \mathscr{D}}\big),$$
$$\lim GVH = \big(GM, (Gq_D)_{D \in \mathscr{D}}\big).$$

On the other hand each object HD has the form

$$HD = (UHD, \alpha_D, VHD)$$

where $\alpha: FU \Rightarrow GV$ was defined in 1.6.2. Considering the natural transformation $\alpha * H: FUH \Rightarrow GVH$ (see 1.3.4), we deduce the existence of a corresponding factorization $h: FL \longrightarrow GM$ between the two limits. It is now straightforward to check that (L, h, M) together with the projections

$$(p_D, q_D): (L, h, M) \longrightarrow (UHD, \alpha_D, VHD)$$

is the limit of H. $\square$

Corollary 2.16.2 *If $\mathscr{C}$ is a complete category and $F: \mathscr{C} \longrightarrow \mathsf{Set}$ is a limit preserving functor, the category $\mathsf{Elts}(F)$ is complete and the forgetful functor $\phi_F: \mathsf{Elts}(F) \longrightarrow \mathscr{C}$ is limit preserving.*

Proof   With 1.6.4 in mind and using its notation, apply the previous result to $1: 1 \longrightarrow \mathsf{Set}$ and $F: \mathscr{C} \longrightarrow \mathsf{Set}$. $\square$

Another interesting example of a comma category is the category $\mathscr{C}/I$, for some fixed object $I \in \mathscr{C}$. Indeed considering the one-point category $\mathbf{1}$ (see 1.6.4) and the functor $\Delta_I: \mathbf{1} \longrightarrow \mathscr{C}$; $\Delta_I(*) = I$, $\mathscr{C}/I$ is just the comma category $(1_{\mathscr{C}}, \Delta_I)$ where $1_{\mathscr{C}}$ is the identity functor on $\mathscr{C}$. It should be noticed that Δ_I does not, in general, preserve limits or colimits. Indeed in $\mathbf{1}$ one has $* \times * = *$ and $* \amalg * = *$, but generally $I \times I \ncong I$ and $I \amalg I \ncong I$. Nevertheless we have the following result.

Proposition 2.16.3 *Consider a category $\mathscr{C}$ and a fixed object $I \in \mathscr{C}$.*

(1) If $\mathscr{C}$ is complete, $\mathscr{C}/I$ is complete.
(2) If $\mathscr{C}$ is cocomplete, $\mathscr{C}/I$ is cocomplete.

Proof   Let us first assume $\mathscr{C}$ is complete. Consider a non-empty family of objects $(f_k: C_k \longrightarrow I)_{k \in K}$ in the category $\mathscr{C}/I$. In $\mathscr{C}$, the diagram constituted by all those morphisms f_k has a limit given by an object L and morphisms $p_k: L \longrightarrow C_k$, $p: L \longrightarrow I$. It is an obvious matter to check that

$$\big((L, p), (p_k)_{k \in K}\big)$$

is the product of the original family in $\mathscr{C}/I$. On the other hand the empty product in $\mathscr{C}/I$, i.e. the terminal object, is just the identity on I.

Now consider two objects $f: C \longrightarrow I$ and $g: D \longrightarrow I$ of $\mathscr{C}/I$. The equalizer of two morphisms $(\alpha, \beta): (C, f) \rightrightarrows (D, g)$ in $\mathscr{C}/I$ is just the

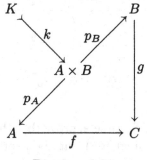

Diagram 2.29

pair $\big((K, f \circ k), k\big)$ where (K, k) is the equalizer of (α, β) in $\mathscr{C}$. By 2.8.1, we conclude that $\mathscr{C}/I$ is complete.

If $\mathscr{C}$ is cocomplete, $\mathscr{D}$ is a small category and $F: \mathscr{D} \longrightarrow \mathscr{C}/I$ is a functor, each FD can be written as a pair (GD, γ_D) where $\gamma_D: D \longrightarrow I$. This immediately induces a functor $G: \mathscr{D} \longrightarrow \mathscr{C}$ whose colimit will be written $(L, (s_D)_{D \in \mathscr{D}})$. Since F takes values in $\mathscr{C}/I$, $(I, (\gamma_D)_{D \in \mathscr{D}})$ is another cocone on G, from which we get a unique factorization $\lambda: L \longrightarrow I$ with the property $\lambda \circ s_D = \gamma_D$. It is immediate that

$$\big((L, \lambda), (s_D)_{D \in \mathscr{D}}\big)$$

is the colimit of F in $\mathscr{C}/I$. □

2.17 Exercises

2.17.1 Consider a category $\mathscr{C}$ with binary products and equalizers. Given two morphisms f, g as in diagram 2.29, prove that the pullback of (f, g) is the equalizer of the pair $(f \circ p_A, g \circ p_B)$.

2.17.2 Consider a functor $F: \mathscr{D} \longrightarrow \mathscr{C}$ and the category of cones on F: its objects are the cones $(M, (r_D)_{D \in \mathscr{D}})$ on F; an arrow between the cones $(M, (r_D)_{D \in \mathscr{D}})$ and $(N, (s_D)_{D \in \mathscr{D}})$ is a morphism $f: M \longrightarrow N$ such that $s_D \circ f = r_D$ for each D. Prove that F has a limit if and only if the functor U from the category of cones on F to the category $\mathscr{C}$, mapping a cone to its vertex, has a colimit.

2.17.3 Consider a functor $F: \mathscr{D} \longrightarrow \mathscr{C}$ and an object $C \in \mathscr{C}$. Write

$$\Delta_C: \mathscr{D} \longrightarrow \mathscr{C}$$

for the constant functor on C (see 1.2.8.e). Prove that a cone on F is just a natural transformation $\Delta_C \Rightarrow F$.

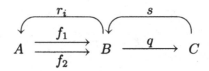

Diagram 2.30

2.17.4 In the category of finite sets, prove that the colimit of the identity functor exists (it is the singleton) but the coproducts which would be needed to construct it via theorem 2.8.1 do not exist.

2.17.5 Consider a category $\mathscr{C}$ with products and a functor $F\colon \mathscr{D} \longrightarrow \mathscr{C}$ where $\mathscr{D}$ is small. Construct α and β as in 2.8.1. Prove that $\mathrm{Ker}\,(\alpha, \beta)$ exists if and only if $\lim F$ exists.

2.17.6 Consider a category $\mathscr{A}$ with products and a functor $F\colon \mathscr{A} \longrightarrow \mathscr{B}$ which preserves products and equalizers. Show that F preserves limits. [Hint: have a look at 2.17.5].

2.17.7 In diagram 2.30, show that (C, q) is the absolute coequalizer of (f_1, f_2) when there exist morphisms s and r_i $(i = 1, \ldots, n)$ such that

$$q \circ f = q \circ g$$
$$q \circ s = 1_C$$
$$s \circ q = f_{i_1} \circ r_1$$
$$f_{i_2} \circ r_1 = f_{i_3} \circ r_2$$
$$\vdots$$
$$f_{i_{2n}} \circ r_n = 1_B$$

where $i_k = 0, 1$. [Hint: if the coequalizer is absolute, apply $\mathscr{C}(C, -)$ to get the existence of s such that $q \circ s = 1_C$; then apply $\mathscr{C}(B, -)$ to get the sequence of r_i's connecting $s \circ q$ and 1_B.]

2.17.8 Prove that a functor $G\colon \mathscr{C} \longrightarrow \mathscr{D}$ is final as long as for each object $D \in \mathscr{D}$, the comma category (Δ_D, G) is connected, where $\Delta_D\colon 1 \longrightarrow \mathscr{D}$ is the (constant) functor on D (see 2.6.7.e for the definition of a connected category). Show that the assumptions of 2.11.2 are stronger than those of the present exercise.

2.17.9 Consider the category $\mathscr{A}$ with two objects $0, 1$ and one single non-identity arrow $0 \longrightarrow 1$. Choose as category $\mathscr{B}$ the poset of diagram 2.31. In $\mathsf{Fun}(\mathscr{A}, \mathscr{B})$, consider the two functors F, G defined by $F(0) = c, F(1) = f$, $G(0) = d, G(1) = g$. Show that their product is

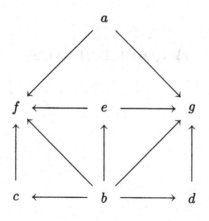

Diagram 2.31

the functor H described by $H(0) = b$, $H(1) = e$. This product is not pointwise since $f \times g$ does not exist in $\mathcal{B}$.

2.17.10 In the situation of 2.8.1, suppose the two products involved in the proof exist as well as the limit $\big(L, (p_D)_{D \in \mathcal{D}}\big)$ of F. This immediately implies the existence of a morphism l such that $p'_D \circ l = p_D$ for each $D \in \mathcal{D}$. Prove that l is the equalizer of α, β.

3

Adjoint functors

3.1 Reflection along a functor

All of us know that considering the monoid $(\mathbb{N}, +)$ of natural numbers, the "best" abelian group associated with it is that of integers, $(\mathbb{Z}, +)$. In fact there exists an obvious forgetful functor

$$U \colon \mathsf{Ab} \longrightarrow \mathsf{Mon} \; ; \; (A, +) \mapsto (A, +)$$

from the category of abelian groups to that of abelian monoids. Given an abelian monoid M, we are looking for a "best" abelian group A such that M can be embedded in UA as a submonoid.

The previous example is in fact somehow misleading, in the sense that it insists too much on "embedding" a monoid in a group. Let us consider a completely different example. There is an obvious embedding functor

$$U \colon \mathsf{Haus} \longrightarrow \mathsf{Top}, \; (X, \mathcal{T}) \mapsto (X, \mathcal{T})$$

from the category of Hausdorff topological spaces and continuous mappings to that of all topological spaces and continuous mappings. A space $(X, \mathcal{T})$ is Hausdorff when the diagonal $\Delta_X \subseteq X \times X$ is closed. Therefore there is a "best" Hausdorff space $(Y, \mathcal{S})$ associated with a topological space $(X, \mathcal{T})$: it is just the quotient of $(X, \mathcal{T})$ by the closure of the diagonal $\overline{\Delta_X} \subseteq X \times X$, which is indeed an equivalence relation on X. This time $(Y, \mathcal{S})$ appears as the "best" Hausdorff quotient of X.

More generally, given an arbitrary functor

$$U \colon \mathscr{A} \longrightarrow \mathscr{B}, \; A \mapsto U(A)$$

between two categories, we can look at the "best" object of $\mathscr{A}$ associated with a given object $B \in \mathscr{B}$. This "best" object $R_B \in \mathscr{A}$ should thus be provided with a "canonical" morphism $\eta_B \colon B \longrightarrow U(R_B)$, which in some cases can be a monomorphism or an epimorphism, but in general is just an arbitrary morphism. But what do "best" and "canonical" mean? Well, just like a limit is the "best" cone associated with a functor, we shall require that any other possibility $B \longrightarrow U(A)$ factors uniquely through the "canonical" choice (R_B, η_B).

Definition 3.1.1 Let $F: \mathcal{A} \longrightarrow \mathcal{B}$ be a functor and B an object of $\mathcal{B}$. A reflection of B along F is a pair (R_B, η_B) where

(1) R_B is an object of $\mathcal{A}$ and $\eta_B: B \longrightarrow F(R_B)$ is a morphism of $\mathcal{B}$,

(2) if $A \in |\mathcal{A}|$ is an object of $\mathcal{A}$ and $b: B \longrightarrow F(A)$ is a morphism of $\mathcal{B}$, there exists a unique morphism $a: R_B \longrightarrow A$ in $\mathcal{A}$ such that $F(a) \circ \eta_B = b$.

Proposition 3.1.2 Let $F: \mathcal{A} \longrightarrow \mathcal{B}$ be a functor and B an object of $\mathcal{B}$. When the reflection of B along F exists, it is unique up to isomorphism.

Proof Consider two reflections (R_B, η_B) and (R'_B, η'_B) of B. By definition, we find morphisms $a: R_B \longrightarrow R'_B$ and $a': R'_B \longrightarrow R_B$ such that $F(a) \circ \eta_B = \eta'_B$ and $F(a') \circ \eta'_B = \eta_B$. From this we deduce immediately that

$$F(a \circ a') \circ \eta'_B = F(a) \circ \eta_B = \eta'_B = F(1_{R'_B}) \circ \eta'_B$$

and, by uniqueness of the factorization, $a \circ a' = 1_{R'_B}$. In an analogous way, we get $a' \circ a = 1_{R_B}$. □

Proposition 3.1.3 Consider a functor $F: \mathcal{A} \longrightarrow \mathcal{B}$ and assume that, for every object $B \in \mathcal{B}$, "the" reflection of B along F exists and such a reflection (R_B, η_B) has been chosen. In that case, there exists a unique functor $R: \mathcal{B} \longrightarrow \mathcal{A}$ satisfying the two properties

(1) $\forall B \in \mathcal{B}$ $R(B) = R_B$,

(2) $(\eta_B: B \longrightarrow FRB)_{B \in \mathcal{B}}$ is a natural transformation.

Proof Considering $b: B \longrightarrow B'$, a morphism of $\mathcal{B}$, the reflection (RB, η_B) of B along F and the pair $(RB', \eta_{B'})$, we deduce the existence of a unique morphism $a: RB \longrightarrow RB'$ such that the right-hand square of diagram 3.1 commutes. We put $R(b) = a$ and it remains to prove that R is a functor.

Consider another morphism $b': B' \longrightarrow B''$ in $\mathcal{B}$. The equalities

$$F(Rb' \circ Rb) \circ \eta_B = FRb' \circ FRb \circ \eta_B = FRb' \circ \eta_{B'} \circ b = \eta_{B''} \circ b' \circ b,$$

$$FR(b' \circ b) \circ \eta_B = \eta_{B''} \circ b' \circ b,$$

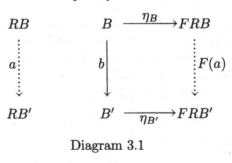

Diagram 3.1

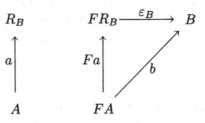

Diagram 3.2

indicate that $Rb' \circ Rb = R(b' \circ b)$, by uniqueness of the factorization. In the same way one proves that F preserves identities. ☐

Definition 3.1.4 *A functor* $R: \mathcal{B} \longrightarrow \mathcal{A}$ *is left adjoint to the functor* $F: \mathcal{A} \longrightarrow \mathcal{B}$ *when there exists a natural transformation* $\eta: 1_{\mathcal{A}} \Rightarrow F \circ R$ *such that for every* $B \in \mathcal{B}$, (RB, η_B) *is a reflection of* B *along* F.

It is an immediate consequence of 3.1.2 that in the situation of 3.1.4, both R and η are defined uniquely up to isomorphism. On the other hand if you allow in your underlying set theory a sufficiently powerful axiom of choice, you can even conclude that a functor $F: \mathcal{A} \longrightarrow \mathcal{B}$ has a left adjoint if and only if each object of $\mathcal{B}$ admits a reflection along F (for each $B \in \mathcal{B}$ choose such a reflection and apply 3.1.3).

The dual notion of "reflection along a functor" is that of "coreflection along a functor $F: \mathcal{A} \longrightarrow \mathcal{B}$"; let us write it explicitly. A coreflection of $B \in \mathcal{B}$ is a pair (R_B, ε_B) where $\varepsilon_B: FR_B \longrightarrow B$ and for every pair (A, b) with $A \in \mathcal{A}$ and $b: FA \longrightarrow B$, there exists a unique morphism $a: A \longrightarrow R_B$ such that $\varepsilon_B \circ F(a) = b$ (see diagram 3.2). In an analogous way a functor $R: \mathcal{B} \longrightarrow \mathcal{A}$ is right adjoint to F when there exists a natural transformation $\varepsilon: F \circ R \Rightarrow 1_{\mathcal{A}}$ such that for each $B \in \mathcal{B}$, (RB, ε_B) is a coreflection of B along F.

We know that an adjoint functor is only defined up to isomorphism. So, in theorem 3.1.5, let us fix a particular functor G left adjoint to F.

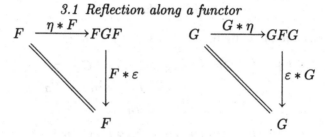

Diagram 3.3

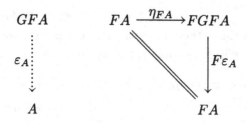

Diagram 3.4

This does not yet fix the natural transformations ε, η of condition (2): these are only determined up to isomorphism (see 3.1.2) even when G has been fixed. See exercise 3.9.1 for an example.

Theorem 3.1.5 *Consider two functors $F: \mathscr{A} \longrightarrow \mathscr{B}$ and $G: \mathscr{B} \longrightarrow \mathscr{A}$. The following conditions are equivalent:*

(1) G is left adjoint to F;

(2) there exist natural transformations $\eta: 1_{\mathscr{B}} \Rightarrow F \circ G$ and $\varepsilon: G \circ F \Rightarrow 1_{\mathscr{A}}$ such that

$$(F * \varepsilon) \circ (\eta * F) = 1_F, \quad (\varepsilon * G) \circ (G * \eta) = 1_G$$

(see diagram 3.3);

(3) there exist bijections

$$\theta_{AB}: \mathscr{A}(GB, A) \cong \mathscr{B}(B, FA)$$

for every object $A \in \mathscr{A}$, $B \in \mathscr{B}$ and those bijections are natural both in A and in B;

(4) F is right adjoint to G.

Proof (1) $\Rightarrow$ (2). The existence of the natural transformation η is just part of the definition of left adjointness. Now consider the reflection (GFA, η_A) of $FA \in \mathscr{B}$ as in diagram 3.4; the identity on FA gives rise to a unique factorization ε_A and we have already $F\varepsilon_A \circ \eta_{FA} = 1_{FA}$. We must prove that $\varepsilon: GF \Rightarrow 1_{\mathscr{A}}$ is natural. Choosing a morphism

$a\colon A \longrightarrow A'$ we have, by naturality of $\eta\colon 1_{\mathscr{A}} \Rightarrow FG$,

$$F(\varepsilon_{A'} \circ GFa) \circ \eta_{FA} = F\varepsilon_{A'} \circ FGFa \circ \eta_{FA}$$
$$= F\varepsilon_{A'} \circ \eta_{FA'} \circ Fa$$
$$= Fa,$$
$$F(a \circ \varepsilon_A) \circ \eta_{FA} = Fa \circ F\varepsilon_A \circ \eta_{FA}$$
$$= Fa,$$

from which $\varepsilon_{A'} \circ GFa = a \circ \varepsilon_A$, by uniqueness of the factorization.

It remains to check the commutativity of the second triangle. Given $B \in \mathscr{B}$ and the reflection (RB, η_B) of B, we have

$$F(\varepsilon_{GB} \circ G\eta_B) \circ \eta_B = F\varepsilon_{GB} \circ FG\eta_B \circ \eta_B$$
$$= F\varepsilon_{GB} \circ \eta_{FGB} \circ \eta_B$$
$$= \eta_B$$
$$= F(1_{GB}) \circ \eta_B,$$

from which $\varepsilon_{GB} \circ G\eta_B = 1_{GB}$ by uniqueness of the factorization.

$(2) \Rightarrow (3)$. Given a morphism $a\colon GB \longrightarrow A$, we define $\theta_{A,B}(a)$ as the composite

$$Fa \circ \eta_B \colon B \longrightarrow FGB \longrightarrow FA.$$

Conversely, given a morphism $b\colon B \longrightarrow FA$, we define $\tau_{A,B}(b)$ as the composite

$$\varepsilon_A \circ Gb \colon GB \longrightarrow GFA \longrightarrow A.$$

It is an immediate consequence of the triangular equalities in (2) and the naturality of η, ε that $\theta_{A,B}$ and $\tau_{A,B}$ are mutual inverses. For example

$$(\tau_{A,B} \circ \theta_{A,B})(a) = \tau_{A,B}(Fa \circ \eta_B)$$
$$= \varepsilon_A \circ G(Fa \circ \eta_B)$$
$$= \varepsilon_A \circ GFa \circ G\eta_B$$
$$= a \circ \varepsilon_{GB} \circ G\eta_B$$
$$= a.$$

To prove the naturality in A, choose a morphism $f\colon A \longrightarrow A'$ in $\mathscr{A}$. One has

$$(\mathscr{B}(B, Ff) \circ \theta_{A,B})(a) = (Ff \circ \theta_{AB})(a) = Ff \circ Fa \circ \eta_B,$$
$$(\theta_{A',B} \circ \mathscr{A}(GB, f))(a) = \theta_{A',B}(f \circ a) = F(f \circ a) \circ \eta_B,$$

$$\mathscr{A}(GB,GB)\xrightarrow[\cong]{\theta_{GB,B}}\mathscr{B}(B,FGB)$$

$$\mathscr{A}(GB,a)\Big\downarrow \qquad\qquad \Big\downarrow\mathscr{B}(B,Fa)$$

$$\mathscr{A}(GB,A)\xrightarrow[\theta_{A,B}]{\cong}\mathscr{B}(B,FA)$$

Diagram 3.5

from which follows the equality of those terms. In an analogous way, one proves the naturality in B.

(3) $\Rightarrow$ (1). We prove that given $B \in \mathscr{B}$, $(GB, \theta_{GB,B}(1_{GB}))$ is a reflection of B along F. Given $A \in \mathscr{A}$ and $b\colon B \longrightarrow FA$, b has the form $\theta_{A,B}(a)$ for a unique morphism $a\colon GB \longrightarrow A$. The naturality of the bijections $\theta_{A,B}$ (see diagram 3.5) implies

$$\begin{aligned}
(Fa \circ \theta_{GB,B})(1_{GB}) &= (\mathscr{B}(B,Fa) \circ \theta_{GB,B})(1_{GB})\\
&= (\theta_{A,B} \circ \mathscr{A}(GB,a))(1_{GB})\\
&= \theta_{A,B}(a)\\
&= b.
\end{aligned}$$

On the other hand if $a'\colon GB \longrightarrow A$ is another morphism of $\mathscr{A}$ such that $Fa' \circ \theta_{GB,B}(1_{GB}) = b$, we have

$$\begin{aligned}
\theta_{A,B}(a') &= (\theta_{A,B} \circ \mathscr{A}(GB,a'))(1_{GB})\\
&= (\mathscr{B}(B,Fa') \circ \theta_{GB,B})(1_{GB})\\
&= (Fa' \circ \theta_{GB,B})(1_{GB})\\
&= b\\
&= \theta_{AB}(a),
\end{aligned}$$

from which $a = a'$, since $\theta_{A,B}$ is a bijection.

(4) $\Leftrightarrow$ (3). Working with the dual categories, we have to prove that F^* is left adjoint to G^* if and only if the following natural bijections hold:

$$\mathscr{A}^*(A, G^*B) \cong \mathscr{B}^*(F^*A, B)$$

which is true via the equivalence (1) $\Leftrightarrow$ (3). $\qquad\square$

The equivalence (1) $\Leftrightarrow$ (4) in theorem 3.1.5 shows in particular the autoduality of the notion of adjoint functors. We shall write $G \dashv F$ to indicate the fact that G is left adjoint to F and thus F is right adjoint to G.

Examples 3.1.6

3.1.6.a Consider the category Mon of monoids and monoid homomorphisms; the "underlying set" functor $U\colon \mathsf{Mon} \longrightarrow \mathsf{Set}$ has a left adjoint functor

$$F\colon \mathsf{Set} \longrightarrow \mathsf{Mon}.$$

For a given set X, FX is just the free monoid on X (the monoid of finite sequences of elements of X, where composition is just concatenation); given a mapping $f\colon X \longrightarrow Y$ between sets, $Ff\colon FX \longrightarrow FY$ is obtained by applying f to each element of a finite sequence in X. There is an obvious mapping $\eta_X\colon X \longrightarrow UFX$ applying x on the sequence (x) with a single element. Given a monoid $(M, *)$ and a mapping $g\colon X \longrightarrow M$, we get the required unique factorization $h\colon FX \longrightarrow (M, *)$ by defining $h(x_1, \dots, x_n) = g(x_1) * \dots * g(x_n)$; the empty sequence is just mapped to the unit element of M.

3.1.6.b Consider the category Gr of groups and group homomorphisms; the "underlying set" functor $U\colon \mathsf{Gr} \longrightarrow \mathsf{Set}$ has a left adjoint functor

$$F\colon \mathsf{Set} \longrightarrow \mathsf{Gr}.$$

For a given set X, we consider first the free monoid M constructed on $X \amalg X$; for clarity, we write x^+ and x^- for the two copies in $X \amalg X$ of an element $x \in X$. The free group FX on X is the quotient of M by the equivalence relation generated by $(x^+, x^-) \approx (\ )$, where $(\ )$ denotes the empty sequence; the composition is induced by that of M. We define a mapping from X to FX by sending the element $x \in X$ onto $[(x^+)] \in FX$, the equivalence class of the sequence consisting only of x^+. It is straightforward to verify that we have got a reflection of X along U.

3.1.6.c Consider the category Ab of abelian groups and group homomorphisms; the "underlying set" functor $U\colon \mathsf{Ab} \longrightarrow \mathsf{Set}$ has a left adjoint functor

$$F\colon \mathsf{Set} \longrightarrow \mathsf{Ab}.$$

For a given set X, just put

$$FX = \bigl\{ (z_x)_{x \in X} \,\big|\, z_x \in \mathbb{Z};\ \{x \mid z_x \neq 0\} \text{ is finite } \bigr\}.$$

In other words FX is just the coproduct of X copies of the abelian group $(\mathbb{Z}, +)$ (see 2.2.4.f). The canonical morphism $X \longrightarrow FX$ maps the element x_0 to the sequence $(z_x)_{x \in X}$ where $z_{x_0} = 1$ and the other components are just 0. Given an abelian group $(A, +)$ and a mapping $f: X \longrightarrow A$, the required unique factorization $g: FX \longrightarrow (A, +)$ is given by $g((z_x)_{x \in X}) = \sum_{x \in X} z_x f(x)$, which makes sense since only finitely many z_x's are non-zero.

3.1.6.d Consider the category Rng of commutative rings with a unit; the "underlying set" functor $U: \mathsf{Rng} \longrightarrow \mathsf{Set}$ has a left adjoint functor

$$F: \mathsf{Set} \longrightarrow \mathsf{Rng}.$$

For a given set X, FX is just the ring of polynomials $\mathbb{Z}[x_1, \ldots, x_n, \ldots]$ where the x_i's are the various elements of X. The canonical mapping $X \longrightarrow FX$ maps an element x to the polynomial x. Given another ring $(A, +, \times)$ and a mapping $f: X \longrightarrow A$, the required factorization $FX \longrightarrow (A, +, \times)$ is given by $g(p) = p(f(x_1), \ldots, f(x_n), \ldots)$ for every polynomial p.

3.1.6.e Consider two commutative rings with unit R, S and a ring homomorphism $f: R \longrightarrow S$. Each S-module M can be seen as an R-module, via the multiplication $r \cdot m = f(r) \cdot m$, where the left-hand side is thus an R-scalar multiplication and the right-hand side is an S-scalar multiplication. This induces immediately a functor $U: \mathsf{Mod}_S \longrightarrow \mathsf{Mod}_R$, called the "extension of scalars". The functor U has both a left and a right adjoint functor. The left adjoint functor is given by

$$\mathsf{Mod}_R \longrightarrow \mathsf{Mod}_S, \quad N \mapsto S \otimes_R N,$$

and the right adjoint functor is given by

$$\mathsf{Mod}_R \longrightarrow \mathsf{Mod}_S, \quad N \mapsto \mathsf{Lin}_R(S, N),$$

with obvious definitions on the morphisms. Now $S \otimes_R N$ is an S-module via the multiplication $s(s' \otimes n) = (ss') \otimes n$; on the other hand the multiplication on $\mathsf{Lin}_R(S, N)$ is given by $(sf)(s') = f(ss')$. The following isomorphisms are well-known to hold:

$$\mathsf{Lin}_S(S \otimes_R N, M) \cong \mathsf{Lin}_R(N, UM),$$
$$\mathsf{Lin}_S(M, \mathsf{Lin}_R(S, N)) \cong \mathsf{Lin}_R(UM, N),$$

for fixed modules $N \in \mathsf{Mod}_R$ and $M \in \mathsf{Mod}_S$; they easily imply the existence of the two adjunctions.

3.1.6.f Consider a fixed set I and the functor $- \times I: \mathsf{Set} \longrightarrow \mathsf{Set}$ on the category of sets. This functor has a right adjoint functor given by

$$(-)^I: \mathsf{Set} \longrightarrow \mathsf{Set}.$$

Indeed for two sets X, Y, the isomorphism

$$\mathsf{Set}(X \times I, Y) \cong \mathsf{Set}(X, Y^I)$$

holds, since it just means that $Y^{X \times I} \cong (Y^I)^X$. Together with the naturality of those bijections, this implies the required result.

3.1.6.g Consider the category Cat of small categories and functors. For a given small category $\mathscr{B}$, the functor $- \times \mathscr{B} \colon \mathsf{Cat} \longrightarrow \mathsf{Cat}$ admits

$$\mathsf{Fun}(\mathscr{B}, -) \colon \mathsf{Cat} \longrightarrow \mathsf{Cat}$$

as a right adjoint functor (see exercise 1.11.8).

3.1.6.h Consider the category Top of topological spaces and continuous mappings and the full subcategory Haus of Hausdorff spaces. The inclusion functor $i \colon \mathsf{Haus} \longrightarrow \mathsf{Top}$ has a left adjoint functor

$$H \colon \mathsf{Top} \longrightarrow \mathsf{Haus}.$$

$H(X, \mathcal{T})$ is just the quotient of $(X, \mathcal{T})$ by the equivalence relation obtained as the closure of the diagonal $\Delta_X \subseteq X \times X$, and the canonical morphism $(X, \mathcal{T}) \longrightarrow H(X, \mathcal{T})$ is just the quotient morphism. (Remember a space is Hausdorff precisely when its diagonal is closed.)

3.1.6.i Consider the category Top_* of pointed topological spaces. An object is a pair $\big((X, \mathcal{T}), x_0\big)$ where $(X, \mathcal{T})$ is a topological space and $x_0 \in X$ is the choice of a "base point in X"; an arrow is just a continuous function mapping the base point to the base point. To avoid too heavy a notation, we shall omit writing the topology explicitly. Given a pointed space (X, x_0), we define $\Omega(X, x_0)$ to be the space of its loops provided with the compact open topology, i.e.

$$\Omega(X, x_0) = \big\{ f \colon (S^1, p) \longrightarrow (X, x_0) \,\big|\, f \in \mathsf{Top}_* \big\}$$

where p is an arbitrary fixed point on the circle S^1; the basic open subsets of $\Omega(X, x_0)$ are given by

$$[K, U] = \big\{ f \in \Omega(X, x_0) \,\big|\, f(K) \subseteq U \big\}$$

where K runs through the compact subsets of S^1 and U runs through the open subsets of X. This construction extends easily to a functor

$$\Omega \colon \mathsf{Top}_* \longrightarrow \mathsf{Top}_*$$

acting on the arrows simply by composition. Now let us define a second functor

$$\Sigma \colon \mathsf{Top}_* \longrightarrow \mathsf{Top}_*$$

called the "suspension functor". Given a pointed space (X, x_0), $\Sigma(X, x_0)$ is obtained as a quotient $X \times S^1 / \approx$ of the topological product $X \times S^1$, where the quotient identifies to a single base point all the pairs (x, p) and (x_0, s). Again Σ extends obviously to a functor. It is a classical result in topology that Σ is left adjoint to Ω.

3.1.6.j Consider the category Top of topological spaces and the "underlying set" functor U: Top $\longrightarrow$ Set. This functor U has both a left adjoint functor M and a right adjoint functor R. Given a set X, LX is just the set X provided with the indiscrete topology, while RX is the same set X provided with the indiscrete topology.

3.1.6.k Consider the category Cat of small categories and the forgetful functor Ob: Cat $\longrightarrow$ Set which maps a small category $\mathscr{C}$ to its set of objects. Ob has both a left adjoint functor L and a right adjoint functor R; for a given set X, LX is the discrete category with X as a set of objects and RX is the category with X as a set of objects and one single arrow from each object to each object.

3.1.6.l Consider an arbitrary category $\mathscr{C}$ and a fixed object $C \in \mathscr{C}$. The singleton set $\{*\}$ admits a reflection along the representable functor

$$\mathscr{C}(C, -) \colon \mathscr{C} \longrightarrow \mathsf{Set};$$

this is just the pair (C, i_C) where $i_C \colon \{*\} \longrightarrow \mathscr{C}(C, C)$ is the mapping "picking up" the identity on C. Indeed giving a mapping

$$\varphi \colon \{*\} \longrightarrow \mathscr{C}(C, D)$$

is just picking up an arrow $f \in \mathscr{C}(C, D)$ and clearly $\mathscr{C}(C, f) \circ i_C = \varphi$, where $i_C \colon \{*\} \longrightarrow \mathscr{C}(C, C)$ picks up the identity on $\mathscr{C}$.

3.1.6.m In the spirit of 1.2.6.b, consider two partially ordered sets A, B viewed as categories and two preorder preserving mappings

$$f \colon A \longrightarrow B, \quad g \colon B \longrightarrow A.$$

Viewed as functors, g is left adjoint to f when one has $(g \circ f)(a) \leq a$ for every $a \in A$ and $b \leq (f \circ g)(b)$ for every $b \in B$. Indeed, the naturality and the commutativity conditions of 3.1.5.(2) are automatically satisfied since in a poset, every diagram is commutative. These conditions yield immediately $f \circ g \circ f = f$ and $g \circ f \circ g = g$. Indeed applying f to $(g \circ f)(a) \leq a$ yields $(f \circ g \circ f)(a) \leq f(a)$ while putting $b = f(a)$ yields $f(a) \leq (f \circ g \circ f)(a)$; the other relation is analogous. A situation of adjunction between posets is also called a "Galois connection".

$$\mathscr{A}(GB, HD) \xrightarrow[\cong]{\theta_{HD,B}} \mathscr{B}(B, FHD)$$

$$\mathscr{A}(1_{GB}, Hd) \Big\downarrow \qquad\qquad \Big\downarrow \mathscr{B}(1_B, FHd)$$

$$\mathscr{A}(GB, HD') \xrightarrow[\theta_{HD',B}]{\cong} \mathscr{B}(B, FHD')$$

Diagram 3.6

3.2 Properties of adjoint functors

Proposition 3.2.1 *Consider the following situation:*

$$\mathscr{A} \underset{F}{\overset{G}{\rightleftarrows}} \mathscr{B} \underset{H}{\overset{K}{\rightleftarrows}} \mathscr{C}$$

where F, G, H, K are functors, with G left adjoint to F and K left adjoint to H. In this case $G \circ K$ is left adjoint to $H \circ F$.

Proof Just consider the canonical bijections

$$\mathscr{A}(GKC, A) \cong \mathscr{B}(KC, FA) \cong \mathscr{C}(C, HFA)$$

for objects $A \in \mathscr{A}$ and $C \in \mathscr{C}$ (see 3.1.5). $\qquad\qquad\qquad\Box$

Proposition 3.2.2 *If the functor $F: \mathscr{A} \longrightarrow \mathscr{B}$ has a left adjoint, F preserves all limits which turn out to exist in $\mathscr{A}$.*

Proof Write $G: \mathscr{B} \longrightarrow \mathscr{A}$ for a left adjoint to F. Consider a category $\mathscr{D}$ and a functor $H: \mathscr{D} \longrightarrow \mathscr{A}$; suppose $(L, (p_D)_{D \in \mathscr{D}})$ is a limit of H. We must prove that $(FL, (Fp_D)_{D \in \mathscr{D}})$ is a limit of $F \circ H$. Clearly $(Fp_D)_{D \in \mathscr{D}}$ is a cone as image of a cone, so it suffices to prove the universal property.

Consider a cone $(B, (q_D)_{D \in \mathscr{D}})$ on $F \circ H$. By adjointness (see 3.1.5), the morphism $q_D: B \longrightarrow FHD$ corresponds to a morphism $r_D: GB \longrightarrow HD$ in $\mathscr{A}$. Given an arrow $d: D \longrightarrow D'$ in $\mathscr{D}$, the naturality of the bijections $\theta_{A,B}$ (see diagram 3.6, notation of 3.1.5) implies that

$$
\begin{aligned}
r_{D'} &= \theta_{HD',B}^{-1}(q_{D'}) \\
&= \theta_{HD',B}^{-1}(FHd \circ q_D) \\
&= \left(\theta_{HD',B}^{-1} \circ \mathscr{B}(1_B, FHd)\right)(q_D) \\
&= \left(\mathscr{A}(1_{GB}, Hd) \circ \theta_{HD,B}^{-1}\right)(q_D) \\
&= \mathscr{A}(1_{GB}, Hd)(r_D) \\
&= Hd \circ r_D.
\end{aligned}
$$

So $(GB, (r_D)_{D \in \mathscr{D}})$ is a cone on H and we get in $\mathscr{A}$ a unique factorization $r: GB \longrightarrow L$ such that for each $D \in \mathscr{D}$, $p_D \circ r = r_D$. This arrow r corresponds via θ_{LB} to a morphism $s = \theta_{L,B}(r): B \longrightarrow FL$ in $\mathscr{B}$ and using again the naturality of the bijections $\theta_{A,B}$ have $Fp_D \circ s = q_D$. Since the $\theta_{A,B}$ are bijective, s is unique with that property. $\qquad \square$

Given a category $\mathscr{C}$ and a small category $\mathscr{D}$, let us consider the functor

$$\Delta: \mathscr{C} \longrightarrow \mathsf{Fun}(\mathscr{D}, \mathscr{C})$$

where given a morphism $f: C \longrightarrow C'$ in $\mathscr{C}$, ΔC is the constant functor on C (see 1.2.8.e) and Δf is the constant natural transformation on f (see 1.3.6.d).

Proposition 3.2.3 *A category $\mathscr{C}$ is cocomplete if and only if, for every small category $\mathscr{D}$, the corresponding functor*

$$\Delta: \mathscr{C} \longrightarrow \mathsf{Fun}(\mathscr{D}, \mathscr{C})$$

has a left adjoint.

Proof Consider a functor $F: \mathscr{D} \longrightarrow \mathscr{C}$. A pair (C, α), where $C \in |\mathscr{C}|$ and $\alpha: F \Rightarrow \Delta C$ is a natural transformation, is just a cocone on F. Thus a reflection of F is just a universal such cocone, i.e. a colimit of F. $\qquad \square$

Finally let us observe how a given adjunction generates many other adjunctions. Given three categories $\mathscr{A}, \mathscr{B}, \mathscr{C}$ and a functor $F: \mathscr{A} \longrightarrow \mathscr{B}$, we write

$$F_*: \mathsf{Fun}(\mathscr{C}, \mathscr{A}) \longrightarrow \mathsf{Fun}(\mathscr{C}, \mathscr{B}), \quad H \mapsto F \circ H,$$

for the functor acting by composition with F. To avoid size problems, we had better suppose that $\mathscr{C}$ is small.

Proposition 3.2.4 *Consider a functor $F: \mathscr{A} \longrightarrow \mathscr{B}$ with a left adjoint $G: \mathscr{B} \longrightarrow \mathscr{A}$. If $\mathscr{C}$ is any small category, $G_*: \mathsf{Fun}(\mathscr{C}, \mathscr{B}) \longrightarrow \mathsf{Fun}(\mathscr{C}, \mathscr{A})$ is itself left adjoint to $F_*: \mathsf{Fun}(\mathscr{C}, \mathscr{A}) \longrightarrow \mathsf{Fun}(\mathscr{C}, \mathscr{B})$.*

Proof Let us write $\eta: 1_{\mathscr{B}} \Rightarrow F \circ G$ and $\varepsilon: G \circ F \Rightarrow 1_{\mathscr{A}}$ for the two natural transformations describing the adjunction (see 3.1.5). Given functors $K: \mathscr{C} \longrightarrow \mathscr{A}$ and $H: \mathscr{C} \longrightarrow \mathscr{B}$, we have corresponding natural transformations

$$\eta * H: H \Rightarrow F \circ G \circ H = F_* G_* H,$$

$$\varepsilon * K: G_* F_* K = GFK \Rightarrow K.$$

This yields two natural transformations

$$\bar{\eta}: 1_{\mathrm{Fun}(\mathscr{C},\mathscr{B})} \Rightarrow F_*G_*, \quad \bar{\varepsilon}: G_*F_* \xrightarrow{\hspace{2cm}} 1_{\mathrm{Fun}(\mathscr{C},\mathscr{A})}.$$

Indeed, given a natural transformation $\alpha: H \Rightarrow H'$, the naturality of $\bar{\eta}$ reduces to the relation $FG\alpha \circ \eta_H = \eta_{H'} \circ \alpha$, which holds by naturality of η. An analogous argument holds for $\bar{\varepsilon}$. Finally the triangular identities satisfied by η, ε immediately imply the corresponding identities for $\bar{\eta}$, $\bar{\varepsilon}$.

$\square$

3.3 The adjoint functor theorem

This section is devoted to the proof of one of the most important results in this book: the adjoint functor theorem (see 3.3.3).

Given a functor $F: \mathscr{A} \longrightarrow \mathscr{B}$ and an object $B \in \mathscr{B}$, we shall consider the functor

$$\mathscr{B}(B, F-): \mathscr{A} \xrightarrow{\hspace{2cm}} \mathsf{Set}$$

and its category of elements (see 1.6.4), which we shall write as $\mathscr{E}_B$, for the sake of brevity. We also write $\phi_B: \mathscr{E}_B \longrightarrow \mathscr{A}$ for the corresponding forgetful functor.

Proposition 3.3.1 *Consider a functor* $F: \mathscr{A} \longrightarrow \mathscr{B}$ *between arbitrary categories. The following conditions are equivalent:*

(1) the object $B \in \mathscr{B}$ *has a reflection along* F;

(2) the functor $\phi_B: \mathscr{E}_B \longrightarrow \mathscr{A}$ *has a limit which is preserved by* F.

It should be noticed that no assumption is made of the completeness of $\mathscr{A}$. Moreover, completeness of $\mathscr{A}$ would not imply the existence of a limit for ϕ_B, since $\mathscr{E}_B$ is in general not a small category.

Proof (1) $\Rightarrow$ (2). Given the object $B \in \mathscr{B}$, consider its reflection (L, α) along F. By 2.11.5, (2) will be proved if we show that (L, α) is initial in $\mathscr{E}_B$. For each object $(A, b) \in \mathscr{E}_B$, b is a morphism $b: B \longrightarrow FA$ in $\mathscr{B}$ and by 3.1.1, we get a unique morphism $p_{(A,b)}: L \longrightarrow A$ in $\mathscr{A}$ such that $Fp_{(A,b)} \circ \alpha = b$, i.e. a unique morphism $p_{(A,b)}: (L, \alpha) \longrightarrow (A, b)$ in $\mathscr{E}_B$.

(2) $\Rightarrow$ (1). Let us now consider the limit $\left(L, p_{(A,b)}\right)_{(A,b)\in\mathscr{E}_B}$ of ϕ_B and the corresponding limit $(FL, Fp_{(A,b)})$ of $F \circ \phi_B$. For every $(A, b) \in \mathscr{E}_B$, define $r_{(A,b)} = b$. This produces a cone on $F \circ \phi_B$, just by definition of $\mathscr{E}_B$, and therefore a unique factorization $\alpha: B \longrightarrow FL$ such that for each $(A, b) \in \mathscr{E}_B$, $Fp_{(A,b)} \circ \alpha = r_{(A,b)} = b$. We shall prove that (L, α) is the reflection of B along F.

Given an object $A \in \mathscr{A}$ and a morphism $b: B \longrightarrow FA$, we already have a factorization $p_{(A,b)}: L \longrightarrow A$ such that $Fp_{(A,b)} \circ \alpha = b$. We must prove its uniqueness.

First of all observe that $(L, \alpha) \in \mathscr{E}_B$. The relation $Fp_{(A,b)} \circ \alpha = b$ indicates that $p_{(A,b)}: (L, \alpha) \longrightarrow (A, b)$ is a morphism of $\mathscr{E}_B$, from which the relation $p_{(A,b)} \circ p_{(L,\alpha)} = p_{(A,b)}$ follows. By definition of the limit of ϕ_B, this implies $p_{(L,\alpha)} = 1_L$. Now if $p: L \longrightarrow A$ is such that $Fp \circ \alpha = b$, then $p: (L, \alpha) \longrightarrow (A, b)$ is a morphism of $\mathscr{E}_B$. Therefore we deduce $p_{(A,b)} = p \circ p_{(L,\alpha)} = p$. $\qquad\qquad \square$

We introduce now the famous "solution set condition", which is a key ingredient for proving the adjoint functor theorem. This is really the first time in this book that "smallness conditions" play a definitely fundamental role.

Definition 3.3.2 (Solution set condition)
A functor $F: \mathscr{A} \longrightarrow \mathscr{B}$ satisfies the solution set condition with respect to an object $B \in \mathscr{B}$ when there exists a set $S_B \subseteq |\mathscr{A}|$ of objects such that

$$\forall A \in \mathscr{A} \;\; \forall b: B \to FA \;\; \exists A' \in S_B \;\; \exists a: A' \to A \;\; \exists b': B \to FA' \;\; F(a) \circ b' = b.$$

The fact that B admits a reflection (R_B, α_B) along F implies that $S_B = \{R_B\}$ can be chosen as solution set, with $b' = \alpha_B$ and moreover a unique. So the solution set condition is a much weaker requirement than the existence of a reflection. In particular observe that when $\mathscr{A}$ is small, the solution set condition is automatically satisfied for every object $B \in \mathscr{B}$: just choose $S_B = |\mathscr{A}|$.

Theorem 3.3.3 (Adjoint functor theorem)
Consider a complete category $\mathscr{A}$ and a functor $F: \mathscr{A} \longrightarrow \mathscr{B}$. The following conditions are equivalent.

(1) F has a left adjoint functor.

(2) The following conditions hold:

    (a) F preserves small limits;

    (b) F satisfies the solution set condition for every object $B \in \mathscr{B}$.

Proof (1) implies (2) by 3.2.2 and the observation following 3.3.1. Conversely consider the full subcategory $\mathscr{S}_B$ of $\mathscr{E}_B$ whose objects are the pairs (A, b) with $A \in S_B$; this category $\mathscr{S}_B$ is small. To conclude the proof, it suffices to show that the inclusion $\mathscr{S}_B \subseteq \mathscr{E}_B$ is a final functor (see 2.11.1). To do this we use proposition 2.11.3. The functors F and $\mathscr{B}(B, -)$ preserve small limits, by assumption and 2.9.4. So $\mathscr{B}(B, F-)$

is limit preserving and $\mathscr{E}_B$ is complete (see 2.16.2). The solution set condition can be reformulated as the fact that, given $(A, b) \in \mathscr{E}_B$, there exist $(A', b') \in \mathscr{S}_B$ and a morphism $a \colon (A', b') \longrightarrow (A, b)$; this is precisely the requirement of 2.11.3.   □

The subsequent results of this section indicate several sufficient conditions implying the assumptions of the adjoint functor theorem. To achieve this, we use freely some notions which will only be introduced in chapter 4.

Theorem 3.3.4 (Special adjoint functor theorem)
Consider a functor $F \colon \mathscr{A} \longrightarrow \mathscr{B}$ and suppose the following conditions are satisfied:

(1) $\mathscr{A}$ is complete;
(2) F preserves small limits;
(3) $\mathscr{A}$ is well-powered;
(4) $\mathscr{A}$ has a cogenerating family.
Under these conditions, F has a left adjoint functor.

Proof By 3.3.3, it suffices to prove the solution set condition for every fixed object $B \in \mathscr{B}$. To do this, consider a cogenerating family $(G_i)_{i \in I}$ of $\mathscr{A}$ and an object $B \in \mathscr{B}$. Define

$$S_B = \left\{ S \,\middle|\, S \text{ is a subobject of } \prod_{i \in I} G_i^{\mathscr{B}(B, FG_i)} \right\}.$$

By $G_i^{\mathscr{B}(B, FG_i)}$ we mean the product of as many copies of G_i as there are elements in $\mathscr{B}(B, FG_i)$; on the other hand the definition of S_B must be understood as the choice of one specific monomorphism

$$S \rightarrowtail \prod_{i \in I} G_i^{\mathscr{B}(B, FG_i)}$$

for each isomorphism class. We shall prove S_B satisfies the requirements of 3.3.2.

Let us consider $A \in \mathscr{A}$ and $b \colon B \longrightarrow FA$. We must find $S \in S_B$ and $a \colon S \longrightarrow A$, $b' \colon B \longrightarrow FS$ such that $Fa \circ b' = b$. We refer to diagram 3.7 for the notation. By 4.5.2 we have a monomorphism α such that $p_f \circ \alpha = f$ for every $i \in I$ and $f \in \mathscr{A}(A, G_i)$. Just by definition of a product, there is also a morphism β such that $p_f \circ \beta = p_{Ff \circ b}$ for every $i \in I$ and $f \in \mathscr{A}(A, G_i)$. Pulling back α along β, we obtain a subobject $S \in S_B$ and a morphism $a \colon Q \longrightarrow A$. Applying F which preserves pullbacks and products, we get an analogous diagram 3.8 in $\mathscr{B}$. There is a

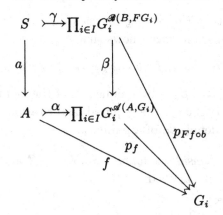

Diagram 3.7

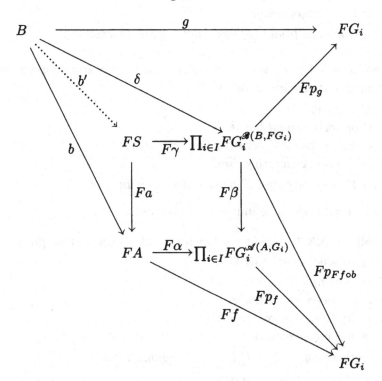

Diagram 3.8

morphism δ such that $Fp_g \circ \delta = g$ for each $i \in I$ and $g \in \mathscr{B}(B, FG_i)$. A straightforward diagram chase shows that

$$Fp_f \circ F\alpha \circ b = Ff \circ b = Fp_{Ffob} \circ \delta = Fp_f \circ F\beta \circ \delta$$

from which $F\alpha \circ b = F\beta \circ \delta$ by uniqueness of the factorization through a product. The existence of the required factorization b' follows at once from this and the definition of a pullback. $\qquad\square$

Corollary 3.3.5 *Consider a functor $F \colon \mathscr{A} \longrightarrow \mathscr{B}$ and suppose the following conditions are satisfied:*

(1) $\mathscr{A}$ is complete;
(2) F preserves small limits;
(3) $\mathscr{A}$ is well-powered;
(4) $\mathscr{A}$ has a cogenerator.

Under these conditions, F has a left adjoint functor. $\qquad\square$

Corollary 3.3.6 *Consider a functor $F \colon \mathscr{A} \longrightarrow \mathscr{B}$ and suppose the following conditions are satisfied:*

(1) $\mathscr{A}$ is complete;
(2) F preserves small limits;
(3) $\mathscr{A}$ has a strong generating family;
(4) $\mathscr{A}$ has a cogenerating family.

Under these conditions, F has a left adjoint functor.

Proof By 4.5.15, $\mathscr{A}$ is indeed well-powered. $\qquad\square$

Corollary 3.3.7 *Consider a functor $F \colon \mathscr{A} \longrightarrow \mathscr{B}$ and suppose the following conditions are satisfied:*

(1) $\mathscr{A}$ is complete;
(2) F preserves small limits;
(3) $\mathscr{A}$ has a strong generator;
(4) $\mathscr{A}$ has a cogenerator.

Under these conditions, F has a left adjoint functor. $\qquad\square$

As a first application of these existence theorems, let us observe that:

Proposition 3.3.8 *Suppose the category $\mathscr{A}$ satisfies the following conditions:*

(1) $\mathscr{A}$ is complete;
(2) $\mathscr{A}$ is well-powered;
(3) $\mathscr{A}$ has a cogenerating family.

Under these conditions, $\mathscr{A}$ is cocomplete as well.

Proof   Applying proposition 3.2.3 and using its notation, it suffices to prove that the functor Δ preserves limits. But this is obvious from 2.15.2. □

Examples 3.3.9

3.3.9.a   The inclusion Ab $\hookrightarrow$ Gr of the category of abelian groups in the category of groups has a left adjoint functor. Indeed the conditions of 3.3.7 are satisfied, with $\mathbb{Z}$ a strong generator and $\mathbb{Q}/\mathbb{Z}$ a cogenerator (see chapter 4).

3.3.9.b   The "unit ball" functor $B\colon \mathsf{Ban}_1 \longrightarrow \mathsf{Set}$ from the category of Banach spaces and linear contractions to the category of sets has a left adjoint functor. Indeed the conditions of 3.3.7 are satisfied (see 2.9.10.c) with $\mathbb{R}$ both a strong generator and a cogenerator (see chapter 4).

3.3.9.c   The inclusion Comp $\hookrightarrow$ Top of the category of compact Hausdorff spaces in the category of all topological spaces has a left adjoint (called the "Stone-Čech compactification"). Indeed the conditions of 3.3.7 are satisfied (see 2.1.7.g and 2.4.6.f) with the singleton as a strong generator and the unit interval $[0,1]$ as a cogenerator (see chapter 4).

3.3.9.d   Consider a functor $F\colon \mathscr{A} \longrightarrow \mathscr{B}$ between small categories $\mathscr{A}$ and $\mathscr{B}$. Consider the corresponding functor

$$F^*\colon \mathsf{Fun}(\mathscr{B}, \mathsf{Set}) \longrightarrow \mathsf{Fun}(\mathscr{A}, \mathsf{Set})$$

obtained by composition with F. By the dual of 2.15.2, the categories $\mathsf{Fun}(\mathscr{B}, \mathsf{Set})$ and $\mathsf{Fun}(\mathscr{A}, \mathsf{Set})$ are cocomplete and the functor F^* preserves colimits. The category $\mathsf{Fun}(\mathscr{B}, \mathsf{Set})$ is certainly co-well-powered: indeed, by corollary 2.15.3, an epimorphism $p\colon F \Rightarrow G$ is such that each $p_B\colon FB \longrightarrow GB$ is surjective; since each set FB has just a set of quotients and $|\mathscr{B}|$ itself is a set, there is just a set of epimorphisms $p\colon F \Rightarrow \bullet$ of domain F. On the other hand the representable functors constitute a generating family for $\mathsf{Fun}(\mathscr{B}, \mathsf{Set})$ (see 4.5.17.b). By the dual of 3.3.4, F^* has a right adjoint functor. This example is an instance of what is called a "Kan extension" (see 3.7.2).

3.3.9.e   If $f\colon A \longrightarrow B$ is a functor between two posets, f has a left adjoint as long as A is complete and f preserves infima. Indeed, the solution set condition is now obvious: it suffices to take as solution set the set of all elements of A (see 2.8.6).

3.4 Fully faithful adjoint functors

Two categories $\mathscr{A}$, $\mathscr{B}$ are isomorphic when there exists a pair of functors $F, G: \mathscr{A} \xrightarrow{\longleftarrow} \mathscr{B}$ with the properties $F \circ G = 1_{\mathscr{B}}$, $G \circ F = 1_{\mathscr{A}}$. Imposing a condition like $FG(B) = B$ on the objects seems in opposition with the global spirit of category theory, where things are defined up to isomorphism. It would therefore sound more reasonable to get interested in situations where identities are replaced by isomorphisms $F \circ G \cong 1_{\mathscr{B}}$, $G \circ F \cong 1_{\mathscr{A}}$. This is the essence of the notion of an "equivalence of categories", which will be presented at the end of this section.

Proposition 3.4.1 *Consider two functors $G \dashv F: \mathscr{A} \xrightarrow{\longleftarrow} \mathscr{B}$, with G left adjoint to F and $\eta: 1_{\mathscr{B}} \Rightarrow FG$, $\varepsilon: GF \Rightarrow 1_{\mathscr{A}}$ the two corresponding natural transformations (see 3.1.5). The following conditions are equivalent:*

(1) F is full and faithful;

(2) ε is an isomorphism.

Under these conditions, $\eta * F$ and $G * \eta$ are isomorphisms as well.

Proof (1) $\Rightarrow$ (2). The morphism $\eta_{FA}: FA \longrightarrow FGFA$ has the form $F\alpha_A$ for some morphism $\alpha_A: A \longrightarrow GFA$, because F is full. From the relation $F\varepsilon_A \circ F\alpha_A = F\varepsilon_A \circ \eta_{FA} = 1_{FA}$ and the faithfulness of F, we deduce $\varepsilon_A \circ \alpha_A = 1_A$. To prove the equality $\alpha_A \circ \varepsilon_A = 1_{GFA}$, it suffices to notice that

$$F(\alpha_A \circ \varepsilon_A) \circ \eta_{FA} = \eta_{FA} \circ F\varepsilon_A \circ \eta_{FA} = \eta_{FA} = F(1_{GFA}) \circ \eta_{FA},$$

since (GFA, η_{FA}) is the reflection of FA along F.

(2) $\Rightarrow$ (1). Consider the composite

$$\mathscr{A}(A, A') \xrightarrow{\mathscr{A}(\varepsilon_A, 1_{A'})} \mathscr{A}(GFA, A') \xrightarrow{\theta_{A', FA}} \mathscr{B}(FA, FA').$$

We know that each individual mapping is a bijection. The composite bijection maps an arrow $a: A \longrightarrow A'$ to

$$\theta_{A', FA}(a \circ \varepsilon_A) = F(a \circ \varepsilon_A) \circ \eta_{FA} = Fa \circ F\varepsilon_A \circ \eta_{FA} = Fa$$

(see 3.1.5), so it is just the action of F.

Now if conditions (1), (2) are satisfied, for all objects $A \in \mathscr{A}$, $B \in \mathscr{B}$ one has $F\varepsilon_A \circ \eta_{FA} = 1_{FA}$ and $\varepsilon_{GB} \circ G\eta_B = 1_{GB}$. Since ε_A and ε_{GB} are isomorphisms, so are η_{FA} and $G\eta_B$. $\qquad\square$

Proposition 3.4.2 *Consider a functor $F: \mathscr{A} \longrightarrow \mathscr{B}$ with both a left adjoint functor G and a right adjoint functor H. If one of those adjoint functors is full and faithful, so is the other adjoint functor.*

Proof   Let us write $\varepsilon: GF \Rightarrow 1_{\mathscr{A}}$, $\eta: 1_{\mathscr{B}} \Rightarrow FG$, $\alpha: FH \Rightarrow 1_{\mathscr{B}}$ and $\beta: 1_{\mathscr{A}} \Rightarrow HF$ for the canonical natural transformations of the adjunctions (see 3.1.5). Let us assume H is full and faithful and let us prove the same holds for G (the converse implication holds by duality). Using 3.4.1, we must prove that η is an isomorphism whenever α is. Assuming α is an isomorphism, we consider the composite

$$FG \xrightarrow{\ FG\alpha^{-1}\ } FGFH \xrightarrow{\ F\varepsilon H\ } FH \xrightarrow{\ \alpha\ } 1_{\mathscr{B}}$$

and prove it is inverse to η. Indeed

$$\alpha \circ F\varepsilon_H \circ FG\alpha^{-1} \circ \eta = \alpha \circ F\varepsilon_H \circ \eta_{FG} \circ \alpha^{-1} = \alpha \circ \alpha^{-1} = 1.$$

On the other hand, the relation $\alpha_F \circ F\beta = 1_F$ implies $F\beta = \alpha_F^{-1}$ and therefore, by naturality of α, ε and by 3.1.5,

$$\begin{aligned}
\eta \circ \alpha \circ F\varepsilon_H \circ FG\alpha^{-1} &= FG\alpha \circ FH\eta \circ F\varepsilon H \circ FG\alpha^{-1} \\
&= FG\alpha \circ F\varepsilon HGF \circ FGFH\eta \circ FG\alpha^{-1} \\
&= FG\alpha \circ F\varepsilon HFG \circ FG\alpha^{-1}FG \circ FG\eta \\
&= FG\alpha \circ F\varepsilon HFG \circ FGF\beta G \circ FG\eta \\
&= F1_G \circ F1_G \\
&= 1_{FG}. \qquad \qquad \square
\end{aligned}$$

Proposition 3.4.3 *Given a functor* $F: \mathscr{A} \longrightarrow \mathscr{B}$, *the following conditions are equivalent:*

(1) *F is full and faithful and has a full and faithful left adjoint G;*

(2) *F has a left adjoint G and the two canonical natural transformations of the adjunction $\eta: 1_{\mathscr{B}} \Rightarrow FG$, $\varepsilon: GF \Rightarrow 1_{\mathscr{A}}$ are isomorphisms;*

(3) *there exists a functor $G: \mathscr{B} \longrightarrow \mathscr{A}$ and two arbitrary natural isomorphisms $1_{\mathscr{B}} \cong FG$, $GF \cong 1_{\mathscr{A}}$;*

(4) *F is full and faithful and each object $B \in \mathscr{B}$ is isomorphic to an object of the form FA, $A \in \mathscr{A}$;*

(5) *the dual condition of (1);*

(6) *the dual condition of (2).*

Proof   Since conditions (3) and (4) are autodual, it suffices to prove the equivalence of (1) to (4). (1) $\Rightarrow$ (2) follows from 3.4.2 and (2) $\Rightarrow$ (3) is obvious.

Let us prove (3) $\Rightarrow$ (4). First of all, $B \in \mathscr{B}$ is isomorphic to FGB, with $GB \in \mathscr{A}$, which proves the second assertion. Now since FG is isomorphic to the identity, FG is full and faithful; therefore G is faithful

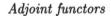

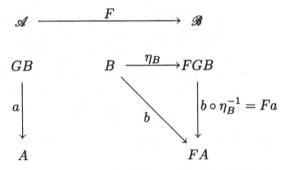

Diagram 3.9

and F is full. In the same way the isomorphism $GF \cong 1_{\mathscr{A}}$ shows that F is faithful and G is full.

To prove (4) $\Rightarrow$ (1), we first show the existence of a left adjoint G to F (see diagram 3.9). Given $B \in \mathscr{B}$, choose an object $GB \in \mathscr{A}$ and an isomorphism

$$\eta_B \colon B \xrightarrow{\;\cong\;} FGB.$$

Given $A \in \mathscr{A}$ and $b \colon B \longrightarrow FA$, the morphism $b \circ \eta_B^{-1} \colon FGB \longrightarrow FA$ has the form Fa for some unique $a \colon GB \longrightarrow A$, which implies immediately that (GB, η_B) is the reflection of B along F. Since each η_B is an isomorphism, the left adjoint functor G is full and faithful (see 3.4.1). $\qquad\square$

We recall that an axiom of choice is needed to deduce the existence of a left adjoint functor from the existence of a reflection for each object (implication (4) $\Rightarrow$ (1)).

But more importantly one should observe that when two isomorphisms $\eta \colon 1_{\mathscr{B}} \Rightarrow FG$ and $\varepsilon \colon GF \Rightarrow 1_{\mathscr{A}}$ are given as in (3), the constructions performed in proving (3) $\Rightarrow$ (4) and (4) $\Rightarrow$ (1) show that G is left adjoint to F, with η as one of the canonical natural transformations of the adjunction; but there is no reason for ε to be the second natural transformation of the adjunction: in other words, the triangles involved in 3.1.5.(2) have no reason to commute (of course, there is some other isomorphism $\varepsilon' \colon GF \Rightarrow 1_{\mathscr{A}}$ constructed from η which makes them commute). See exercise 3.9.4 for a counterexample. By duality, one could clearly choose ε as one of the canonical natural transformations of the adjunction and construct some η' as second natural transformation.

Definition 3.4.4 *A functor $F \colon \mathscr{A} \longrightarrow \mathscr{B}$ which satisfies the conditions*

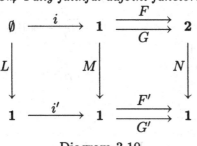

Diagram 3.10

of proposition 3.4.3 is called an *equivalence of categories.*

It is a matter of fact that two equivalent categories share most categorical properties with each other. For example:

Proposition 3.4.5 *Consider an equivalence* $F: \mathscr{A} \longrightarrow \mathscr{B}$ *of categories. If $\mathscr{A}$ is (finitely) complete, so is $\mathscr{B}$.*

Proof Consider $G: \mathscr{B} \longrightarrow \mathscr{A}$, left adjoint to F (see 3.4.3). Given a (finite) small category $\mathscr{D}$ and a functor $H: \mathscr{D} \longrightarrow \mathscr{B}$, the functor GH has a limit since $\mathscr{A}$ is (finitely) complete. Since F preserves limits (see 3.2.2), FGH has a limit. But the limit of FGH is isomorphic to the limit of H, just because FG is isomorphic to $1_{\mathscr{B}}$. $\square$

The proof of the previous proposition emphasizes the fact that equivalent categories have the same properties as far as limits are concerned. One should nevertheless be careful when using equivalences. It is not true that in a construction, *replacing categories by equivalent ones produces equivalent results.* For example write **1** for the discrete category with a single object $*$ and **2** for the category with two objects A,B and just one single isomorphism between them. Consider diagram 3.10 in **Cat**, where $\emptyset$ stands for the empty category, $F(*) = A$ and $G(*) = B$. The horizontal lines are equalizers and the diagram commutes, meaning that $F' \circ M = N \circ F$, $G' \circ M = N \circ G$ and $M \circ i = i' \circ L$. Now M and N are equivalences of categories, but L is not. The trouble comes from the fact that, although M, N are equivalences making the diagram commutative, no choice of the corresponding adjoint equivalences can be made which still makes the diagram commutative. In other words, writing $\mathscr{K}$ for the category $\bullet \underset{\longrightarrow}{\longrightarrow} \bullet$ defining equalizers (see 2.6.7.b), the two functors $\mathscr{K} \underset{\longrightarrow}{\longrightarrow} \textbf{Cat}$ given respectively by F, G and F', G' are not equivalent in the 2-category **Cat**, in the sense of 7.1.2. So replacing categories by equivalent ones when computing a limit in **Cat** does not in general yield equivalent results.

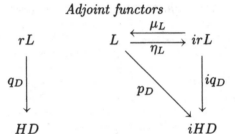

$$\begin{array}{ccc}
rL & L \xrightarrow[\eta_L]{\overset{\mu_L}{\longleftarrow}} irL \\
\Big\downarrow{q_D} & p_D & \Big\downarrow{iq_D} \\
HD & & iHD
\end{array}$$

Diagram 3.11

3.5 Reflective subcategories

When $F\colon \mathscr{A} \longrightarrow \mathscr{B}$ is a full and faithful functor, the corestriction of F to its image $F\colon \mathscr{A} \longrightarrow F(\mathscr{A})$ is obviously an equivalence of categories (see 3.4.3.(4)). Replacing $F(\mathscr{A})$ by the full subcategory $\mathscr{A}' \subseteq \mathscr{B}$ generated by all the objects $B \in \mathscr{B}$ isomorphic to an object of the form FA, one still has an equivalence $F\colon \mathscr{A} \longrightarrow \mathscr{A}'$ (see 3.4.3.(4)). Working with $\mathscr{A}'$ instead of $\mathscr{A}$ enables us to consider a canonical inclusion $\mathscr{A}' \subseteq \mathscr{B}$ instead of a full and faithful functor $F\colon \mathscr{A} \longrightarrow \mathscr{B}$.

Definition 3.5.1 *A full subcategory $\mathscr{A}$ of a category $\mathscr{B}$ is replete when, with every object $A \in \mathscr{A}$, $\mathscr{A}$ also contains every object $B \in \mathscr{B}$ isomorphic to A.*

Definition 3.5.2 *A reflective subcategory of a category $\mathscr{B}$ is a full replete subcategory $\mathscr{A}$ of $\mathscr{B}$ whose inclusion $i\colon \mathscr{A} \longrightarrow \mathscr{B}$ in $\mathscr{B}$ admits a left adjoint $r\colon \mathscr{B} \longrightarrow \mathscr{A}$, called the reflection.*

The fact of choosing $\mathscr{A}$ to be replete in $\mathscr{B}$ is unessential; it is just an easy way to choose a canonical element in the class of all subcategories equivalent to $\mathscr{A}$. This specific choice also makes life easier since most constructions of category theory are only defined up to an isomorphism.

Proposition 3.5.3 *Consider a (finitely) complete category $\mathscr{B}$ and a reflective subcategory $\mathscr{A}$ of $\mathscr{B}$. Under these conditions $\mathscr{A}$ itself is (finitely) complete.*

Proof Take a (finite) small category $\mathscr{D}$, a functor $H\colon \mathscr{D} \longrightarrow \mathscr{A}$ and the limit $\big(L, (p_D)_{D \in \mathscr{D}}\big)$ of $i \circ H$, where $i\colon \mathscr{A} \longrightarrow \mathscr{B}$ is the canonical inclusion and $r\colon \mathscr{B} \longrightarrow \mathscr{A}$ is its left adjoint (see diagram 3.11). Each morphism p_D gives rise to a unique factorization $q_D\colon rL \longrightarrow HD$ in $\mathscr{A}$, with the property $iq_D \circ \eta_L = p_D$ (see 3.1.1). Given $d\colon D \longrightarrow D'$ in $\mathscr{D}$, the relations

$$i(Hd \circ q_D) \circ \eta_L = iHd \circ p_D = p_{D'} = iq_{D'} \circ \eta_L$$

imply $Hd \circ q_D = q_{D'}$, by definition of a reflection (see 3.1.1). So the morphisms $(q_D)_{D \in \mathcal{D}}$ constitute a cone on H, and thus the morphisms $(iq_D)_{D \in \mathcal{D}}$ constitute a cone on iH. This yields a unique factorization $\mu_L \colon irL \longrightarrow L$ such that $p_D \circ \mu_L = iq_D$ for each D. From $p_D \circ \mu_L \circ \eta_L = iq_D \circ \eta_L = p_D$ we deduce $\mu_L \circ \eta_L = 1_L$ by definition of a limit. On the other hand $\eta_L \circ \mu_L \colon irL \longrightarrow irL$ has the form $i(\nu_L)$ for a unique $\nu_L \colon rL \longrightarrow rL$, just because i is full and faithful. From the relations

$$i(\nu_L) \circ \eta_L = \eta_L \circ \mu_L \circ \eta_L = \eta_L = i(1_{rL}) \circ \eta_L$$

we deduce $\nu_L = 1_{rL}$, by definition of a reflection. In particular $\eta_L \circ \mu_L = i(\nu_L) = 1_{irL}$ and η_L is an isomorphism. Since $\mathcal{A}$ is replete in $\mathcal{B}$, this proves that L belongs already to $\mathcal{A}$. The conclusion is then obvious. $\square$

By duality, a coreflective subcategory of a cocomplete category is itself cocomplete. This obvious remark stands here just to emphasize the fact that our next result is by no means dual to 3.5.3.

Proposition 3.5.4 *Consider a (finitely) cocomplete category $\mathcal{B}$ and a reflective subcategory $\mathcal{A}$. Under these conditions, $\mathcal{A}$ itself is (finitely) cocomplete.*

Proof Take a (finite) small category $\mathcal{D}$ and a functor $H \colon \mathcal{D} \longrightarrow \mathcal{A}$; write $\big(L, (s_D)_{D \in \mathcal{D}}\big)$ for the colimit of iH, where $i \colon \mathcal{A} \longrightarrow \mathcal{B}$ is the canonical inclusion. Writing $r \colon \mathcal{B} \longrightarrow \mathcal{A}$ for the reflection, we know already that $\big(rL, (rs_D)_{D \in \mathcal{D}}\big)$ is the colimit of riH (see 3.2.2). But ri is isomorphic to the identity on $\mathcal{A}$ (see 3.4.1), therefore $\big(rL, (rs_D)_{D \in \mathcal{D}}\big)$ is also the colimit of H. $\square$

Definition 3.5.5 *A localization of a category $\mathcal{B}$ with finite limits is a reflective subcategory $\mathcal{A}$ of $\mathcal{B}$ whose reflection preserves finite limits.*

Definition 3.5.6 *An essential localization of a category $\mathcal{B}$ is a reflective subcategory $\mathcal{A}$ of $\mathcal{B}$ whose reflection itself admits a left adjoint.*

A functor with a left adjoint preserves all limits (see 3.2.2). So when $\mathcal{B}$ has finite limits, every essential localization of $\mathcal{B}$ is certainly a localization. Moreover if $i \colon \mathcal{A} \longrightarrow \mathcal{B}$ is the canonical inclusion and $l \dashv r \dashv i$ are the reflection and its left adjoint, the functor l is again full and faithful (see 3.4.2). Therefore the full subcategory $l(\mathcal{A}) \subseteq \mathcal{B}$ is, up to an equivalence, a coreflective subcategory of $\mathcal{B}$. It should be noticed that l has in general no reason at all to coincide with the canonical inclusion i of $\mathcal{A}$ in $\mathcal{B}$.

Proposition 3.5.7 *Consider a category $\mathscr{B}$ with finite limits and filtered colimits, and a localization $\mathscr{A}$ of $\mathscr{B}$. If in $\mathscr{B}$ finite limits commute with filtered colimits, the same property holds in $\mathscr{A}$.*

Proof We use the notation of section 2.13 and consider a functor

$$F\colon \mathscr{C} \times \mathscr{D} \longrightarrow \mathscr{A},$$

with $\mathscr{C}$ filtered and $\mathscr{D}$ finite. We write Lim and Colim for limits and colimits in $\mathscr{B}$ and correspondingly, lim and colim for limits and colimits in $\mathscr{A}$. Using the construction of limits and colimits in $\mathscr{A}$ as described in 3.5.3, 3.5.4, the result is proved by the following isomorphisms, where $i\colon \mathscr{A} \longrightarrow \mathscr{B}$ is the canonical inclusion and $r\colon \mathscr{B} \longrightarrow \mathscr{A}$ is the reflection. We recall that i, r preserve finite limits while r preserves colimits (see 3.2.2).

$$\begin{aligned}
\lim_{D\in\mathscr{D}}\mathrm{colim}_{C\in\mathscr{C}} &\cong \lim_{D\in\mathscr{D}} r\big(\mathrm{Colim}_{C\in\mathscr{C}} iF(C,D)\big) \\
&\cong r\mathrm{Lim}_{D\in\mathscr{D}} ir\big(\mathrm{Colim}_{C\in\mathscr{C}} iF(C,D)\big) \\
&\cong rir\mathrm{Lim}_{D\in\mathscr{D}} \mathrm{Colim}_{C\in\mathscr{C}} iF(C,D) \\
&\cong rir\mathrm{Colim}_{C\in\mathscr{C}} \mathrm{Lim}_{D\in\mathscr{D}} iF(C,D) \\
&\cong ri\mathrm{colim}_{C\in\mathscr{C}} r\mathrm{Lim}_{D\in\mathscr{D}} iF(C,D) \\
&\cong ri\mathrm{colim}_{C\in\mathscr{C}} \lim_{D\in\mathscr{D}} riF(C,D) \\
&\cong \mathrm{colim}_{C\in\mathscr{C}} \lim_{D\in\mathscr{D}} F(C,D). \qquad \square
\end{aligned}$$

3.6 Epireflective subcategories

In this section, we pay special attention to the reflective subcategories $i\colon \mathscr{A} \hookrightarrow \mathscr{B}$ with the property that, given an object B of $\mathscr{B}$ and its reflection (rB, η_B), the canonical morphism $\eta_B\colon B \longrightarrow irB$ is an epimorphism. While studying this particular topic, we shall freely use some notions which will only be introduced in chapter 4.

Definition 3.6.1 *Consider a category $\mathscr{B}$ and a reflective subcategory $\mathscr{A}$ of $\mathscr{B}$, with corresponding adjunction $r \dashv i\colon \mathscr{A} \overset{\longleftarrow}{\longrightarrow} \mathscr{B}$. The reflection r is an epireflection when, for every object $B \in \mathscr{B}$, the universal morphism $\eta_B\colon B \longrightarrow irB$ is an epimorphism.*

Proposition 3.6.2 *Consider a category $\mathscr{B}$ in which every morphism can be factored as an epimorhism followed by a strong monomorphism. For a reflective subcategory $r \dashv i\colon \mathscr{A} \overset{\longleftarrow}{\longrightarrow} \mathscr{B}$ of $\mathscr{B}$, the following conditions are equivalent:*

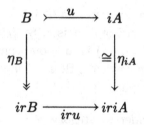

Diagram 3.12

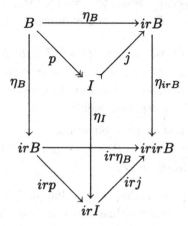

Diagram 3.13

(1) the reflection is an epireflection;

(2) given a strong monomorphism $u: B \rightarrowtail iA$ in $\mathscr{B}$, with $B \in \mathscr{B}$ and $A \in \mathscr{A}$, the object B belongs to $\mathscr{A}$.

Proof (1) $\Rightarrow$ (2). The consideration of the commutative square in diagram 3.12, where η_{iA} is an isomorphism and u is a strong monomorphism, indicates that η_B is a strong monomorphism as well (see 4.3.6). Since η_B is by assumption an epimorphism, it is an isomorphism (see 4.3.6 again). Since $ir(B) \in \mathscr{A}$, and $B \cong ir(B)$, $B \in \mathscr{A}$ because $\mathscr{A}$ is replete in $\mathscr{B}$ (see 3.5.2).

(2) $\Rightarrow$ (1). Given an object $B \in \mathscr{B}$, let us consider the canonical morphism $\eta_B: B \longrightarrow ir(B)$ and its strong-mono–epi factorization

$$B \xrightarrow{\;p\;} I \rightarrowtail \xrightarrow{\;j\;} ir(B).$$

Let us consider diagram 3.13. By 3.4.1, $\eta_{ir(B)}$ and $r\eta_B$ are isomorphisms. $r(B) \in \mathscr{A}$, thus $I \in \mathscr{A}$ by assumption; so again by 3.4.1, η_I is an isomorphism. Since $ir(\eta_B)$ is an isomorphism, $ir(p)$ is a strong monomorphism. Since p is an epimorphism and η_I is an isomorphism, $ir(p)$ is an epimorphism; thus it is an isomorphism. So $ir(p)$ and $ir(\eta_B)$ are both isomorphisms, from which $ir(j)$ is an isomorphism as well. Finally η_I, $\eta_{ir(B)}$ and $ir(j)$ are isomorphisms, thus j is an isomorphism and η_B is isomorphic to the epimorphism p.   □

Definition 3.6.3 *Consider a category $\mathscr{B}$ and a reflective subcategory $\mathscr{A}$ of $\mathscr{B}$, with corresponding adjunction $r \dashv i\colon \mathscr{A} \underset{\longrightarrow}{\overset{\longleftarrow}{}} \mathscr{B}$. The reflection r is a strong epireflection when, for every object $B \in \mathscr{B}$, the universal morphism $\eta_B\colon B \longrightarrow ir B$ is a strong epimorphism.*

Proposition 3.6.4 *Consider a category $\mathscr{B}$ in which every morphism factors as a strong epimorphism followed by a monomorphism. For a reflective subcategory $\mathscr{A}$ of $\mathscr{B}$, with corresponding adjunction $r \dashv i\colon \mathscr{A} \underset{\longrightarrow}{\overset{\longleftarrow}{}} \mathscr{B}$, the following conditions are equivalent:*

(1) the reflection is a strong epireflection;
(2) given a monomorphism $u\colon B \rightarrowtail iA$ in $\mathscr{B}$, with $B \in \mathscr{B}$ and $A \in \mathscr{A}$, the object B belongs to $\mathscr{A}$.

Proof   This perfectly analogous to that of 3.6.2, replacing "strong monomorphism" by monomorphism and "epimorphism" by "strong epimorphism".   □

By analogy, one could define the notions of "monoreflection" (each η_B is a monomorphism) or "strong monoreflection" (each η_B is a strong monomorphism). It should be clear that these notions are by no means dual to those of epireflection or strong epireflection. With the notation of 3.6.1, the dual notion of "being an epireflection" is "being a *coreflection with each $\eta_B\colon ir B \longrightarrow B$ a monomorphism*".

3.7 Kan extensions

In example 3.3.9.d, we considered a functor $F\colon \mathscr{A} \longrightarrow \mathscr{B}$ between two small categories $\mathscr{A}, \mathscr{B}$ and the corresponding functor

$$F^*\colon \mathsf{Fun}(\mathscr{B}, \mathsf{Set}) \longrightarrow \mathsf{Fun}(\mathscr{A}, \mathsf{Set})$$

obtained by composition with F. From the cocompleteness of Set and various properties of the category $\mathsf{Fun}(\mathscr{B}, \mathsf{Set})$, strongly depending on the fact of working with the category Set, we proved the existence of a right adjoint to F^*. In this section, we shall replace Set by an arbitrary cocomplete category... and prove that F^* always has a left adjoint!

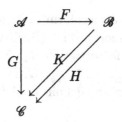

Diagram 3.14

Definition 3.7.1 *Consider two functors $F: \mathscr{A} \longrightarrow \mathscr{B}$ and $G: \mathscr{A} \longrightarrow \mathscr{C}$. The left Kan extension of G along F, if it exists, is a pair (K, α) where*

- *$K: \mathscr{B} \longrightarrow \mathscr{C}$ is a functor,*
- *$\alpha: G \Rightarrow K \circ F$ is a natural transformation,*

satisfying the following universal property: if (H, β) is another pair with

- *$H: \mathscr{B} \longrightarrow \mathscr{C}$ a functor,*
- *$\beta: G \Rightarrow H \circ F$ a natural transformation,*

there exists a unique natural transformation $\gamma: K \Rightarrow H$ satisfying the equality $(\gamma * F) \circ \alpha = \beta$ (see diagram 3.14).

We shall use the notation $\mathrm{Lan}\,_F G$ to denote the left Kan extension of G along F. The notation $\mathrm{Ran}\,_F G$ is used for the dual notion of right Kan extension.

When $\mathscr{A}$ and $\mathscr{B}$ are small, we can consider the functor

$$F^*: \mathsf{Fun}(\mathscr{B}, \mathscr{C}) \longrightarrow \mathsf{Fun}(\mathscr{A}, \mathscr{C})$$

acting by composition with F. The existence of the left Kan extension of G along F means exactly the existence of a reflection (K, α) for G along F^*.

Theorem 3.7.2 *Consider two functors $F: \mathscr{A} \longrightarrow \mathscr{B}$ and $G: \mathscr{A} \longrightarrow \mathscr{C}$, with $\mathscr{A}$ small and $\mathscr{C}$ cocomplete. Under these conditions, the left Kan extension of G along F exists.*

Proof We use the same notation as in 3.7.1. To define K, consider first a fixed object $B \in \mathscr{B}$. We write $\mathscr{E}_B$ for the category of elements of the contravariant functor $\mathscr{B}(F-, B): \mathscr{A} \longrightarrow \mathsf{Set}$ (see 1.6.4) and $\phi_B: \mathscr{E}_B \longrightarrow \mathscr{A}$ for the corresponding forgetful functor. Now $\mathscr{E}_B$ is small since $\mathscr{A}$ is. We write $\left(KB, (s^B_{(A,b)})_{(A,b) \in \mathscr{E}_B}\right)$ for the colimit of $G \circ \phi_B$. This defines K on the objects.

Given a morphism $f: B \longrightarrow B'$ in $\mathscr{B}$ and an object $(A, b) \in \mathscr{E}_B$, the pair $(A, f \circ b)$ is an object of $\mathscr{E}_{B'}$. The family $\left(s^{B'}_{(A, f \circ b)}\right)_{(A,b) \in \mathscr{E}_B}$ is a cocone on $G \circ \phi_B$ because a morphism $a: (A, b) \longrightarrow (A', b')$ in $\mathscr{E}_B$ immediately gives rise to the morphism $a: (A, f \circ b) \longrightarrow (A', f \circ b')$ in $\mathscr{E}_{B'}$. Therefore we get a unique factorization $Kf: KB \longrightarrow KB'$ through the colimit KB with $Kf \circ s^B_{(A,b)} = s^{B'}_{(A,f \circ b)}$. This defines K on the arrows. The uniqueness condition in the definition of Kf easily implies that K is indeed a functor.

To define α, we must construct a morphism $\alpha_A: GA \longrightarrow KFA$ for each object $A \in \mathscr{A}$. It suffices to define $\alpha_A = s^{FA}_{(A,1_{FA})}$. Let us prove the naturality of α. Given a morphism $a: A \longrightarrow A'$ in $\mathscr{A}$, we have indeed

$$KFa \circ s^{FA}_{(A,1_{FA})} = s^{FA'}_{(A,Fa)} = s^{FA'}_{(A',1_{FA'})} \circ Ga,$$

where the first equality holds by definition of KFa and the second equality holds because $a: (A, Fa) \longrightarrow (A', 1_{FA'})$ is a morphism of $\mathscr{E}_{FA'}$.

Now consider a functor $H: \mathscr{B} \longrightarrow \mathscr{C}$ together with a natural transformation $\beta: G \Rightarrow H \circ F$. To construct $\gamma: K \Rightarrow H$, let us fix an object $B \in \mathscr{B}$. For each $(A, b) \in \mathscr{E}_B$ consider the composite

$$(G \circ \phi_B)(A, b) = GA \xrightarrow{\beta_A} HFA \xrightarrow{Hb} HB.$$

Those morphisms constitute a cone on $G \circ \phi_B$, just by definition of $\mathscr{E}_B$ and naturality of β. So we get a unique factorization $\gamma_B: KB \longrightarrow HB$ through the colimit KB yielding $\gamma_B \circ s^B_{(A,b)} = Hb \circ \beta_A$. To prove the naturality of γ, consider a morphism $f: B \longrightarrow B'$ in $\mathscr{B}$ and an object $(A, b) \in \mathscr{E}_B$. The relations

$$\begin{aligned} Hf \circ \gamma_B \circ s^B_{(A,b)} &= Hf \circ Hb \circ \beta_A \\ &= H(f \circ b) \circ \beta_A \\ &= \gamma_{B'} \circ s^{B'}_{(A,f \circ b)} \\ &= \gamma_{B'} \circ Kf \circ s^B_{(A,b)} \end{aligned}$$

imply the required identity $Hf \circ \gamma_B = \gamma_{B'} \circ Kf$, by definition of a colimit.

The condition $(\gamma * F) \circ \alpha = \beta$ reduces to the equality $\gamma_{FA} \circ \alpha_A = \beta_A$, which is just the relation

$$\gamma_{FA} \circ s^{FA}_{(A,1_{FA})} = H(1_{FA}) \circ \beta_A$$

appearing in the definition of γ_{FA}. The uniqueness of γ with such a property reduces easily to the uniqueness condition in the definition of γ_{FA}. $\qquad\square$

A left Kan extension is called "pointwise" when each $\mathrm{Lan}_F G(B)$ can be computed by the colimit formula of theorem 3.7.2. See exercise 3.9.7 for an example of a non-pointwise Kan extension.

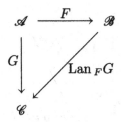

Diagram 3.15

It should be noticed that, given a Kan extension, the triangle of diagram 3.15 does not in general commute (see exercise 3.8.7). But it "commutes up to isomorphism" in a special case of interest, as attested by the next result. Exercise 3.9.5 shows that the assumption in 3.7.3 is by no means necessary.

Proposition 3.7.3 *Consider a full and faithful functor* $F: \mathscr{A} \longrightarrow \mathscr{B}$ *with* $\mathscr{A}$ *a small category. Let* $\mathscr{C}$ *be a cocomplete category. Given a functor* $G: \mathscr{A} \longrightarrow \mathscr{C}$, *the canonical natural transformation* $G \Rightarrow (\operatorname{Lan}_F G) \circ F$ *is an isomorphism.*

Proof   We use the notation of 3.7.2. Given an object $A \in \mathscr{A}$, let us prove that the pair $(A, 1_{FA})$ is a terminal object for $\mathscr{E}_{FA}$. Given $(A', b) \in \mathscr{E}_{FA}$, we have a morphism $b: FA' \longrightarrow FA$ which, by fullness of F, has the form Fa for some $a: A' \longrightarrow A$. In particular we obtain a morphism $a: (A', b) \longrightarrow (A, 1_{FA})$ in $\mathscr{E}_{FA}$. Given another morphism of this kind $a': (A', b) \longrightarrow (A, 1_{FA})$, the relation $Fa' \circ 1_{FA} = b$ implies $Fa' = Fa$, thus $a' = a$ by faithfulness of F. Applying 2.11.5 we conclude that

$$KFA \cong \operatorname{colim} G \circ \phi_{FA} \cong G \circ \phi_{FA}(A, 1_{FA}) \cong GA. \qquad \square$$

Proposition 3.7.4 *Consider categories* $\mathscr{A}, \mathscr{B}, \mathscr{C}, \mathscr{D}$ *with* $\mathscr{A}, \mathscr{B}$ *small. Consider functors* $F: \mathscr{A} \longrightarrow \mathscr{B}$, $G: \mathscr{A} \longrightarrow \mathscr{C}$, $K: \mathscr{B} \longrightarrow \mathscr{C}$, $L: \mathscr{C} \longrightarrow \mathscr{D}$, $R: \mathscr{D} \longrightarrow \mathscr{C}$ *with* L *left adjoint to* R *and* $K = \operatorname{Lan}_F G$. *Under these conditions* $L \circ \operatorname{Lan}_F G = \operatorname{Lan}_F LG$ *(see diagram 3.16).*

Proof   Applying 3.7.1 and 3.2.4, we get the following bijections, for every functor $H: \mathscr{B} \longrightarrow \mathscr{D}$.

$$\operatorname{Nat}(L \circ \operatorname{Lan}_F G, H) \cong \operatorname{Nat}(\operatorname{Lan}_F G, RH)$$
$$\cong \operatorname{Nat}(G, RHF)$$
$$\cong \operatorname{Nat}(LG, HF)$$
$$\cong \operatorname{Nat}(\operatorname{Lan}_F LG, H).$$

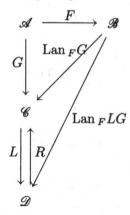

Diagram 3.16

So $L \circ \mathrm{Lan}_F G \cong \mathrm{Lan}_F LG$, by putting successively $H = L \circ \mathrm{Lan}_F G$, $H = \mathrm{Lan}_F LG$.   □

We have already seen that the existence of an adjoint functor or a Kan extension can be reduced to the existence of a limit or a colimit (see 3.3.1 and 3.7.2). On the other hand the existence of a limit or a Kan extension can be reduced to that of an adjoint functor (see 3.2.3 and 3.7.1). We shall complete the picture by proving that the existence of a limit or an adjoint functor can be reduced to that of a Kan extension.

Proposition 3.7.5 *Consider a functor $G: \mathscr{A} \longrightarrow \mathscr{C}$, with $\mathscr{A}$ a small category. Write $\mathbf{1}$ for the category with a single object and a single arrow, and $F: \mathscr{A} \longrightarrow \mathbf{1}$ for the corresponding functor. The functor G has a colimit if and only if the left Kan extension $\mathrm{Lan}_F G$ of G along F exists.*

Proof   Going back to the proof of 3.7.2 and writing $*$ for the unique object of $\mathbf{1}$, we observe that $\mathscr{E}_*$ is just the category $\mathscr{A}$, so that when $\mathrm{colim}\,G$ exists, so does $\mathrm{colim}\,G \circ \phi_*$ and $\mathrm{Lan}_F G$ can be constructed pointwise as in 3.7.2.

Conversely if $\mathrm{Lan}_F G$ exists, it is just the choice of an object $L \in \mathscr{C}$ and a natural transformation $G \Rightarrow \Delta_L$, i.e. a cocone on G with vertex L. The universality of this cocone is just the universality of $\mathrm{Lan}_F G$.   □

Proposition 3.7.6 *Consider a functor $F: \mathscr{A} \longrightarrow \mathscr{B}$ between small categories. The following conditions are equivalent:*

(1) F has a right adjoint G;

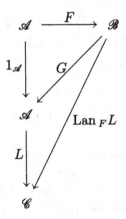

Diagram 3.17

(2) Lan $_F 1_{\mathscr{A}}$ exists and, for every functor $L: \mathscr{A} \longrightarrow \mathscr{C}$, the isomorphism
$L \circ$ Lan $_F 1_{\mathscr{A}} \cong$ Lan $_F L$ holds;

(3) Lan $_F 1_{\mathscr{A}}$ exists and the isomorphism $F \circ$ Lan $_F 1_{\mathscr{A}} \cong$ Lan $_F F$ holds.
(see diagram 3.17.)

Proof (1) $\Rightarrow$ (2). Let us write $\eta: 1_{\mathscr{A}} \Rightarrow GF$ and $\varepsilon: FG \Rightarrow 1_{\mathscr{B}}$ for the natural transformations describing the adjunction (see 3.1.5). We immediately get a natural transformation $\alpha = L * \eta: L \Rightarrow LGF$. Given a functor $H: \mathscr{B} \longrightarrow \mathscr{C}$ and a natural transformation $\beta: L \Rightarrow HF$, let us consider the composite

$$\gamma = (H * \varepsilon) \circ (\beta * G): LG \Rightarrow HFG \Rightarrow H.$$

The relation $(\gamma * F) \circ \alpha = \beta$ follows from the naturality of β and the triangular equalities of the adjunction (see 3.1.5):

$$\begin{aligned}
(\gamma * F) \circ \alpha &= (H * \varepsilon * F) \circ (\beta * G * F) \circ (L * \eta) \\
&= (H * \varepsilon * F) \circ (H * F * \eta) \circ \beta \\
&= H\big((\varepsilon * F) \circ (F * \eta)\big) \circ \beta \\
&= H(1_p) \circ \beta \\
&= \beta.
\end{aligned}$$

Moreover, if $\gamma': LG \Rightarrow H$ satisfies $(\gamma' * F) \circ \alpha = \beta$, we compute immediately that

$$\begin{aligned}
\gamma &= (H * \varepsilon) \circ (\beta * G) \\
&= (H * \varepsilon) \circ (\gamma' * F * G) \circ (L * \eta * G)
\end{aligned}$$

$$= \gamma' \circ (L * G * \varepsilon) \circ (L * \eta * G)$$
$$= \gamma' \circ L\big((G * \varepsilon) \circ (\eta * G)\big)$$
$$= \gamma' \circ L(1_G)$$
$$= \gamma'.$$

This proves that $(L \circ G, L * \eta) = \operatorname{Lan}_F L$ for every functor L.

(2) $\Rightarrow$ (3) is obvious.

(3) $\Rightarrow$ (1) Let us write $\operatorname{Lan}_F 1_{\mathscr{A}} = (G, \eta)$. By assumption $\operatorname{Lan}_F F = (F \circ G, F * \eta)$. Considering the functor $1_{\mathscr{B}} \colon \mathscr{B} \longrightarrow \mathscr{B}$ and the identity natural transformation $F = 1_{\mathscr{B}} \circ F$, the definition of $\operatorname{Lan}_F F$ yields a natural transformation $\varepsilon \colon F \circ G \Rightarrow 1_{\mathscr{B}}$ with the property $(\varepsilon * F) \circ (F * \eta) = 1_F$. It remains to prove that $(G * \varepsilon) \circ (\eta * G) = 1_G$ (see 3.1.5) which, by the uniqueness condition in the definition of $\operatorname{Lan}_F 1_{\mathscr{A}}$, is equivalent to proving

$$\Big(\big((G * \varepsilon) \circ (\eta * G)\big) * F\Big) \circ \eta = (1_G * F) \circ \eta.$$

Indeed, using the naturality of η and the triangular equality which is already proved, we get

$$\Big(\big((G * \varepsilon) \circ (\eta * G)\big) * F\Big) \circ \eta = (G * \varepsilon * F) \circ (\eta * G * F) \circ \eta$$
$$= (G * \varepsilon * F) \circ (G * F * \eta) \circ \eta$$
$$= G\big((\varepsilon * F) \circ (F * \eta)\big) \circ \eta$$
$$= G(1_F) \circ \eta$$
$$= \eta. \qquad \square$$

3.8 Tensor product of set-valued functors

Proposition 3.8.1 *Consider a small category $\mathscr{A}$ and two functors*

$$F \colon \mathscr{A}^* \longrightarrow \mathsf{Set}, \quad G \colon \mathscr{A} \longrightarrow \mathsf{Set}.$$

Write

$$Y \colon \mathscr{A} \longrightarrow \mathsf{Fun}(\mathscr{A}^*, \mathsf{Set}), \quad Y^* \colon \mathscr{A}^* \longrightarrow \mathsf{Fun}(\mathscr{A}, \mathsf{Set})$$

for the covariant Yoneda embeddings

$$Y(A) = \mathscr{A}(-, A), \quad Y^*(A) = \mathscr{A}(A, -).$$

Under these considions, there is a bijection

$$(\operatorname{Lan}_{Y^*} F)(G) \cong (\operatorname{Lan}_Y G)(F).$$

Proof We can apply the constructions of 3.7.2 to produce these two objects. More precisely we consider the category $\mathscr{C}$ defined as follows.

- *Objects:* pairs (A, α) where $\alpha \colon \mathscr{A}(A, -) \Rightarrow G$ is a natural transformation.
- *Arrows:* an arrow $f \colon (A, \alpha) \longrightarrow (A', \alpha')$. is a morphism $f \colon A' \longrightarrow A$ in $\mathscr{A}$ such that $\alpha \circ \mathscr{A}(f, -) = \alpha'$

We also take the obvious forgetful functor $\phi \colon \mathscr{C} \longrightarrow \mathscr{A}^*$ and consider the category $\mathscr{D}$ defined as follows.

- *Objects:* pairs (B, β) where $\beta \colon \mathscr{A}(-, B) \Rightarrow F$ is a natural transformation.
- *Arrows:* an arrow $g \colon (B, \beta) \longrightarrow (B', \beta')$ is a morphism $g \colon B \longrightarrow B'$ in $\mathscr{A}$ such that $\beta' \circ \mathscr{A}(-, g) = \beta$.

Likewise we take the obvious forgetful functor $\psi \colon \mathscr{D} \longrightarrow \mathscr{A}$. We must prove that, as objects of **Set**,

$$\operatorname{colim}_{(A,\alpha)}(F \circ \phi)(A, \alpha) \cong \operatorname{colim}_{(B,\beta)}(G \circ \psi)(B, \beta).$$

Let us recall (see 2.15.6) that we have the following isomorphisms:

$$G = \operatorname{colim}_{(A,\alpha)}(Y^* \circ \phi)(A, \alpha),$$
$$F = \operatorname{colim}_{(B,\beta)}(Y \circ \psi)(B, \beta).$$

Applying 2.15.2 and 2.12.1, we get

$$\begin{aligned}
(\operatorname{Lan}_{Y^*} F)(G) &\cong \operatorname{colim}_{(A,\alpha)} F(A) \\
&\cong \operatorname{colim}_{(A,\alpha)} \Big(\big(\operatorname{colim}_{(B,\beta)} \mathscr{A}(-, B)\big)(A) \Big) \\
&\cong \operatorname{colim}_{(A,\alpha)} \operatorname{colim}_{(B,\beta)} \mathscr{A}(A, B) \\
&\cong \operatorname{colim}_{(B,\beta)} \operatorname{colim}_{(A,\alpha)} \mathscr{A}(A, B) \\
&\cong \operatorname{colim}_{(B,\beta)} \Big(\big(\operatorname{colim}_{(A,\alpha)} \mathscr{A}(A, -)\big)(B) \Big) \\
&\cong \operatorname{colim}_{(B,\beta)} G(B) \\
&\cong (\operatorname{Lan}_Y G)(F). \qquad \qquad \square
\end{aligned}$$

Under the conditions of 3.8.1, one often writes $F \otimes G$ for the set

$$(\operatorname{Lan}_{Y^*} F)(G) \cong F \otimes G \cong (\operatorname{Lan}_Y G)(F).$$

This interchange formula will play a key role in the theories of left exact and flat functors (see chapter 6).

3.9 Exercises

3.9.1 Consider the category $\mathscr{C}$ with a single object $*$ and just two arrows: the identity 1 on $*$ and a morphism f such that $f \circ f = 1$. Prove that f determines a natural transformation $\varphi \colon 1_\mathscr{C} \Rightarrow 1_\mathscr{C}$. The identity functor $1_\mathscr{C}$ on $\mathscr{C}$ is left adjoint to itself and the corresponding natural transformations $\eta \colon 1_\mathscr{C} \Rightarrow 1_\mathscr{C} \circ 1_\mathscr{C}$, $\varepsilon \colon 1_\mathscr{C} \circ 1_\mathscr{C} \Rightarrow 1_\mathscr{C}$ can both be chosen to be the identity; but they can also both be chosen to be the transformation φ.

3.9.2 Prove that a functor $F \colon \mathscr{A} \longrightarrow \mathscr{B}$ has a left adjoint functor if and only if for every object $B \in \mathscr{B}$, the functor

$$\mathscr{B}(B, F-) \colon \mathscr{A} \longrightarrow \mathsf{Set}$$

is representable.

3.9.3 Consider the category Idem whose objects are the pairs (X, v) of a set X provided with an idempotent endomorphism $v \colon X \longrightarrow X$, $v \circ v = v$. A morphism of Idem $(X, v) \longrightarrow (Y, w)$ is just a mapping $f \colon X \longrightarrow Y$ satisfying $w \circ f = f \circ v$. There is a canonical full embedding of the category Set of sets and mappings in Idem:

$$i \colon \mathsf{Set} \longrightarrow \mathsf{Idem}, \quad X \mapsto (X, 1_X).$$

Consider now the functor determined by

$$j \colon \mathsf{Idem} \longrightarrow \mathsf{Set}, \quad (X, v) \mapsto \left\{ x \in X \,\middle|\, v(x) = x \right\}.$$

Prove that j is both left and right adjoint to i.

3.9.4 In exercise 3.9.1, observe the existence of a pair of functors $1_\mathscr{C}$, $1_\mathscr{C} \colon \mathscr{C} \overleftrightarrow{} \mathscr{C}$ and a pair of natural isomorphisms

$$\mathrm{id} \colon 1_\mathscr{C} \Rightarrow 1_\mathscr{C} \circ 1_\mathscr{C}, \quad \varphi \colon 1_\mathscr{C} \circ 1_\mathscr{C} \longrightarrow 1_\mathscr{C},$$

expressing the fact that $1_\mathscr{C}$ is an equivalence of categories (see 3.4.3.(3)). Prove that id, φ do not exhibit the adjunction between $1_\mathscr{C}$ and $1_\mathscr{C}$ (see 3.4.3.(2)).

3.9.5 Consider two functors $F \colon \mathscr{A} \longrightarrow \mathscr{B}$ and $G \colon \mathscr{A} \longrightarrow \mathscr{C}$, with the property

$$\forall A, A' \in \mathscr{A} \quad \forall f, f' \colon A \rightrightarrows A' \quad Gf = Gf' \Rightarrow Ff = Ff'.$$

Suppose $\mathscr{A}$, $\mathscr{B}$ are small, $\mathscr{C}$ is cocomplete and F is full. Under these conditions $\mathrm{Lan}_G F$ exists and the isomorphism $G \cong \mathrm{Lan}_G F \circ F$ holds.

3.9.6 Consider a functor $F: \mathscr{A} \longrightarrow \mathscr{B}$ between small categories. For each object $A \in \mathscr{A}$ the Kan extension $\operatorname{Lan}_F \mathscr{A}(A, -)$ exists and is given by $\mathscr{B}(FA, -)$. [Hint: apply the Yoneda lemma.] The equality

$$\operatorname{Lan}_F \mathscr{A}(A, -) \circ F = \mathscr{A}(A, -)$$

holds precisely when F is full and faithful.

3.9.7 Consider the category **1** with a single object A and a single arrow 1_A, the category **2** with two distinct objects A, B and just the identity arrows $1_A, 1_B$, and the category $\vec{\mathbf{2}}$ with two distinct objects A, B, the identity arrrows $1_A, 1_B$ and an additional arrow $f: B \longrightarrow A$. Define the functors $F: \mathbf{1} \longrightarrow \vec{\mathbf{2}}$ and $G: \mathbf{1} \longrightarrow \mathbf{2}$ by $FA = A$, $GA = A$. Observe that the only two functors from $\vec{\mathbf{2}}$ to **2** are the constant functors Δ_A and Δ_B. Check that Δ_A is the left Kan extension $\operatorname{Lan}_F G$. With the notation of 3.7.2, prove that $\mathscr{E}_B$ is an empty category and $\Delta_A(B)$ is not isomorphic to the colimit of $G \circ \phi_B$. So the Kan extension $\operatorname{Lan}_F G$ is not pointwise.

4

Generators and projectives

4.1 Well-powered categories

Given a set X, each of us knows the notion of "subset of X". Our best approximation to this is, up to now, the notion of "monomorphism with codomain X". But for example $f: \{\pi\} \longrightarrow \mathbb{N}$, $f(\pi) = 0$ is a monomorphism in Set... but is not formally a "subset" of $\mathbb{N}$. In fact a monomorphism $f: A \longrightarrow B$ in Set defines a subset of B, namely the subset $f(A)$, which as a set is isomorphic to A via f.

Definition 4.1.1 *Consider a category $\mathscr{A}$ and an object $A \in \mathscr{A}$. Two monomorphisms $f: R \rightarrowtail A$ and $g: S \rightarrowtail A$ are equivalent when there exists an isomorphism $\tau: R \xrightarrow{\cong} S$ such that $g \circ \tau = f$. An equivalence class of monomorphisms with codomain A is called a subobject of A. The dual notion is that of a "quotient of A".*

When the category $\mathscr{A}$ is not small, there can be a proper class of subobjects of a given object $A \in \mathscr{A}$.

Definition 4.1.2 *A category $\mathscr{A}$ is well-powered when the subobjects of every object constitute a set.*

It is most often obvious to observe that a category is well-powered:

(1) in the category Set of sets, the subobjects of a set X are in bijection with the subsets of X;
(2) in the category Gr of groups, the subobjects of a group G are in bijection with the subgroups of G;
(3) in the category Top of topological spaces, the subobjects of a topological space $(X, \mathcal{T})$ are in bijection with the pairs $(Y, \mathcal{S})$, where Y is a subset of X and $\mathcal{S}$ is a topology finer than the topology on Y induced by $\mathcal{T}$.

And so on .... It is thus obvious, in all those cases, that the category is well-powered. Examples of non-well-powered categories are often quite unnatural. For instance observe that in a partially ordered set (or class), each arrow is of course a monomorphism, since you can never find two distinct arrows between two specified objects. Therefore in a partially ordered class with top element 1, the subobjects of 1 are in bijection with the elements of the class.

4.2 Intersection and union

Given an object A of a category $\mathscr{C}$, let us consider the class $\mathsf{Mono}(A)$ of all monomorphisms with codomain A. A monomorphism $r\colon R \rightarrowtail A$ is smaller than a monomorphism $s\colon S \rightarrowtail A$ when there exists a (mono) morphism $t\colon R \rightarrowtail S$ such that $s \circ t = r$. Observe that t can be chosen an isomorphism precisely when, in addition, s is smaller than r. Indeed if there exists $t'\colon S \rightarrowtail R$ such that $r \circ t' = s$, we have $r \circ t' \circ t = r$ and $s \circ t \circ t' = s$, from which $t' \circ t = 1_R$ and $t \circ t' = 1_S$ since r and s are monomorphisms. Performing the quotient on $\mathsf{Mono}(A)$ which identifies isomorphic monomorphisms, we obtain a partial order on the class $\mathsf{Sub}(A)$ of subobjects of A. We recall that $\mathscr{C}$ is well-powered when, for each $A \in \mathscr{C}$, $\mathsf{Sub}(A)$ is a set (see 4.1.2).

Since $\mathsf{Sub}(A)$ is a partially ordered class, it makes sense to consider the existence of the infimum or the supremum of a family of subobjects. Let us make clear that by "family", we always mean a set-indexed family. Except when some confusion could arise, we speak freely of the monomorphism $r\colon R \rightarrowtail A$ or the corresponding subobject, without emphasizing the difference.

Definition 4.2.1 *Consider an object A of a category $\mathscr{C}$. By the intersection of a family of subobjects of A, we mean their infimum in $\mathsf{Sub}(A)$. By the union of a family of subobjects of A, we mean their supremum in $\mathsf{Sub}(A)$.*

Proposition 4.2.2 *Consider an object A of a category $\mathscr{C}$ and suppose $\mathsf{Sub}(A)$ is a set. The following conditions are equivalent:*
(1) the intersection of every family of subobjects of A exists;
(2) the union of every family of subobjects of A exists.

Proof In a poset, it is well-known that

$$\sup_{i\in I} s_i = \inf\{s \mid \forall i \in I \ \ s_i \le s\},$$
$$\inf_{i\in I} s_i = \sup\{s \mid \forall i \in I \ \ s \le s_i\}. \qquad \square$$

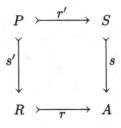

Diagram 4.1

Proposition 4.2.3 *In a category $\mathscr{C}$ with pullbacks, the intersection of two subobjects of the same object $A \in \mathscr{C}$ always exists and is given by their pullback.*

Proof If r, s are monomorphisms, so are r', s' obtained by pullback (see 2.5.3) as in diagram 4.1. Thus $r \circ s' = s \circ r' \colon P \rightarrowtail A$ is another subobject of A and, by definition of a pullback, it is obviously the infimum of r, s in $\mathsf{Sub}(A)$. □

The previous result generalizes easily to the case of arbitrary intersections:

Proposition 4.2.4 *In a complete category, the intersection of every family of subobjects of a fixed object always exists.*

Proof Given first a non-empty family $(s_i \colon S_i \rightarrowtail A)_{i \in I}$ of subobjects, we compute the limit $\bigl(L, (p_i)_{i \in I}\bigr)$ of the diagram constituted of the various morphisms s_i. All the composites $s_i \circ p_i \colon L \longrightarrow A$ are equal by definition of a limit and, since the set I of indices is not empty, this effectively gives us a morphism $s \colon L \longrightarrow A$. This morphism s is a monomorphism because, given $x, y \colon X \rightrightarrows L$ with $s \circ x = s \circ y$, we have $s_i \circ p_i \circ x = s \circ x = s \circ y = s_i \circ p_i \circ y$ for every $i \in I$, and thus $p_i \circ x = p_i \circ y$ since s_i is a monomorphism. But then $x = y$ (see 2.6.4). Just by definition of a limit, the subobject $s \colon L \rightarrowtail A$ is the intersection of the subobjects $(s_i)_{i \in I}$.

It remains to consider the case of an empty set of indices, i.e. to prove the existence of a biggest subobject of A. This is just the identity on A. □

Corollary 4.2.5 *In a complete and well-powered category, the intersection and the union of every family of subobjects of a fixed object always exist.*

Proof By 4.2.4 and 4.2.2. □

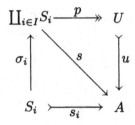

Diagram 4.2

At this stage one should avoid a classical mistake. Computing the union of two subobjects is by no means a problem dual to that of computing their intersection. Dualizing 4.2.3 tells us something about the poset of quotients of A (epimorphisms with domain A), not about unions in $\mathsf{Sub}(A)$. In the same way let us observe that in 4.2.5 the existence of unions is by no means related to any assumption on colimits: it relies on the formal formulas used in 4.2.2. In particular a finite version of 4.2.5 does not hold: a finitely complete and well-powered category certainly admits finite intersections of subobjects (see 4.2.3), but not in general finite unions of subobjects. Finite unions have been constructed in 4.2.5 using possibly infinite intersections. For a counterexample, just consider a $\wedge$-semi-lattice with a top element which is not a lattice.

Anticipating the results of section 4.4, let us give a construction of the union of subobjects which applies very widely.

Proposition 4.2.6 *In a category with (finite) coproducts and strong-epi–mono factorizations, the union of a (finite) family of subobjects always exists.*

Proof   Consider subobjects $s_i \colon S_i \rightarrowtail A$, the corresponding factorization $s \colon \coprod_{i \in I} S_i \longrightarrow A$ through the coproduct and the canonical morphisms $\sigma_i \colon S_i \longrightarrow \coprod_{i \in I} S_i$ of the coproduct. Let us write $s = u \circ p$ for the strong-epi–mono factorization of s. This yields the commutative diagram 4.2. In particular each s_i factors through u, proving $S_i \subseteq U$.

Next consider $t \colon T \rightarrowtail A$, another subobject through which each s_i factors as $s_i = t \circ t_i$ (see diagram 4.3). There is a unique factorization τ through the coproduct, such that $\tau \circ \sigma_i = t_i$. Since

$$t \circ \tau \circ \sigma_i = t \circ t_i = s_i = s \circ \sigma_i,$$

we deduce that $t \circ \tau = s$, by definition of a coproduct. Since t is a monomorphism, by 4.4.5.(3) u factors through t. $\square$

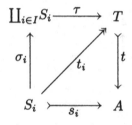

Diagram 4.3

4.3 Strong epimorphisms

Among epimorphisms, let us point out some particularly interesting classes.

Definition 4.3.1 *In a category, an epimorphism is called regular when it is the coequalizer of a pair of arrows.*

Definition 4.3.2 *An epimorphism $f: A \longrightarrow B$ in a category is called extremal when it does not factor through any proper subobject of B; i.e. given $f = i \circ p$ with i a monomorphism, i is necessarily an isomorphism.*

Let us first observe some obvious properties.

Proposition 4.3.3 *In a category $\mathscr{C}$*

(1) *every regular epimorphism is extremal,*
(2) *if a composite $f \circ g$ is an extremal epimorphism, f itself is an extremal epimorphism,*
(3) *a morphism which is both a monomorphism and an extremal epimorphism is an isomorphism.*

Proof   Let $f = \text{Coker}\,(u, v)$ and $f = i \circ p$, with i a monomorphism. From $i \circ p \circ u = f \circ u = f \circ v = i \circ p \circ v$ one deduces $p \circ u = p \circ v$, since i is a monomorphism. Thus there exists a unique factorization j through $f = \text{Coker}\,(u, v)$, such that $j \circ f = p$ (see diagram 4.4). From $i \circ j \circ f = i \circ p = f$, we deduce $i \circ j = 1_B$ since f is an epimorphism. From $i \circ j \circ i = 1_B \circ i = i$, one deduces $j \circ i = 1_I$ since i is a monomorphism.

Now if $f \circ g$ is an extremal epimorphism, f is an epimorphism by 1.8.2. If $f = i \circ p$ with i a monomorphism, $f \circ g = i \circ p \circ g$ and i is an isomorphism because $f \circ g$ is an extremal epimorphism.

Finally if $f: A \longrightarrow B$ is both a monomorphism and an extremal epimorphism, from $f = f \circ 1_A$ we deduce that f is an isomorphism.   □

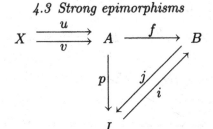

Diagram 4.4

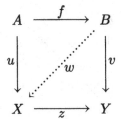

Diagram 4.5

Proposition 4.3.4 *In a category $\mathscr{A}$, suppose that every morphism which is both a monomorphism and an epimorphism is necessarily an isomorphism. In those conditions, every epimorphism is extremal.*

Proof If $f = i \circ p$ with f an epimorphism and i a monomorphism, i is both a monomorphism and an epimorphism (see 1.8.2), thus it is an isomorphism. □

The notion of regular epimorphism will prove to be crucial (see for example chapter 2 of volume 2). On the other hand, the notion of extremal epimorphism coincides most often with the more sophisticated notion of "strong epimorphism" and this notion plays a key role in category theory.

Definition 4.3.5 *In a category $\mathscr{A}$, an epimorphism $f: A \longrightarrow B$ is called a strong epimorphism when, for every commutative square $z \circ u = v \circ f$ as in diagram 4.5, with $z: X \longrightarrow Y$ a monomorphism, there exists a (unique) arrow $w: B \longrightarrow X$ such that $w \circ f = u$, $z \circ w = v$.*

The uniqueness condition in 4.3.5 is of course redundant since by assumption, f is an epimorphism and z is a monomorphism. Here are the key properties of strong epimorphisms:

Proposition 4.3.6 *In a category $\mathscr{A}$,*

(1) the composite of two strong epimorphisms is a strong epimorphism,

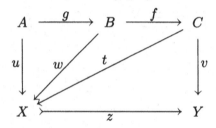

Diagram 4.6

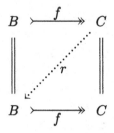

Diagram 4.7

(2) *if a composite $f \circ g$ is a strong epimorphism, f is a strong epimorphism,*

(3) *a morphism which is both a monomorphism and a strong epimorphism is an isomorphism,*

(4) *every regular epimorphism is strong,*

(5) *every strong epimorphism is extremal.*

Proof If f, g are strong epimorphisms, in diagram 4.6 choose $z \circ u = v \circ f \circ g$ with z a monomorphism. Since g is a strong epimorphism, we find w such that $w \circ g = u$ and $z \circ w = vf$. Further since f is a strong epimorphism, we find t such that $t \circ f = w$ and $z \circ t = v$. It is obvious that t is the required factorization.

Now suppose that $f \circ g$ is a strong epimorphism in diagram 4.6 and choose $z \circ w = v \circ f$, with z a monomorphism. Putting $u = w \circ g$, one gets a factorization t such that $t \circ f \circ g = u$, $z \circ t = v$ since $f \circ g$ is a strong epimorphism. From $z \circ t \circ f = v \circ f = z \circ w$ one deduces $t \circ f = w$ since z is a monomorphism. It is obvious that t is the required factorization.

If f is both a monomorphism and a strong epimorphism, considering diagram 4.7 we find r such that $r \circ f = 1_B$, $f \circ r = 1_C$. Thus f is an isomorphism.

If $f = \text{Coker}\,(a, b)$ and $z \circ w = v \circ f$ with z a monomorphism, consider diagram 4.8. From $z \circ w \circ a = v \circ f \circ a = v \circ f \circ b = z \circ w \circ b$, we

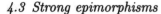

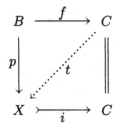

Diagram 4.8

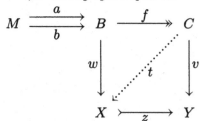

Diagram 4.9

deduce $w \circ a = w \circ b$ since z is a monomorphism. Therefore we get some factorization t of w through $f = \mathsf{Coker}\,(a, b)$. From $w = t \circ f$ we deduce $z \circ t \circ f = z \circ w = v \circ f$ and thus $z \circ t = v$ since f is an epimorphism (see 2.4.3).

Finally if f is a strong epimorphism and $f = i \circ p$ with i a monomorphism, consider diagram 4.9. There exists a unique t yielding $t \circ f = p$, $i \circ t = 1_C$. By the dual of 1.9.3, i is an isomorphism.   □

Strong epimorphisms play a particularly nice role in finitely complete categories. This is due to the following results:

Proposition 4.3.7 *Consider a finitely complete category $\mathscr{A}$.*

(1) *Let a morphism $f \colon A \longrightarrow B$ satisfy the diagonal property of 4.3.5, i.e. given $z \circ u = v \circ f$ with z a monomorphism, there exists a unique w such that $z \circ w = v$, $w \circ f = u$. Under these conditions, f is an epimorphism and thus a strong epimorphism.*

(2) *Let a morphism $f \colon A \longrightarrow B$ be such that given any factorization $f = i \circ p$ with i a monomorphism, i must be an isomorphism. Under these conditions f is an epimorphism, thus an extremal epimorphism.*

(3) *An epimorphism is extremal if and only if it is strong.*

Proof   Take a morphism $f \colon A \longrightarrow B$ satisfying the "diagonal condition". If $u \circ f = v \circ f$, consider diagram 4.10 where $k = \mathsf{Ker}\,(u, v)$ and

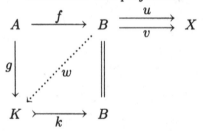

Diagram 4.10

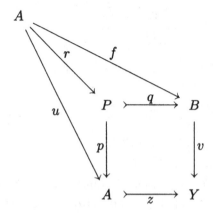

Diagram 4.11

g is the unique factorization of f through $k = \mathsf{Ker}\,(u,v)$. Since k is a monomorphism (see 2.4.3), there exists by assumption a unique w such that $g = w \circ f$ and $k \circ w = 1_B$. Thus k is an epimorphism (see 1.8.3) and from $u \circ k = v \circ k$, we deduce $u = v$. So f is an epimorphism.

Now if f does not factor through any proper subobject of B, choose again $u \circ f = v \circ f$ and $k = \mathsf{Ker}\,(u,v)$. The morphism f factors through $k = \mathsf{Ker}\,(u,v)$ as $f = k \circ g$, thus the monomorphism k is an isomorphism by assumption. From $u \circ k = v \circ k$ we conclude that $u = v$. So f is an epimorphism.

Finally suppose $f \colon A \longrightarrow B$ is an extremal epimorphism. If $z \circ u = v \circ f$ with z a monomorphism, let us consider diagram 4.11 where the square is a pullback. There is a unique factorization r such that $q \circ r = f$, $p \circ r = u$. Since z is a monomorphism, q is a monomorphism as well (see 2.5.3). Since f is an extremal epimorphism, q is an isomorphism. Therefore $p \circ q^{-1} \colon B \longrightarrow A$ is the expected factorization. It is unique since f is an epimorphism and z is a monomorphism. The converse implication has been proved in 4.3.6.(5). $\qquad\qquad \square$

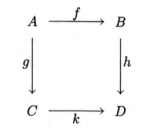

Diagram 4.12

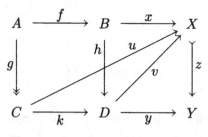

Diagram 4.13

Another major property of strong or regular epimorphisms is attested by the following result (compare with the dual of 2.5.3).

Proposition 4.3.8 *In a category* $\mathscr{C}$,

(1) strong epimorphisms are stable under pushout,

(2) regular epimorphisms are stable under pushout,

as long as the required pushouts exist. In other words, in a pushout square as in diagram 4.12, if g is a strong epimorphism, h is a strong epimorphism as well; if g is a regular epimorphism, h is a regular epimorphism as well.

Proof   If g is a strong epimorphism, consider the commutative diagram 4.13 where z is a monomorphism. We first find u such that $u \circ g = x \circ f$ and $z \circ u = y \circ k$. Since $u \circ g = x \circ f$ and the original square is a pushout, we get a unique v such that $v \circ h = x$ and $v \circ k = u$. Notice that h is an epimorphism, since g is (see 2.5.3). Thus from $z \circ v \circ h = z \circ x = y \circ h$ we deduce $z \circ v = y$. The uniqueness of v follows from the fact that z is a monomorphism (or h an epimorphism).

Now if g is a regular epimorphism, consider diagram 4.14 where $g = $ Coker (m, n). Let us prove that $h = $ Coker $(f \circ m, f \circ n)$. First of all

$$h \circ f \circ m = k \circ g \circ m = k \circ g \circ n = h \circ f \circ n.$$

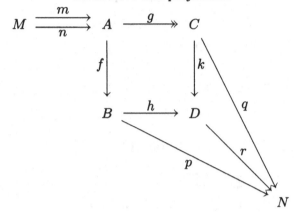

Diagram 4.14

Next consider p such that $p \circ f \circ m = p \circ f \circ n$. Since $g = \mathsf{Coker}\,(m, n)$, there exists a unique q such that $q \circ g = p \circ f$ and since the square is a pushout, this yields a unique r such that $r \circ h = p$, $r \circ k = q$. This morphism r is also unique for just the property $r \circ h = p$ since h is an epimorphism (see 2.5.3). $\qquad\square$

Like epimorphisms, the notions of strong or regular epimorphism are preserved by right adjoint functors (see 3.2.2 and 2.9.3).

Proposition 4.3.9 *Let* $F: \mathcal{A} \longrightarrow \mathcal{B}$ *be a functor admitting a left adjoint functor* $G: \mathcal{B} \longrightarrow \mathcal{A}$. *The functor* F *preserves strong monomorphisms and regular monomorphisms, and the functor* G *preserves strong epimorphisms and regular epimorphisms.*

Proof By 3.1.5, 3.2.2 and 2.9.3, it suffices to prove the statement concerning strong epimorphisms. Let us write

$$\eta: 1_{\mathcal{B}} \Rightarrow FG, \quad \varepsilon: GF \Rightarrow 1_{\mathcal{A}}$$

for the canonical natural transformations of the adjunction.

Given a strong epimorphism f in $\mathcal{B}$ and a diagram $v \circ Gf = z \circ u$ in $\mathcal{A}$ with z a monomorphism, consider the situation of diagram 4.15. F preserves limits (see 3.2.2), thus Fz is still a monomorphism (see 2.9.3). By naturality of η,

$$Fv \circ \eta_B \circ f = Fv \circ FGf \circ \eta_A = Fz \circ Fu \circ \eta_A.$$

So the second square is commutative, from which there is a unique w such that $w \circ f = Fu \circ \eta_A$, $Fz \circ w = Fv \circ \eta_B$. Therefore we get (see

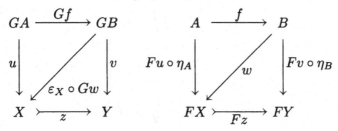

Diagram 4.15

3.1.5)

$$\varepsilon_X \circ Gw \circ Gf = \varepsilon_X \circ GFu \circ G\eta_A = u \circ \varepsilon_{GA} \circ G\eta_A = u,$$

$$z \circ \varepsilon_X \circ Gw = \varepsilon_Y \circ GFz \circ Gw = \varepsilon_Y \circ GFv \circ G\eta_B$$

$$= v \circ \varepsilon_{GB} \circ G\eta_B = v. \qquad \Box$$

Examples 4.3.10

4.3.10.a In the categories Set of sets, Gr of groups, Ab of abelian groups, Mod_R of R-modules, the epimorphisms $f\colon A \longrightarrow B$ are exactly the surjective homomorphisms (see 1.8.5). So B is the quotient of A by the equivalence relation identifying a, a' when $f(a) = f(a')$. In other words, f is in each case the coequalizer of its kernel pair (see section 2.5) and so every epimorphism is regular. In particular every epimorphism is also strong and extremal (see 4.3.3).

In the same categories, every monomorphism is regular, thus strong. In the category Set, a monomorphism $f\colon A \rightarrowtail B$ is the equalizer of its characteristic mapping

$$\varphi\colon B \longrightarrow \{0,1\}, \quad \varphi(b) = \left\{ \begin{array}{ll} 1 & \text{if } b \in f(A) \\ 0 & \text{if } b \notin f(A) \end{array} \right.$$

and the constant mapping $\Delta_1\colon B \longrightarrow \{0,1\}$ on 1. In the category Gr and with the notation of 1.8.5.d, the amalgamation property applied in the case $G = H$ indicates precisely that $K \subseteq G$ is the equalizer of the two canonical inclusions $G \rightrightarrows G *_K G$. In the category Ab, a monomorphism $f\colon A \rightarrowtail B$ is the equalizer of the quotient map $p\colon B \twoheadrightarrow B/f(A)$ and the zero map $0\colon B \longrightarrow B/f(A)$.

4.3.10.b In the category Top of topological spaces and continuous mappings, a morphism $f\colon A \longrightarrow B$ can be factored through its image $f(A) \subseteq B$, provided with the induced topology. But if R is the kernel pair of f,

$$R = \left\{ (a, a') \in A \times A \,\middle|\, f(a) = f(a') \right\},$$

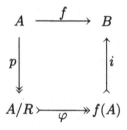

Diagram 4.16

$f(A)$ is a bijection with the quotient $A \longrightarrow A/R$, the bijection being just

$$\varphi \colon A/R \longrightarrow f(A), \quad [a] \mapsto f(a).$$

Providing A/R with the quotient topology, φ is thus continuous. This gives us a factorization $f = i \circ \varphi \circ p$ as in diagram 4.16, with φ a bijection, thus both a monomorphism and an epimorphism. Since **Top** is (finitely) complete and cocomplete, p is an extremal, thus strong, epimorphism because A/R is provided with the quotient topology. In the same way i is a strong monomorphism because $f(A)$ is provided with the induced topology. Observe that in general $\varphi \circ p$ is not a strong epimorphism and $i \circ \varphi$ is not a strong monomorphism.

4.3.10.c   In the category **Rng** of commutative rings with unit, every epimorphism $f \colon A \longrightarrow B$ factors through its image $f(A)$. Therefore the epimorphism is extremal (or strong) precisely when $B = f(A)$, thus when f is surjective. Recall that in **Rng**, there are non-surjective epimorphisms, thus there exist in **Rng** epimorphisms which are not strong (see 1.8.5.f).

4.3.10.d   In the category **Haus** of Hausdorff spaces and continuous mappings, a monomorphism (i.e. an injection) $f \colon A \longrightarrow B$ is extremal precisely when A is provided with the topology induced by that of B. But the kernel $k \colon K \longrightarrow A$ of a pair $g, h \colon A \rightrightarrows B$ can be presented as the pullback of diagram 4.17, where Δ is the diagonal. Since B is Hausdorff, the diagonal is closed in $B \times B$ and therefore K is closed in A. Thus a non-closed subspace $A \longrightarrow B$ with the induced topology is an example of an extremal (or strong) but non-regular monomorphism.

4.3.10.e   In the category **Ban₁** of Banach spaces and linear contractions, monomorphisms coincide with injections (see 1.7.7.f). The equalizer of two morphisms $f, g \colon A \rightrightarrows B$ is just

$$\operatorname{Ker}(f, g) = \{a \in A \,|\, f(a) = g(a)\} = (f - g)^{-1}(0)$$

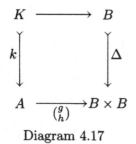

Diagram 4.17

provided with the induced norm; this is indeed a Banach space since the subspace $\mathsf{Ker}\,(f,g)$ is closed in A as inverse image of $\{0\}$ under the continuous mapping $f - g$. Conversely, given a closed linear subspace $K \subseteq A$, this space K is a Banach space and moreover A/K provided with the quotient norm is again a Banach space. Therefore $K \subseteq A$ is the equalizer of the two morphisms $0, p\colon A \rightrightarrows A/K$ where p is the canonical projection. Since in Ban_1 isomorphisms are just isometries (see 1.9.6.f), the regular monomorphisms are the isometric injections.

Observe that in Ban_1, strong monomorphisms coincide with regular monomorphisms. Indeed choose a strong monomorphism $f\colon K \rightarrowtail A$ and consider the Banach space $\overline{f(K)} \subseteq A$. The factorization

$$f\colon K \rightarrowtail \overline{f(K)}$$

is a strong monomorphism since f is (see 4.3.6), but it is also an epimorphism (see 1.8.5.g): therefore it is an isomorphism. Since isomorphisms are isometric (see 2-1.9.6.f), $f\colon K \rightarrowtail A$ is an isometric injection, thus a regular monomorphism.

Notice also that every strong epimorphism in Ban_1 is surjective on the unit balls, from which in particular, every strong epimorphism is surjective. Indeed if $f\colon A \twoheadrightarrow B$ is a strong epimorphism and $b \in B$, $\|b\| \leq 1$, consider diagram 4.18 where

$$g\colon \mathbb{R} \longrightarrow B, \quad g(r) = rb.$$

The existence of the factorization h implies $b = g(1) = f\big(h(1)\big)$ with $\|h(1)\| \leq 1$. Thus f is surjective on the unit balls. Conversely, if a linear contraction $f\colon A \longrightarrow B$ is surjective on the unit balls, it is a strong epimorphism. Indeed f is surjective and we get a linear isomorphism

$$\varphi\colon A/\mathsf{Ker}\,f \xrightarrow{\ \cong\ } B$$

between Banach spaces. Via this isomorphism, the unit ball of $A/\mathsf{Ker}\,f$ is mapped into the unit ball of B because f is a linear contraction. This

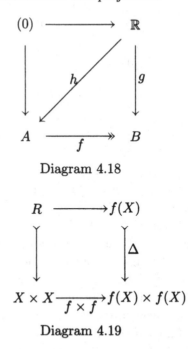

Diagram 4.18

Diagram 4.19

yields a bijection between the unit balls, since f is surjective on the unit balls. Thus via the linear isomorphism φ, both Banach spaces have the same unit ball, hence are isomorphic. So f is a quotient map, thus a regular and strong epimorphism. This proves that in Ban_1, strong epimorphisms are regular and coincide with those linear contractions which are surjective on the unit balls.

4.3.10.f   In the category Comp of compact Hausdorff spaces, every epimorphism $f\colon X \longrightarrow\!\!\!\!\!\rightarrow Y$ is regular. To prove this, consider the set theoretical image $f(X) \subseteq Y$ provided with the (Hausdorff) topology induced by that of Y. Since f is continuous and X is compact, $f(X)$ is compact Hausdorff as well. In the category Set of sets, the surjection $f\colon X \longrightarrow\!\!\!\!\!\rightarrow f(X)$ is the coequalizer of its kernel pair (see 2.5.7) and this kernel pair R can be obtained via the pullback of diagram 4.19, where Δ is the diagonal. Since $f(X)$ is Hausdorff, Δ is closed, thus $R = (f \times f)^{-1}(\Delta)$ is closed in $X \times X$. Since X is compact Hausdorff, $X \times X$ is compact Hausdorff and thus R, closed in it, is compact Hausdorff. So writing p_1, p_2 for the two projections, $p_1, p_2\colon R \rightrightarrows X$ is the kernel pair of f in Comp. Writing $p\colon X \longrightarrow\!\!\!\!\!\rightarrow X/R$ for the topological quotient of X by R, X/R is still compact as a continuous image of the compact space X and it is Hausdorff because R is a closed equivalence

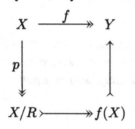

Diagram 4.20

relation. We get the commutative diagram 4.20 in Top where the factorization $X/R \longrightarrow f(X)$ is bijective since in Set, $f(X)$ is the coequalizer of p_1, p_2. But the continuous bijection $X/R \longrightarrow f(X)$ is defined between compact Hausdorff spaces; it is thus a homeomorphism. Finally since f is an epimorphism, $f(X)$ must be dense in Y (the argument of 1.8.5.c applies since a disjoint union of two compact Hausdorff spaces is compact Hausdorff and a Hausdorff quotient of a compact Hausdorff space is compact Hausdorff). But $f(X)$ is closed in Y, since it is compact. Thus $f(X) = Y$ and the epimorphism f is homeomorphic to the coequalizer p.

4.3.10.g If $\mathscr{C}$ is a small category, every epimorphism in the category Fun$(\mathscr{C}, \text{Set})$ of set-valued functors is regular. Indeed, an epimorphism $\alpha: F \Rightarrow G$ is such that each $\alpha_C: FC \longrightarrow GC$ is an epimorphism in Set (see 2.15.3), i.e. the coequalizer of its kernel pair (see 2.5.7). Since kernel pairs and coequalizers in Fun$(\mathscr{C}, \text{Set})$ are computed pointwise (see 2.15.2), α is itself the coequalizer of its kernel pair and therefore is regular, strong and extremal (see 4.3.6,3).

4.4 Epi–mono factorizations

The following notion, even if a little bit strange from a categorical point of view, turns out to be quite useful in the applications.

Definition 4.4.1 *A category $\mathscr{C}$ is finitely well-complete when*

(1) *$\mathscr{C}$ is finitely complete,*
(2) *given an object $C \in \mathscr{C}$, the intersection of an arbitrary class of subobjects of C always exists.*

To understand the unifying role of definition 4.4.1, it suffices to observe that

Proposition 4.4.2

(1) A complete and well-powered category is finitely well-complete.

(2) A finitely complete category where every object has just finitely many subobjects is finitely well-complete.

Proof See 4.2.3 and 4.2.4.   □

Proposition 4.4.3 *In a finitely well-complete category, every morphism f factors as $f = i \circ p$, where i is a monomorphism and p is a strong epimorphism.*

Proof   Given $f: A \longrightarrow B$, consider all the possible factorizations $f = i_k \circ p_k$, with i_k a monomorphism. Compute the intersection $i = I \rightarrowtail B$ of all the monomorphisms i_k. The compatible family $p_k: A \longrightarrow I_k$ of morphisms factors through the limit I of the various I_k, from which $p: A \longrightarrow I$ such that $i \circ p = f$.

   If p admits a factorization $p = j \circ q$ with j a monomorphism, then $f = i \circ p = i \circ j \circ q$ with $i \circ j$ a monomorphism. As subobjects of B, one has of course $i \circ j \leq i$... since $i \circ j$ factors through i via j ! But f factors through $i \circ j$, so $i \circ j$ is one of the i_k's and thus $i \leq i \circ j$. Finally i and $i \circ j$ are isomorphic subobjects, i.e. j is an isomorphism. Therefore p is a strong epimorphism (see 4.3.7.(2)).   □

Definition 4.4.4 *A category $\mathscr{C}$ has strong-epi–mono factorizations when every morphism f of $\mathscr{C}$ factors as $f = i \circ p$, with p a strong epimorphism and i a monomorphism. The monomorphism i is also called the "image" of f.*

Proposition 4.4.5 *Let $\mathscr{C}$ be a category with strong-epi–mono factorizations.*

(1) The strong-epi–mono factorization of an arrow is unique up to an isomorphism.

(2) The strong-epi–mono factorizations are natural in the sense that given the commutative outer rectangle of diagram 4.21, with (i, p) and (j, q) strong-epi–mono factorizations, there exists a unique morphism h making the whole diagram commute.

(3) If $h = i \circ p$ is a strong-epi–mono factorization and $h = k \circ r$, where k is a monomorphism, there exists a unique t such that $r = t \circ p$, $i = k \circ t$ (see diagram 4.22).

Proof   Given $i \circ p = i' \circ p'$ with p, p' strong epimorphisms and i, i' monomorphisms, consider diagram 4.23. There exists u such that $u \circ p =$

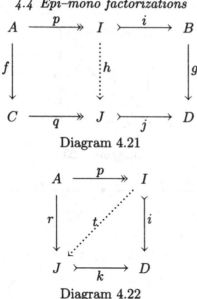

Diagram 4.21

Diagram 4.22

p', $i' \circ u = i$ because p is a strong epimorphism and i' is a monomorphism. There exists v such that $v \circ p' = p$ and $i \circ v = i'$ because p' is a strong epimorphism and i is a monomorphism. Therefore $i \circ v \circ u \circ p = i' \circ p' = i \circ p$, thus $v \circ u = 1_I$ since p is an epimorphism and i is a monomorphism. In the same way $u \circ v = 1_{I'}$.

Let us now consider the situation of point (2). The uniqueness of h is obvious since p is an epimorphism and j is a monomorphism. To prove the existence of h, consider diagram 4.24 where $v \circ u$ and $l \circ k$ are strong-epi–mono factorizations of $q \circ f$ and $g \circ i$. Then $(j \circ v) \circ u$ and $l \circ (k \circ p)$ are both the strong-epi–mono factorization of the global composite $j \circ q \circ f = g \circ i \circ p$. Therefore we get an isomorphism s making diagram 4.24 commutative and h is just $v \circ s^{-1} \circ k$.

For point (3), factor r as $r = l \circ s$ with l a monomorphism and s a strong epimorphism. Putting $f = 1_A$, $g = 1_D$, $q = s$, $j = k \circ l$ in point (2) yields a unique h such that $h \circ p = s$, $k \circ l \circ h = i$, as in diagram 4.25. Putting $t = l \circ h$ yields $t \circ p = l \circ h \circ p = l \circ s = r$, $k \circ t = k \circ l \circ h = i$. Such a morphism t is unique since k is a monomorphism. □

Proposition 4.4.6 *Let $\mathscr{C}$ be a category with pullbacks and epi–strong-mono factorizations. Given a morphism $f \colon A \longrightarrow B$ and writing* $\mathrm{Str}(A)$, $\mathrm{Str}(B)$ *for the posets of strong subobjects, the inverse image functor*

$$f^{-1} \colon \mathrm{Str}(B) \longrightarrow \mathrm{Str}(A)$$

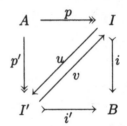

Diagram 4.23

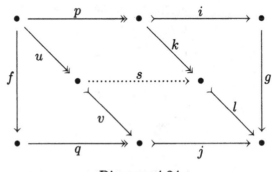

Diagram 4.24

is right adjoint to the direct strong image functor

$$f^{+1} \colon \mathrm{Str}(A) \longrightarrow \mathrm{Str}(B).$$

The same conclusion holds replacing "strong" by "regular".

Proof By 4.3.8, pulling back along f maps the elements of $\mathrm{Str}(B)$ to those of $\mathrm{Str}(A)$. On the other hand, given a subobject $a \colon A' \rightarrowtail A$ in $\mathrm{Str}(A)$ we consider in diagram 4.26 the epi–strong-mono factorization of $f \circ a$, existing by assumption, and this defines f^{+1}. By definition of a pull-back, if $B' \subseteq B$ is in $\mathrm{Str}(B)$ and $f^{+1}(A') \subseteq B'$, one gets $A' \subseteq f^{-1}(B')$. Conversely if $A' \subseteq f^{-1}(B')$, consider the epi–strong-mono factorization of diagram 4.27. Now I is a strong subobject of B' and by 4.3.6, a strong subobject of B as well. By uniqueness of the factorization (see 4.4.5), $I = f^{+1}(A')$; in particular $f^{+1}(A') \subseteq B'$. $\qquad\square$

4.5 Generators

Let us make clear again that, when mentioning a *family* indexed by I, we assume I to be a set.

Definition 4.5.1 *Let $\mathscr{C}$ be a category. A family $(G_i)_{i \in I}$ of objects of $\mathscr{C}$ is called a family of generators when, given any two parallel morphisms*

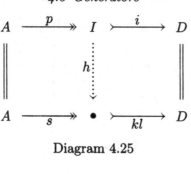

Diagram 4.25

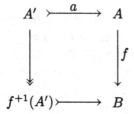

Diagram 4.26

$u, v: A \rightrightarrows B$ in $\mathscr{C}$,

$$\forall i \in I \;\; \forall g: G_i \longrightarrow A \quad u \circ g = v \circ g \Rightarrow u = v.$$

When the family is reduced to a single element $\{G\}$, we say that G is a generator of $\mathscr{C}$.

Some people prefer the term "separator" instead of generator, which is very sensible. Nevertheless generators are important because of the following property which indicates that, in good cases, every object can be recaptured as "a quotient of a coproduct of generators"; this result justifies the terminology "generator".

Proposition 4.5.2 *Let $\mathscr{C}$ be a category with coproducts and $(G_i)_{i \in I}$ a family of objects of $\mathscr{C}$. The following conditions are equivalent:*

(1) *$(G_i)_{i \in I}$ is a family of generators;*

(2) *for every object $C \in \mathscr{C}$, the unique morphism*

$$\gamma_C: \coprod_{i \in I, \, f \in \mathscr{C}(G_i, C)} (\text{domain of } f) \longrightarrow C$$

such that $\gamma_C \circ s_f = f$ is an epimorphism.

Proof For the sake of brevity, we shall often write $\coprod G_i$ to indicate the coproduct of the statement; we write s_f for the canonical morphism of the coproduct corresponding to $f \in \mathscr{C}(G_i, C)$, $i \in I$.

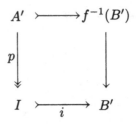

Diagram 4.27

Suppose $(G_i)_{i \in I}$ is a family of generators and choose $u, v \colon C \rightrightarrows D$ such that $u \circ \gamma_C = v \circ \gamma_C$. Then $u \circ f = u \circ \gamma_C \circ s_f = v \circ \gamma_C \circ s_f = v \circ f$ for every $i \in I$ and every $f \in \mathscr{C}(G_i, C)$. Thus $u = v$.

Conversely, suppose that $u \circ f = v \circ f$ for every $i \in I$ and every $f \in \mathscr{C}(G_i, C)$. Then $u \circ \gamma_C \circ s_f = u \circ f = v \circ f = v \circ \gamma_C \circ s_f$ and thus $u \circ \gamma_C = v \circ \gamma_C$, by the property of a coproduct. Finally $u = v$ since γ_C is an epimorphism.   □

Definition 4.5.3 *Let $\mathscr{C}$ be a category with coproducts and $(G_i)_{i \in I}$ a family of objects of $\mathscr{C}$. The family $(G_i)_{i \in I}$ is a strong family of generators when, for every object $C \in \mathscr{C}$, the morphism γ_C of 4.5.2 is a strong epimorphism. $(G_i)_{i \in I}$ is a regular family of generators when, for every object $C \in \mathscr{C}$, the morphism γ_C of 4.5.2 is a regular epimorphism. When the family is reduced to a single element $\{G\}$, we say that G is a strong or a regular generator, according to the case.*

Let us observe that when $(G_i)_{i \in I}$ is a regular family of generators, for every object C, $\gamma_C = \mathsf{Coker}\,(u, v)$ for some pair (u, v) of morphisms. Since γ_P is an epimorphism, one clearly has $\gamma_C = \mathsf{Coker}\,(u \circ \gamma_P, v \circ \gamma_P)$:

$$\coprod_{j,g} G_j \xrightarrow{\ \gamma_P\ } P \mathrel{\substack{\xrightarrow{u} \\[-4pt] \xrightarrow[v]{}}} \coprod_{i,f} G_i \xrightarrow{\ \gamma_C\ } C.$$

Thus C has been presented as the coequalizer of two morphisms defined between two coproducts of the generators G_i. But if the objects of the diagram presenting C are coproducts of the generators, nothing can be said about the two morphisms $u \circ \gamma_P$, $v \circ \gamma_P$ between them. There is no reason why they should be determined by morphisms $G_i \longrightarrow G_j$ between the generators. This indicates the interest of the next notion.

Definition 4.5.4 *Let $\mathscr{C}$ be a category and $(G_i)_{i \in I}$ a family of objects of $\mathscr{C}$. Let us write $\mathscr{G}$ for the full subcategory of $\mathscr{C}$ generated by the G_i's and $\mathscr{G}/C$ for the full subcategory of $\mathscr{C}/C$ generated by the objects of*

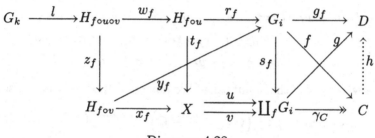

Diagram 4.28

the form $f: G_i \longrightarrow C$. The family $(G_i)_{i\in I}$ is a dense family of generators when for every object $C \in \mathscr{C}$, the colimit of the functor

$$\Gamma^C: \mathscr{G}/C \longrightarrow \mathscr{C}, \quad (f: G_i \longrightarrow C) \mapsto G_i,$$

is precisely $(C, (f)_{f\in\mathscr{G}/C})$. When the family is reduced to a single element $\{G\}$, G is called a dense generator.

Let us study the relations between dense and regular generators.

Proposition 4.5.5 *In a category with coproducts, every dense family of generators is regular and every regular family of generators is strong.*

Proof   The first statement is an immediate consequence of the construction of the colimit from coproducts and coequalizers, as described in 2.8.1 (see exercise 2.17.10). The rest follows from 3.3.6.   □

Proposition 4.5.6 *Let $\mathscr{C}$ be a category with pullbacks and universal coproducts (see section 2.14). Given a family $(G_i)_{i\in I}$ of objects of $\mathscr{C}$, the following conditions are equivalent:*

(1) $(G_i)_{i\in I}$ is a regular family of generators;

(2) $(G_i)_{i\in I}$ is a dense family of generators.

Proof   (2) $\Rightarrow$ (1) has been proved in 4.5.5. Conversely let us consider diagram 4.28 where γ_C has been defined in 4.5.2. Writing s_f for the canonical morphisms of the coproduct, we thus have $\gamma_C \circ s_f = f$. With the notation of 4.5.4, this is of course a cocone on Γ^C and we must prove it is a colimit cocone. So we choose another cocone $(g_f)_f$ on Γ^C. By definition of a coproduct, we get immediately a unique g such that $g \circ s_f = g_f$ for each $f \in |\mathscr{G}/C|$. By assumption, γ_C is regular; let us write $\gamma_C = \mathsf{Coker}\,(u, v)$. (H_{fou}, r_f, t_f) is the pullback of u, s_f; (H_{fov}, x_f, y_f) is the pullback of v, s_f; (H_{fouov}, z_f, w_f) is the pullback of x_f, t_f.

Let us observe first that for every morphism $l: G_k \longrightarrow H_{fouov}$, whatever the indices, one has

$$f \circ r_f \circ w_f \circ l = \gamma_C \circ s_f \circ r_f \circ w_f \circ l = \gamma_C \circ u \circ t_f \circ w_f \circ l$$
$$= \gamma_C \circ u \circ x_f \circ z_f \circ l = \gamma_C \circ v \circ x_f \circ z_f \circ l$$
$$= \gamma_C \circ s_f \circ y_f \circ z_f \circ l = f \circ y_f \circ z_f \circ l.$$

Let us write m for this composite. We have obtained two morphisms

$$r_f \circ w_f \circ l, \; y_f \circ z_f \circ l : \left(G_k \xrightarrow{\;m\;} C \right) \rightrightarrows \left(G_i \xrightarrow{\;f\;} C \right)$$

in $\mathscr{G}/C$ and, since the $(g_f)_{f \in I}$ constitute a cocone on Γ^C,

$$g_f \circ (r_f \circ w_f \circ l) = g \circ m = g_f \circ (y_f \circ z_f \circ l),$$

which implies

$$g \circ u \circ x_f \circ z_f \circ l = g \circ u \circ y_f \circ w_f \circ l = g \circ s_f \circ r_f \circ w_f \circ l$$
$$= g_f \circ r_f \circ w_f \circ l = g_f \circ y_f \circ z_f \circ l$$
$$= g \circ s_f \circ y_f \circ z_f \circ l = g \circ v \circ x_f \circ z_f \circ l.$$

Since this is valid for every k, l and the G_k's are a family of generators, $g \circ u \circ x_f \circ z_f = g \circ v \circ x_f \circ z_f$. But by universality of coproducts, $(X, (t_f)_f)$, $(X, (x_f)_f)$, $(H_{f \circ u}, (w_f)_f)$ and $(H_{f \circ v}, (z_f)_f)$ are coproducts as well. Therefore $g \circ u = g \circ v$ and we get a unique h such that $h \circ \gamma_C = g$, since $\gamma_C = \mathsf{Coker}\,(u, v)$.

Observe that $h \circ f = h \circ \gamma_C \circ s_f = g \circ s_f = g_f$, thus h is indeed a good factorization. On the other hand it is unique since γ_C is an epimorphism and $\left(\coprod_{i,f} G_i, (s_f)_f \right)$ is a coproduct. $\qquad\square$

Now comes a "functorial" approach to generators. For this we need some terminology.

Definition 4.5.7

(1) A family of functors $(F_i \colon \mathscr{A} \longrightarrow \mathscr{B}_i)_{i \in I}$ is *collectively faithful* when given morphisms $f, g \colon A \rightrightarrows A'$ in $\mathscr{A}$

$$\big(\forall i \in I \; F_i(f) = F_i(g) \big) \Rightarrow (f = g).$$

(2) A family of functors $(F_i \colon \mathscr{A} \longrightarrow \mathscr{B}_i)_{i \in I}$ *collectively reflects isomorphisms* when, given a morphism $f \colon A \longrightarrow A'$ in $\mathscr{A}$,

$$\big(\forall i \in I \; F_i(f) \text{ is an isomorphism} \big) \Rightarrow (f \text{ is an isomorphism}).$$

Proposition 4.5.8 *Let $\mathscr{C}$ be a category and $(G_i)_{i \in I}$ a family of objects in $\mathscr{C}$. The following conditions are equivalent:*

(1) $(G_i)_{i \in I}$ *is a family of generators;*

(2) *the functors $\mathscr{C}(G_i, -) \colon \mathscr{C} \longrightarrow \mathsf{Set}$ are collectively faithful.*

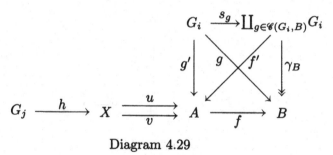

Diagram 4.29

Proof Collective faithfulness of the functors $\mathscr{C}(G_i, -)$ means that given $f, g: A \rightrightarrows A'$ in $\mathscr{A}$,

$$\left(\forall i \in I \ \forall h_i: G_i \longrightarrow A \ \ f \circ h = g \circ h\right) \Rightarrow \left(f = g\right).$$

This is precisely the definition of $(G_i)_{i \in I}$ being a family of generators. $\qquad \square$

Corollary 4.5.9 *Let $\mathscr{C}$ be a category and $G \in \mathscr{C}$. The following conditions are equivalent:*

(1) G is a generator;

(2) the functor $\mathscr{C}(G, -): \mathscr{C} \longrightarrow$ Set is faithful. $\qquad \square$

Proposition 4.5.10 *Let $\mathscr{C}$ be a finitely complete category with coproducts. Given a family $(G_i)_{i \in I}$ of objects of $\mathscr{C}$, the following conditions are equivalent:*

(1) $(G_i)_{i \in I}$ is a strong family of generators;

(2) the functors $\mathscr{C}(G_i, -): \mathscr{C} \longrightarrow$ Set collectively reflect isomorphisms.

Proof Let $(G_i)_{i \in I}$ be a strong family of generators. Take $f: A \longrightarrow B$ such that $\mathscr{C}(G_i, f)$ is an isomorphism for each i. We use the notation of 4.5.3 and consider diagram 4.29. For every $i \in I$ and $g: G_i \longrightarrow B$, one has $g = \gamma_B \circ s_g \in \mathscr{C}(G_i, B)$. Since $\mathscr{C}(G_i, f): \mathscr{C}(G_i, A) \longrightarrow \mathscr{C}(G_i, B)$ is a bijection, there is a unique g' such that $\gamma_B \circ s_g = g = f \circ g'$. Doing this for all $i \in I$, $g \in \mathscr{C}(G_i, B)$, one gets a factorization f' through the coproduct, with $f' \circ s_g = g'$. From

$$f \circ f' \circ s_g = f \circ g' = \gamma_B \circ s_g$$

we deduce $f \circ f' = \gamma_B$ by definition of a coproduct. Since γ_B is a strong epimorphism, f is a strong epimorphism as well (see 4.3.6).

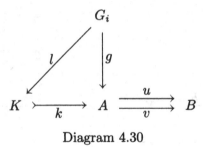

Diagram 4.30

To prove that f is a monomorphism, choose u, v such that $f \circ u = f \circ v$. For all $j \in I$ and $h: G_j \longrightarrow X$, one has $f \circ u \circ h = f \circ v \circ h$, i.e.

$$\mathscr{C}(G_j, f)(u \circ h) = \mathscr{C}(G_j, f)(v \circ h).$$

Since $\mathscr{C}(g_j, f)$ is bijective, $u \circ h = v \circ h$ and since $(G_j)_{j \in I}$ is a family of generators, $u = v$. So f is both a monomorphism and a strong epimorphism: it is an isomorphism (see 4.3.6).

Conversely suppose the functors $\mathscr{C}(G_i, -)$ collectively reflect isomorphisms. If $\gamma_C = j \circ p$ with j a monomorphism, $\mathscr{C}(G_i, j)$ is injective for every i by 2.9.4. Moreover, given $g: G_i \longrightarrow C$, one has

$$g = \gamma_C \circ s_g = j \circ p \circ s_g = \mathscr{C}(G_i, j)(p \circ s_g),$$

proving that $\mathscr{C}(G_i, j)$ is surjective as well. Therefore $\mathscr{C}(G_i, j)$ is bijective for each index i and j is an isomorphism by assumption. By 4.3.7, γ_C is an extremal and thus strong epimorphism. $\qquad\square$

Corollary 4.5.11 *Let $\mathscr{C}$ be a finitely complete category with coproducts and $G \in \mathscr{C}$. The following conditions are equivalent:*

(1) G is a strong generator;

(2) the functor $\mathscr{C}(G, -): \mathscr{C} \longrightarrow$ Set reflects isomorphisms. $\qquad\square$

Let us now observe that

Proposition 4.5.12 *Let $\mathscr{C}$ be a category with equalizers and $(G_i)_{i \in I}$ a family of objects of $\mathscr{C}$. If the functors $\mathscr{C}(G_i, -)$ collectively reflect isomorphisms, $(G_i)_{i \in I}$ is a family of generators.*

Proof Consider u, v such that for every $i \in I$ and $g: G_i \longrightarrow A$, $u \circ g = v \circ g$ (see diagram 4.30). Putting $k = \mathsf{Ker}\,(u, v)$, with each g is associated a unique l such that $k \circ l = g$. This means precisely that each $\mathscr{C}(G_i, k)$ is bijective, thus k is an isomorphism. From $u \circ k = v \circ k$ one deduces $u = v$. $\qquad\square$

Proposition 4.5.12 shows, together with 4.5.10, that the following definition is compatible with the terminology of 4.5.3. Some authors just take 4.5.13 as the definition of a strong family of generators, dropping the assumption on the existence of finite limits which stands here just to ensure compatibility with 4.5.3.

Definition 4.5.13 Let $\mathscr{C}$ be a category (with finite limits). A family $(G_i)_{i \in I}$ of objects of $\mathscr{C}$ is a strong family of generators when the family of functors $\mathscr{C}(G_i, -) \colon \mathscr{C} \longrightarrow$ Set collectively reflects isomorphisms. When the family is reduced to a single object $\{G\}$, G is called a strong generator.

Proposition 4.5.14 Consider a category $\mathscr{C}$, a family $(G_i)_{i \in I}$ of objects of $\mathscr{C}$ and the corresponding full subcategory $\mathscr{G}$. The following conditions are equivalent:

(1) $(G_i)_{i \in I}$ is a dense family of generators;
(2) the functor $\Gamma \colon \mathscr{C} \longrightarrow \mathsf{Fun}(\mathscr{G}^*, \mathsf{Set}); C \mapsto \mathscr{C}(-, C)$ is full and faithful.

Proof   Let us observe that the functor $\Gamma \colon \mathscr{C} \longrightarrow \mathsf{Fun}(\mathscr{G}^*, \mathsf{Set})$ takes as value at $C \in \mathscr{C}$ the functor

$$\Gamma C \colon \mathscr{G}^* \longrightarrow \mathsf{Set}, \quad G_i \mapsto \mathscr{C}(G_i, C),$$

which is just the restriction to $\mathscr{G}^*$ of the representable functor

$$\mathscr{C}(-, C) \colon \mathscr{G}^* \longrightarrow \mathsf{Set}, \quad D \mapsto \mathscr{C}(D, C).$$

But let us make clear that $\mathscr{C}(-, C)$, when restricted to $\mathscr{G}^*$, is by no means a representable functor (a functor represented by an object of $\mathscr{G}$). Moreover we should avoid considering the "category" $\mathsf{Fun}(\mathscr{C}^*, \mathsf{Set})$ since $\mathscr{C}$ is not small in general.

Let us suppose first that $(G_i)_{i \in I}$ is a dense family of generators. Given $f \colon C \longrightarrow D$ in $\mathscr{C}$, we have a natural transformation

$$\mathscr{C}(-, f) \colon \mathscr{C}(-, C) \Rightarrow \mathscr{C}(-, D)$$

between the corresponding representable functors; it restricts to a natural transformation

$$\Gamma f \colon \Gamma C \Rightarrow \Gamma D.$$

Conversely given a natural transformation $\alpha \colon \Gamma C \Rightarrow \Gamma D$, for every $i \in I$ and every $g \colon G_i \longrightarrow C$ we get a morphism $\alpha_{G_i}(g) \colon G_i \longrightarrow D$. Those morphisms give rise to a cocone on $\mathscr{G}/C$, just by naturality of α. Therefore we get a unique factorization $\theta(\alpha) \colon C \longrightarrow D$ through the corresponding colimit C, such that $\theta(\alpha) \circ g = \alpha_{G_i}(g)$ for every $i \in I$ and $g \colon G_i \longrightarrow C$.

Let us prove that $\theta\Gamma(f) = f$. Indeed for every $g\colon G_i \longrightarrow C$

$$\theta\Gamma(f) \circ g = (\Gamma f)_{G_i}(g) = \mathscr{C}(G_i, f)(g) = f \circ g,$$

from which $\theta\Gamma(f) = f$ since the $(G_i)_{i \in I}$ are a family of generators (see 4.5.5 and 4.5.2). On the other hand, given $\alpha\colon \Gamma^C \Rightarrow \Gamma^D$, $\Gamma\theta(\alpha) = \alpha$ since, given $g\colon G_i \longrightarrow C$,

$$\bigl(\Gamma\theta(\alpha)\bigr)_{G_i}(g) = \mathscr{C}\bigl(G_i, \theta(\alpha)\bigr)(g) = \theta(\alpha) \circ g = \alpha_{G_i}(g).$$

Conversely let us suppose that Γ is full and faithful. The category $\mathsf{Elts}\bigl(\Gamma(C)\bigr)$ (see 1.6.4) is just the category $\mathscr{G}/C$, by definition of $\Gamma(C)$. Therefore writing $Y\colon\mathscr{G} \longrightarrow \mathsf{Fun}(\mathscr{G}^*, \mathsf{Set})$ for the covariant Yoneda embedding $Y(G) = \mathscr{G}(-, G)$, and $\phi\colon\mathscr{G}/C \longrightarrow \mathscr{G}$ for the forgetful functor $\phi(G_i \xrightarrow{f} C) = G_i$, we know that the colimit of the functor

$$\mathscr{G}/C \xrightarrow{\ \phi\ } \mathscr{G} \xrightarrow{\ Y\ } \mathsf{Fun}(\mathscr{G}^*, \mathsf{Set})$$

is precisely $\bigl(\Gamma C, (\Gamma g)_{g \in \mathscr{G}/C}\bigr)$ (see 2.15.6). Observe that this composite is precisely equal to

$$\mathscr{G}/C \xrightarrow{\ \phi\ } \mathscr{C} \xrightarrow{\ \Gamma\ } \mathsf{Fun}(\mathscr{G}^*, \mathsf{Set}).$$

Since Γ is full and faithful, it reflects colimits (see 2.9.9). Therefore

$$\bigl(C, (g)_{g \in \mathscr{G}/C}\bigr)$$

is indeed the colimit of $\phi\colon\mathscr{G}/C \longrightarrow \mathscr{C}$. $\square$

Here is an interesting consequence of the existence of generators.

Proposition 4.5.15 *If a category $\mathscr{C}$ with finite limits possesses a strong family $(G_i)_{i \in I}$ of generators, $\mathscr{C}$ is well-powered.*

Proof   Given an object $C \in \mathscr{C}$ and the class $\mathsf{Mono}(C)$ of monomorphisms with codomain C, consider the mapping in Set

$$\alpha\colon \mathsf{Mono}(C) \longrightarrow \coprod_{i \in I} \mathscr{C}(G_i, C) = \bigl\{(g, i) \,\big|\, i \in I \ ; \ g \in \mathscr{C}(G_i, C)\bigr\},$$

$$\alpha(s\colon S \rightarrowtail C) = \bigl\{(g, i) \,\big|\, i \in I; \ g\colon G_i \longrightarrow C; \ \exists h\colon G_i \longrightarrow S \ \ s \circ h = g\bigr\}.$$

Let us prove that when $\alpha(s\colon S \rightarrowtail C) = \alpha(r\colon R \rightarrowtail C)$, r and s define the same subobject of C. This will prove that α factors as an injection

$$\mathsf{Sub}(C) \rightarrowtail \coprod_{i \in I} \mathscr{C}(G_i, C)$$

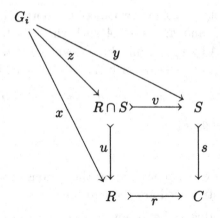

Diagram 4.31

through the class $\mathsf{Sub}(C)$ of subobjects of C. Since the coproduct is a set, $\mathsf{Sub}(C)$ will be a set as well.

Thus we suppose $\alpha(s) = \alpha(r)$ and we consider diagram 4.31, where the square is a pullback. Given a morphism $x\colon G_i \longrightarrow R$, the morphism $r \circ x\colon G_i \longrightarrow C$ factors through r, thus $r \circ x \in \alpha(r)$. Then $r \circ x \in \alpha(s)$ and we find y such that $r \circ x = s \circ y$. By definition of a pullback, we get z such that $u \circ z = x$, $v \circ z = y$. Thus $\mathscr{C}(G_i, u)$ is surjective and, since u is a monomorphism, it is injective as well (see 2.9.4). By assumption, u is then an isomorphism (see 4.5.13). In the same way v is an isomorphism and r, s are isomorphic monomorphisms. □

In a special case of interest, the existence of a generator can be reduced to that of a family of generators. The reader should refer to section 1.1 of volume 2 for what concerns zero objects (a zero object is one which is both initial and final).

Proposition 4.5.16 *Let $\mathscr{A}$ be a category with coproducts and a zero object. The following conditions are equivalent:*

(1) $\mathscr{A}$ has a family of generators;

(2) $\mathscr{A}$ has a generator.

Proof (2) $\Rightarrow$ (1) is obvious. Conversely choose a family $(G_i)_{i \in I}$ of generators and consider the coproduct $\left(\coprod_{i \in I} G_i, (s_i)_{i \in I} \right)$ of this family. Let us prove that $\coprod_{i \in I} G_i$ is a generator.

First of all if I is the empty set, any two morphisms $f, g\colon A \rightrightarrows B$ of $\mathscr{A}$ must be equal and thus $\coprod_{i \in I} G_i$ is just the zero object and is a generator.

If I is not empty, consider two distinct morphisms $f, g \colon A \rightrightarrows B$. There exists $i \in I$ and $h \colon G_i \longrightarrow A$ such that $f \circ h \neq g \circ h$. Let us define $k_j \colon G_j \longrightarrow A$ by $k_i = h$ and $k_j = 0$ when $j \neq i$. We get a unique factorization $k \colon \coprod_{j \in I} G_j \longrightarrow A$ such that $k \circ s_j = k_j$. In particular

$$f \circ k \circ s_i = f \circ k_i = f \circ h \neq g \circ h = g \circ k_i = g \circ h \circ s_i,$$

from which $f \circ k \neq g \circ k$. $\qquad\qquad\qquad\qquad\qquad\qquad\qquad\qquad\square$

Examples 4.5.17

These examples show in particular that the various notions of generators which have been introduced are not equivalent.

4.5.17.a In the category Set of sets, the singleton is a dense generator. Indeed writing **1** for the (full) subcategory generated by the singleton, $\mathbf{1}/X$ for a set X is just the discrete category $\mathscr{X}$ with $|\mathscr{X}| = X$. The colimit of the corresponding functor $\mathscr{X} \longrightarrow$ Set is thus the coproduct of X copies of the singleton, i.e. the set X itself.

4.5.17.b More generally if $\mathscr{C}$ is a small category, the representable functors on $\mathscr{C}$ constitute a dense family of generators for the corresponding category Fun$(\mathscr{C}, \text{Set})$ of all set-valued functors on $\mathscr{C}$. This is precisely the content of theorem 2.15.6. By 4.5.5 and 2.15.4, the representable functors also constitute a regular and a strong family of generators.

4.5.17.c In the category Gr of groups or Ab of abelian groups, the group $(\mathbb{Z}, +)$ of integers is a strong generator. Indeed a group homomorphism $f \colon \mathbb{Z} \longrightarrow G$ is such that $f(z) = zf(1)$, thus is entirely determined by $f(1)$. Conversely given $x \in G$, $x = f(1)$ for the group homomorphism $f \colon \mathbb{Z} \longrightarrow G$ defined by $f(z) = zx$. Therefore the functors Gr$(\mathbb{Z}, -)$ or Ab$(\mathbb{Z}, -)$ are isomorphic to the "underlying set functor". Since a bijective group homomorphism is an isomorphism, this forgetful functor reflects isomorphisms. So $\mathbb{Z}$ is a strong generator both in Ab and in Gr.

4.5.17.d Let R be a ring with unit; it is a strong generator in the category Mod$_R$ of right R-modules. Indeed a R-linear morphism $f \colon R \longrightarrow M$ to a module M has the form $f(r) = f(1)r$; conversely given $m \in M$, m has the form $m = f(1)$ for the R-linear mapping defined by $f(r) = mr$. So the functor Mod$_R(R, -)$ can be identified with the underlying set functor. Since a bijective R-linear mapping is an isomorphism, this forgetful functor reflects isomorphisms. Thus R is a strong generator in Mod$_R$.

4.5.17.e In the category Ban$_1$ of real Banach spaces and linear contractions, a morphism $f \colon \mathbb{R} \longrightarrow B$ has the form $f(r) = rf(1)$ with

$\|f(1)\| \leq 1$. Conversely given $b \in B$ with $\|b\| \leq 1$, $b = f(1)$ if $f: \mathbb{R} \longrightarrow B$ is defined by $f(r) = rb$. So the functor $\mathsf{Ban}_1(\mathbb{R}, -)$ can be identified with the unit ball functor. A linear contraction which is bijective on the unit balls is automatically an isomorphism, thus $\mathbb{R}$ is a strong generator in Ban_1.

4.5.17.f In the category Top of topological spaces and continuous mappings, the singleton is a generator. The continuous mappings from the singleton to a space A correspond precisely with the points of A. And two continuous mappings $f, g: A \xrightarrow{\quad} B$ are equal precisely when they coincide on each point of A. The singleton is not a strong generator in Top. Indeed given a continuous mapping $f: A \longrightarrow B$, $\mathsf{Top}(1, f)$ is a bijection precisely when f is a bijection. But a continuous bijection is in general not a homeomorphism.

4.5.17.g In the category Comp of compact Hausdorff spaces, the same argument as in 4.5.17.f shows that a continuous mapping $f: A \longrightarrow B$ is such that $\mathsf{Comp}(1, f)$ is bijective, precisely when f is a continuous bijection. But a continuous bijection between compact Hausdorff spaces is a homeomorphism. Thus applying 4.5.7, we conclude that the singleton is a strong generator in Comp. The singleton is not a dense generator in Comp. Indeed given a compact Hausdorff space X, write **1** for the (full) subcategory of Comp generated by the singleton. The category $\mathbf{1}/X$ is just the discrete category $\mathscr{X}$ with $|\mathscr{X}| = X$. The colimit of the corresponding functor $\mathscr{X} \longrightarrow \mathsf{Comp}$ is thus the coproduct in Comp of X copies of the singleton, i.e. the Stone–Čech compactification $\tilde{X}$ of the set X provided with the discrete topology (see 3.5.4 and 3.3.9.c). But such a space $\tilde{X}$ is disconnected as soon as X has at least two points (see **Kelley**), so it cannot be X itself when X is connected. Thus **1** is not a dense generator in Comp.

In fact the singleton is a regular generator in Comp since, in that category, every epimorphism is regular (see 4.3.10.f). So we have an example of a regular generator which is not dense.

4.5.17.h In the category Cat of small categories and functors, the category $\mathbf{2} = \{X \xrightarrow{z} Y\}$ is a strong generator. Indeed given a small category $\mathscr{C}$, $\mathsf{Cat}(\mathbf{2}, \mathscr{C})$ is just the set of morphisms of $\mathscr{C}$. For a functor $F: \mathscr{C} \longrightarrow \mathscr{D}$, $\mathsf{Cat}(\mathbf{2}, F)$ is a bijection precisely when given a morphism $d: D_1 \longrightarrow D_2$ of $\mathscr{D}$, there exists a unique morphism $c: C_1 \longrightarrow C_2$ of $\mathscr{C}$ such that $F(c) = d$. Choosing d to be an identity, one concludes that F is bijective on the objects. Then fixing the two objects D_1, D_2, one concludes that F is full and faithful, thus an isomorphism. So **2** is indeed

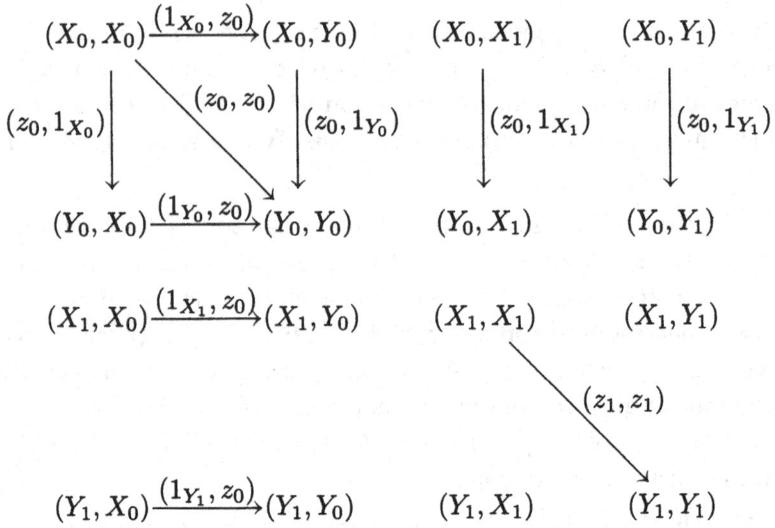

$$(X_0, X_0) \xrightarrow{(1_{X_0}, z_0)} (X_0, Y_0) \qquad (X_0, X_1) \qquad (X_0, Y_1)$$

$$\downarrow (z_0, 1_{X_0}) \qquad \searrow (z_0, z_0) \qquad \downarrow (z_0, 1_{Y_0}) \qquad \downarrow (z_0, 1_{X_1}) \qquad \downarrow (z_0, 1_{Y_1})$$

$$(Y_0, X_0) \xrightarrow{(1_{Y_0}, z_0)} (Y_0, Y_0) \qquad (Y_0, X_1) \qquad (Y_0, Y_1)$$

$$(X_1, X_0) \xrightarrow{(1_{X_1}, z_0)} (X_1, Y_0) \qquad (X_1, X_1) \qquad (X_1, Y_1)$$

$$\searrow (z_1, z_1)$$

$$(Y_1, X_0) \xrightarrow{(1_{Y_1}, z_0)} (Y_1, Y_0) \qquad (Y_1, X_1) \qquad (Y_1, Y_1)$$

Diagram 4.32

a strong generator.

Let us prove that **2** is not a regular generator. Consider the small category $\mathscr{E}$ with a single object $*$ and two arrows $1_*, e$ with $e \circ e = e$. There are just two functors from **2** to $\mathscr{E}$, respectively mapping u to 1_* or e. Using indices $1, 2$ to denote the elements in the two components of the coproduct $\mathbf{2} \amalg \mathbf{2}$, we have to show that the functor

$$Q \colon \mathbf{2} \amalg \mathbf{2} \longrightarrow \mathscr{E}, \quad Q(z_1) = 1_*, \quad Q(z_2) = e$$

is not a regular epimorphism. If Q is regular, it is the coequalizer of its kernel pair (see 2.5.7). This kernel pair $(\mathscr{P}, P_1, P_2)$ is easy to compute; $\mathscr{P}$ is described by diagram 4.32 and P_1, P_2 are the first and second projections. So a functor $G \colon \mathbf{2} \amalg \mathbf{2} \longrightarrow \mathscr{X}$ satisfies $G \circ P_1 = G \circ P_2$ precisely when $G(X_0) = G(Y_0) = G(X_1) = G(Y_1)$ and $G(z_0) = G(1_{X_0})$. Therefore $\mathrm{Coker}\,(P_1, P_2)$ is the functor

$$H \colon \mathbf{2} \amalg \mathbf{2} \longrightarrow \mathscr{N}, \quad H(z_0) = 1_*, \quad H(s_1) = n,$$

where $\mathscr{N}$ is the category with a single object $*$ and infinitely many arrows $1_*, n, n^2, n^3, \ldots, n^k, \ldots$ from $*$ to $*$. Indeed given G such that $G \circ P_1 = G \circ P_2$, the unique required factorization K such that $K \circ H = G$ is given by

$$K \colon \mathscr{N} \longrightarrow \mathscr{X}, \quad K(*) = G(X_0), \quad K(n) = G(z_1).$$

So Q is not the coequalizer of its kernel pair, thus it is not a regular epimorphism.

4.5.17.i If $\mathscr{C}$ is an arbitrary category (not necessarily small), observe that for every object $C \in \mathscr{C}$, the colimit of the functor

$$\Gamma^C\colon \mathscr{C}/C \longrightarrow \mathscr{C}, \quad (f\colon D \longrightarrow C) \mapsto D,$$

is precisely $(C, (f)_{f\in\mathscr{C}/C})$, just because $1_C\colon C =\!=\!= C$ is the terminal object of $\mathscr{C}/C$ (see 2.11.5). In this sense, one could say that *the class of all objects of $\mathscr{C}$ is a dense class of generators*.

Moreover, the class of all representable functors $\mathscr{C}(C, -)\colon \mathscr{C} \longrightarrow \mathsf{Set}$ collectively reflects isomorphisms. Indeed if $f\colon X \longrightarrow Y$ is such that each mapping

$$\mathscr{C}(C, f)\colon \mathscr{C}(C, X) \longrightarrow \mathscr{C}(C, Y)$$

is bijective, putting $C = Y$ yields a unique morphism $g\colon Y \longrightarrow X$ such that $f \circ g = 1_Y$. Thus f is a retraction. It is also a monomorphism since given $u, v\colon Z \rightrightarrows X$ such that $f \circ u = f \circ v$, one gets $\mathscr{C}(Z, f)(u) = \mathscr{C}(Z, f)(v)$ and thus $u = v$. So f is an isomorphism (see 1.9.3). Thus one could say that *the class of all objects of $\mathscr{C}$ is a strong class of generators*.

Finally the class of all representable functors is collectively faithful. If $u, v\colon X \rightrightarrows Y$ are such that $\mathscr{C}(C, u) = \mathscr{C}(C, v)$ for all $C \in \mathscr{C}$, in particular

$$u = \mathscr{C}(X, u)(1_X) = \mathscr{C}(X, v)(1_X) = v.$$

Therefore one could say that *the class of all objects of $\mathscr{C}$ is a class of generators*.

4.6 Projectives

Definition 4.6.1 *An object P of a category $\mathscr{C}$ is projective when, given a strong epimorphism $p\colon X \longrightarrow Y$ and a morphism $f\colon P \longrightarrow Y$, there exists a factorization $g\colon P \longrightarrow X$ such that $p \circ g = f$ (see diagram 4.33).*

One should insist here on the fact that the uniqueness of g is by no means required. Let us also indicate that some authors omit the word "strong" in the definition of projective object.

Here is a functorial description of projectivity.

Proposition 4.6.2 *For an object P of a category $\mathscr{C}$, the following conditions are equivalent:*

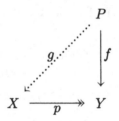

Diagram 4.33

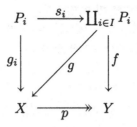

Diagram 4.34

(1) P is projective;

(2) the functor $\mathscr{C}(P, -) \colon \mathscr{C} \longrightarrow$ **Set** preserves epimorphisms.

Proof   With the notation of 4.6.1, the projectivity of P means that $\mathscr{C}(P, p)$ is surjective for every strong epimorphism p.   □

Proposition 4.6.3 *A coproduct of projective objects, when it exists, is again a projective object.*

Proof   If the P_i's are projective, consider diagram 4.34 where p is a strong epimorphism, f is an arbitrary morphism and the s_i's are the canonical morphisms of the coproduct. Since each P_i is projective, we find g_i such that $f \circ s_i = p \circ g_i$. By definition of a coproduct, we get g such that $g \circ s_i = g_i$ for every i. Thus $p \circ g \circ s_i = p \circ g_i = f \circ s_i$, from which $p \circ g = f$.   □

Proposition 4.6.4 *A retract of a projective object is again projective.*

Proof Consider diagram 4.35 where P is projective, $r \circ i = 1_R$ and p is a strong epimorphism. Given $f \colon R \longrightarrow Y$, by projectivity of P we find g such that $p \circ g = f \circ r$. Then $p \circ g \circ i = f \circ r \circ i = f$.   □

Definition 4.6.5 *A category $\mathscr{C}$ has enough projectives when every object is a strong quotient of a projective object.*

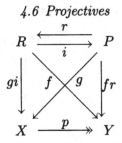

Diagram 4.35

Proposition 4.6.6 *Let $\mathscr{C}$ be a category with coproducts. Suppose $\mathscr{C}$ has a family $(G_i)_{i \in I}$ of strong generators with each G_i projective. Then $\mathscr{C}$ has enough projectives.*

Proof By proposition 4.5.2, for every $C \in \mathscr{C}$ there exists a strong epimorphism $\gamma_C \colon \coprod_j G_j \twoheadrightarrow C$; by 4.6.3 the coproduct is a projective object. □

Here finally is a useful property, in a special case of interest. The reader should refer to section 9.1 for what concerns zero objects (a zero object is one which is both initial and final).

Proposition 4.6.7 *Let $\mathscr{A}$ be a category with coproducts and a zero object. If $(P_i)_{i \in I}$ is a family of objects of $\mathscr{A}$, the following conditions are equivalent:*

(1) the coproduct $\coprod_{i \in I} P_i$ is projective;
(2) for every $i \in I$, P_i is projective.

Proof For a fixed index $j \in I$, we can define $f_j \colon P_i \longrightarrow P_j$ by $f_j = 1_{P_i}$ if $i = j$ and $f_j = 0$ if $i \neq j$. This yields a factorization $p_j \colon \coprod_{i \in I} P_i \longrightarrow P_j$ such that $p_j \circ s_j = 1_{P_j}$ and $p_j \circ s_i = 0$, for $i \neq j$ (the s_i's are the canonical injections of the coproduct). In particular each P_j is a retract of $\coprod_{i \in I} P_i$.

If $\coprod_{i \in I} P_i$ is projective, so is each individual P_j by 4.6.4. If every P_j is projective, so is $\coprod_{i \in I} P_i$ by 4.6.3. □

Examples 4.6.8

4.6.8.a In the category Set of sets, every object is projective. With the notation of 4.6.1, for each $y \in Y$, $p^{-1}(y)$ is non-empty since p is surjective. It suffices then to choose $g(x) \in p^{-1}(f(x))$. Observe that this is exactly the axiom of choice. In particular the singleton is a projective generator (see 4.5.17.a).

4.6.8.b If $\mathscr{C}$ is a small category, let us prove that the representable functors constitute a family of projective generators (see 4.5.17.b). Given

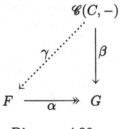

Diagram 4.36

diagram 4.36 in $\mathsf{Fun}(\mathscr{C}, \mathsf{Set})$ with α a strong epimorphism, we know that each α_D,

$$\alpha_D \colon FD \longrightarrow GD \quad (D \in \mathscr{D}),$$

is surjective (see 2.15.3). By the Yoneda lemma, the statement to be proved is equivalent to saying that given $\beta \in G(C)$, there exists an element $\gamma \in F(C)$ such that $\alpha_C(\gamma) = \beta$, which is obvious since α_C is surjective.

4.6.8.c In the category Gr of groups or Ab of abelian groups, $(\mathbb{Z}, +)$ is a projective generator (see 4.5.17.c). Indeed a strong epimorphism $p \colon X \longrightarrow Y$ is a surjection (see 1.8.5.d,e). Given a group homomorphism $f \colon \mathbb{Z} \longrightarrow Y$, choose $x \in p^{-1}(f(1))$ and define $g \colon \mathbb{Z} \longrightarrow X$ by $g(z) = zx$. One has

$$(p \circ g)(z) = p(zx) = zp(x) = zf(1) = f(z).$$

4.6.8.d In the category Mod_R of right R-modules on a ring R with unit, R is a projective generator (see 4.5.17.d). Indeed a strong epimorphism $p \colon X \longrightarrow Y$ is a surjection (see 1.8.5.e). Given a R-linear mapping $f \colon R \longrightarrow Y$, choose $x \in p^{-1}(f(1))$ and define $g \colon R \longrightarrow X$ by $g(r) = rx$. One has

$$(p \circ g)(r) = p(rx) = rp(x) = rf(1) = f(r).$$

4.6.8.e In the category Ban_1 of Banach spaces and linear contractions, $\mathbb{R}$ is a projective generator (see 4.5.17.e). Indeed a strong epimorphism $p \colon X \longrightarrow Y$ is surjective on the unit balls (see 4.3.10.c). A linear contraction $f \colon \mathbb{R} \longrightarrow Y$ has the form $f(r) = rf(1)$, $r \in \mathbb{R}$, with $\|f(1)\| \leq 1$. Choosing $x \in X$ such that $\|x\| \leq 1$ and $p(x) = f(1)$, it suffices to define $g \colon \mathbb{R} \longrightarrow X$ by $g(r) = rx$. Notice that $\mathbb{R}$ does not have the extension property of 4.6.1 with respect to those epimorphisms of Ban_1 which are not surjective.

4.6.8.f In the category Top of topological spaces, the singleton is a projective generator (see 4.5.17.f). Indeed a strong epimorphism

$$p: X \longrightarrow Y$$

is in particular a continuous surjection and given $f: \{*\} \longrightarrow Y$, it suffices to choose $g(*) \in p^{-1}(f(*))$.

4.6.8.g In the category Comp of compact Hausdorff spaces, the singleton is a projective generator (see 4.5.17.g). The argument is the same as in the previous example.

4.7 Injective cogenerators

The notion of cogenerator is dual to that of generator; the notion of injective object is dual to that of projective object. So as far as a theoretical treatment of injective cogenerators is concerned, it suffices to dualize the results of sections 4.5 and 4.6. In particular a category with products and an injective cogenerator always has enough injectives (see 4.6.6).

But the existence of injective cogenerators in concrete examples is generally hard to prove and very often related to the axiom of choice. For that reason we give some emphasis to various of these examples. We focus our attention on the more categorical aspects of the problem, referring freely to some big theorem of algebra, topology, analysis,... when necessary.

Proposition 4.7.1 *In the category* Set *of sets and mappings, the two-point set* $\{0, 1\}$ *is an injective cogenerator.*

Proof Given two distinct mappings $f, g: X \rightrightarrows Y$, there is $x \in X$ such that $f(x) \neq g(x)$. It suffices to choose any mapping $h: Y \longrightarrow \{0, 1\}$ such that $h(f(x)) = 0, h(g(x)) = 1$ and one gets $h \circ f \neq h \circ g$. So $\{0, 1\}$ is a cogenerator.

Given now an injection $f: X \rightarrowtail Y$ and a morphism $g: X \longrightarrow \{0, 1\}$, it suffices to define $h: Y \longrightarrow \{0, 1\}$ by $h(y) = g(x)$ if $y = f(x)$ and $h(y) = 0$ otherwise, to get $h \circ f = g$. $\square$

For the next result, we freely use various notions and results which will be studied in chapters 3 and 5 of volume 3. Let us nevertheless mention right now that the Grothendieck toposes can be exactly characterized as the localizations of the categories $\text{Fun}(\mathscr{C}^*, \text{Set})$ of (contravariant) set-valued functors on a small category $\mathscr{C}$ (see 3.5.5, volume 3). In particular

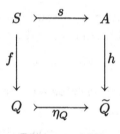

Diagram 4.37

Fun($\mathscr{C}^*$, Set) itself is always a Grothendieck topos. In a topos, every monomorphism is regular and every epimorphism is regular as well.

Proposition 4.7.2 *Every Grothendieck topos has an injective cogenerator.*

Proof A Grothendieck topos $\mathscr{E}$ is in particular a topos (see 3.2.9, volume 3). In $\mathscr{E}$, every epimorphism $f\colon A \longrightarrow B$ is regular and is therefore the coequalizer of its kernel pair $\alpha, \beta\colon P \rightrightarrows A$ (see 3.6.1, volume 3, and 2.5.7). Since P is a subobject of $A \times A$ and $\mathscr{E}$ is well-powered (see 5.3.5, volume 3), there is just a set of equivalence relations on A and thus just a set of quotients of A. Choose now a family $(G_i)_{i \in I}$ of generators of $\mathscr{E}$ (see 3.6.1, volume 3) and, for each index $i \in I$, consider the set $(q_i^k)_{k \in K_i}$ of quotients of the object $G_i \amalg G_i$

$$q_k^i \colon G_i \amalg G_i \longrightarrow Q_k^i$$

(well, we choose an epimorphism in each equivalence class). We consider the coproduct $Q = \amalg_{i \in I, k \in K_i} Q_k^i$. By 5.5.2 and 5.6.1, volume 3, there exist an injective object $\widetilde{Q}$ and a monomorphism $\eta_Q\colon Q \rightarrowtail \widetilde{Q}$ with the property that given a subobject $s\colon S \rightarrowtail A$ and an arbitrary morphism $f\colon S \longrightarrow Q$, there exists a unique morphism $h\colon A \longrightarrow \widetilde{Q}$ such that $h \circ s = \eta_Q \circ f$ (see diagram 4.37). We shall prove that this injective object $\widetilde{Q}$ is also a cogenerator in $\mathscr{E}$.

Consider two distinct morphisms $\alpha, \beta\colon A \rightrightarrows B$ in $\mathscr{E}$. Since the $(G_i)_{i \in I}$ constitute a family of generators, there exist $i_0 \in I$ and $g\colon G_{i_0} \longrightarrow A$ such that $\alpha \circ g \neq \beta \circ g$. Consider diagram 4.38 where $j \circ q_{k_0}^{i_0}$ is the epi–mono factorization of $(\alpha \circ g, \beta \circ g)$ (see 5.9.4, volume 3) and s_1, s_2, $s_{k_0}^{i_0}$ are the canonical inclusions of the coproducts. By the property of $\widetilde{Q}$ we have already mentioned (see 5.5.2, volume 3), there exists a unique morphism $h\colon A \longrightarrow \widetilde{Q}$ such that $h \circ j = \eta_Q \circ s_{k_0}^{i_0}$. Since η_Q and $s_{k_0}^{i_0}$ are

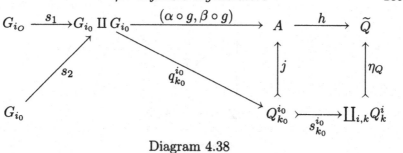

Diagram 4.38

monomorphisms (see 3.4.8, 3.4.10 and 5.5.2, volume 3)

$$\alpha \circ g \neq \beta \circ g \Rightarrow (\alpha \circ g, \beta \circ g) \circ s_1 = \alpha \circ g \neq \beta \circ g = (\alpha \circ g, \beta \circ g) \circ s_2$$
$$\Rightarrow j \circ q_{k_0}^{i_0} \circ s_1 \neq j \circ q_{k_0}^{i_0} \circ s_2$$
$$\Rightarrow q_{k_0}^{i_0} \circ s_1 \neq q_{k_0}^{i_0} \circ s_2$$
$$\Rightarrow \eta_q \circ s_{k_0}^{i_0} \circ q_{k_0}^{i_0} \circ s_1 \neq \eta_Q \circ s_{k_0}^{i_0} \circ q_{k_0}^{i_0} \circ s_2$$
$$\Rightarrow h \circ (\alpha \circ g, \beta \circ g) \circ s_1 \neq h \circ (\alpha \circ g, \beta \circ g) \circ s_2$$
$$\Rightarrow h \circ \alpha \circ g \neq h \circ \beta \circ g$$
$$\Rightarrow h \circ \alpha \neq h \circ \beta.$$

This concludes the proof. $\qquad\square$

Proposition 4.7.3 *The category* Gr *of groups and group homomorphisms does not have any cogenerator.*

Proof Suppose G is a cogenerator in Gr. Recall that a group A is simple when its only normal subgroups are A and (0). If A is a non-zero simple group, the two morphisms $0, 1_A \colon A \rightrightarrows A$ are distinct so that we can find $\alpha \colon A \longrightarrow G$ such that $\alpha \circ 0 \neq \alpha \circ 1_A$, i.e. $\alpha \neq 0$. The kernel of α is a normal subgroup of A which is not A itself, since $\alpha \neq 0$. Since A is simple, this implies $\operatorname{Ker}\alpha = (0)$ and thus α is injective. Therefore the cogenerator G contains as subgroups all the simple groups: this is impossible since there are simple groups of arbitrarily large cardinality (see **Kuroš**). $\qquad\square$

Proposition 4.7.4 *In the category* Ab *of abelian groups and group homomorphisms,* $(\mathbb{Q}/\mathbb{Z}, +)$ *is an injective cogenerator.*

Proof In Ab, all monomorphisms are strong (see 4.3.10.a). We recall that an abelian group A is divisible when given $q \in A$ and $n \in \mathbb{N}$, $n \neq 0$, there exists $b \in A$ such that $nb = a$. Another way to state this property is saying that given $n \in \mathbb{N}$, $n \neq 0$ and a group homomorphism $f \colon n\mathbb{Z} \longrightarrow A$,

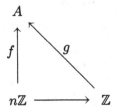

Diagram 4.39

there exists a unique extension $g\colon \mathbb{Z} \longrightarrow A$ (see diagram 4.39). Indeed f is completely determined by the element $a = f(n)$ and finding g is just finding $b = g(1)$ such that $nb = ng(1) = g(n) = f(n) = a$. It is a classical result that, under the axiom of choice, an abelian group is injective iff it is divisible (the so-called Baer criterion for injectivity, depending on the axiom of choice; see **Kuroš**).

Now $\mathbb{Q}/\mathbb{Z}$ is obviously divisible, as given $\left[\frac{a}{b}\right] \in \mathbb{Q}/\mathbb{Z}$ and $0 \neq n \in \mathbb{N}$, $n\left[\frac{a}{nb}\right] = \left[\frac{a}{b}\right]$. Thus $\mathbb{Q}/\mathbb{Z}$ is injective. It is also a cogenerator. Indeed given two distinct group homomorphisms $f, g\colon A \rightrightarrows B$, there exists $a \in A$ such that $f(a) \neq g(a)$. We put $b = f(a) - g(a)$ and consider the subgroup $<b>$ of B generated by b. If b is of order $n \in \mathbb{N}$, $n \neq 0$, i.e.

$$n = \inf\{m \in \mathbb{N} \mid m \neq 0 \ , \ mb = 0\},$$

we define

$$h\colon <b> \longrightarrow \mathbb{Q}/\mathbb{Z}, \quad h(zb) = \left[\frac{z}{n}\right], \quad z \in \mathbb{Z}.$$

Now h is clearly a group homomorphism and $h(b) = \left[\frac{1}{n}\right] \neq 0$. In the case where $nb \neq 0$ for every $0 \neq n \in \mathbb{N}$, we define

$$h\colon <b> \longrightarrow \mathbb{Q}/\mathbb{Z}, \quad h(zb) = \left[\frac{z}{2}\right], \quad z \in \mathbb{Z},$$

and again h is a group homomorphism such that $h(b) = \left[\frac{1}{2}\right] \neq 0$. In both cases, since $\mathbb{Q}/\mathbb{Z}$ is injective, we extend h to a group homomorphism $k\colon B \longrightarrow \mathbb{Q}/\mathbb{Z}$ and we have

$$(k \circ f)(a) - (k \circ g)(a) = k(b) = h(b) \neq 0;$$

thus $k \circ f(a) \neq k \circ g(a)$ and $k \circ f \neq k \circ g$. $\qquad \square$

Proposition 4.7.5 *Let R be a ring with a unit. One gets an injective cogenerator in the category Mod_R of right R-modules by considering*

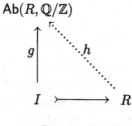

Diagram 4.40

the abelian group $\mathsf{Ab}(R, \mathbb{Q}/\mathbb{Z})$ of group homomorphisms and providing it with the scalar multiplication defined by

$$fr: R \longrightarrow \mathbb{Q}/\mathbb{Z}, \quad s \mapsto f(rs)$$

for $f \in \mathsf{Ab}(R, \mathbb{Q}/\mathbb{Z})$ and $r \in R$.

Proof In Mod_R, all monomorphisms are strong (see 4.3.10.a). It is obvious that we have defined a right R-module structure on $\mathsf{Ab}(R, \mathbb{Q}/\mathbb{Z})$. Consider now the situation of diagram 4.40, where I is a right ideal of R and g is R-linear. Since g is a group homomorphism, it gives rise to another group homomorphism

$$t: I \otimes_{\mathbb{Z}} R \longrightarrow \mathbb{Q}/\mathbb{Z}, \quad t(i \otimes r) = g(i)(r).$$

Given $s \in R$ one observes that, by definition of the R-module structure on $\mathsf{Ab}(R, \mathbb{Q}/\mathbb{Z})$,

$$t(is \otimes r) = g(is)(r) = g(i)(sr) = t(i \otimes sr).$$

Therefore t factors through $I \otimes_R R \cong I$ and we get a group homomorphism:

$$u: I \longrightarrow \mathbb{Q}/\mathbb{Z}, \quad u(i) = t(i \otimes 1).$$

Since $\mathbb{Q}/\mathbb{Z}$ is injective in Ab, there exists some extension v of u (see 4.7.4):

$$v: R \longrightarrow \mathbb{Q}/\mathbb{Z}, \quad v(i) = u(i), \quad i \in I.$$

We can now define the required morphism h:

$$h: R \longrightarrow \mathsf{Ab}(R, \mathbb{Q}/\mathbb{Z}), \quad h(r) = vr.$$

But h is obviously R-linear and given $i \in I$

$$h(i) = vi: R \longrightarrow \mathbb{Q}/\mathbb{Z}$$

with

$$h(i)(r) = (vi)(r) = v(ir) = u(ir) = t(ir \otimes 1) = t(i \otimes r) = g(i)(r),$$

since I is a right ideal and thus $ir \in I$. This proves that $h(i) = g(i)$, thus h is an extension of g. So $\mathsf{Ab}(R, \mathbb{Q}/\mathbb{Z})$ has the extension property with respect to the ideals of R. Under the axiom of choice, the Baer criterion (see **Faith**) tells us that this is precisely the condition for being an injective right R-module.

So $\mathsf{Ab}(R, \mathbb{Q}/\mathbb{Z})$ is an injective R-module and it remains to prove it is a cogenerator. Given two R-linear mappings $f, g: A \underset{\longrightarrow}{\longrightarrow} B$, there is $a \in A$ such that $f(a) \neq g(a)$. We put $b = f(a) - g(a)$ and consider the submodule $< b >$ of B generated by b. Since $b \neq 0$, there exists a group homomorphism $h: < b > \longrightarrow \mathbb{Q}/\mathbb{Z}$ such that $h(b) \neq 0$ (see 4.7.4). Composing with the R-linear mapping

$$k: R \longrightarrow\ < b >, \quad r \mapsto br,$$

we get a group homomorphism $h \circ k: R \longrightarrow \mathbb{Q}/\mathbb{Z}$. Let us define an R-linear mapping by

$$p: < b > \longrightarrow \mathsf{Ab}(R, \mathbb{Q}/\mathbb{Z}), \quad p(br) = (h \circ k)r, \quad r \in R$$

Since $\mathsf{Ab}(R, \mathbb{Q}/\mathbb{Z})$ is an injective R-module, p can be extended to a R-linear mapping

$$q: B \longrightarrow \mathsf{Ab}(R, \mathbb{Q}/\mathbb{Z}), \quad q(b) = p(b).$$

It suffices now to observe that $q(b): R \longrightarrow \mathbb{Q}/\mathbb{Z}$ satisfies

$$\big(q(b)\big)(1) = \big(p(b)\big)(1) = \big(h \circ k\big)(1) = h(b) \neq 0.$$

Therefore $q(b) \neq 0$ or, in other words, $q\big(f(a)\big) \neq q\big(g(a)\big)$ and thus $q \circ f \neq q \circ g$. $\square$

Proposition 4.7.6 *In the category* Ban_1 *of real Banach spaces and linear contractions,* $\mathbb{R}$ *is an injective cogenerator; this cogenerator is regular.*

Proof Let A be a Banach space and $a \in A$, $a \neq 0$. Let us consider the subspace $\mathbb{R}a \subseteq A$. Since $\mathbb{R}$ is finite dimensional (in fact, one dimensional) it is closed and thus a Banach subspace of A. There is an obvious linear mapping

$$f_a: \mathbb{R}a \longrightarrow \mathbb{R}, \quad f_a(ra) = r\|a\|,$$

and $\|f_a\| = 1$ since

$$|f_a(ra)| = |r| \cdot \|a\| = \|ra\|.$$

By the Hahn–Banach theorem (see **Naimark**), there exists a linear extension

$$g_a \colon A \longrightarrow \mathbb{R}, \quad g_a(ra) = f_a(ra)$$

such that $\|g_a\| = \|f_a\|$.

Let us now consider the morphism

$$\gamma_A \colon A \longrightarrow \mathbb{R}^{\mathrm{Ban}_1(A,\mathbb{R})}, \quad \gamma_A(a) = \big(f(a)\big)_{f \in \mathrm{Ban}_1(A,\mathbb{R})}$$

obtained by duality from 4.5.2 applied to the single object $\mathbb{R}$. When $a \in A$, $a \neq 0$, one has $g_a(a) = f_a(a) = \|a\| \neq 0$, thus γ_A is an injection. Moreover $\|\gamma_A(a)\| \leq \|a\|$ since γ_A is a linear contraction and, on the other hand (see 2.1.7.d),

$$\|\gamma_A(a)\| = \sup_f \|f(a)\| \geq \|f_a(a)\| = \|a\|,$$

so that finally $\|\gamma_A(a)\| = \|a\|$. Thus γ_A is an isometric injection, i.e. a regular monomorphism (see 4.3.10.e). This proves that $\mathbb{R}$ is a regular cogenerator (see 4.5.3).

Now if $f \colon A \rightarrowtail B$ is a strong monomorphism, it is an isometric injection (see 4.3.10.a) and $g \colon A \longrightarrow \mathbb{R}$ is a linear contraction, the Hahn–Banach theorem implies the existence of $h \colon B \longrightarrow \mathbb{R}$ with $\|h\| = \|g\|$ and $g = h \circ f$. $\qquad\square$

Proposition 4.7.7 *In the category* Top *of topological spaces and continuous mappings, the two-point space* $\{0,1\}$ *provided with the indiscrete topology is an injective cogenerator.*

Proof If X is a topological space, every mapping $f \colon X \longrightarrow \{0,1\}$ is continuous when $\{0,1\}$ is provided with the indiscrete topology. Therefore the result follows immediately from 4.7.1 and the fact that in Top, (strong) monomorphisms are injective (see 1.7.7.b). $\qquad\square$

Proposition 4.7.8 *In the category* Comp *of compact Hausdorff spaces and continuous mappings, the unit interval* $[0,1]$ *is an injective cogenerator.*

Proof Let us recall that a compact Hausdorff space A is always normal, i.e. two disjoint closed subsets of A can be included in two disjoint open subsets of A. For a normal space A, the famous Urysohn extension theorem (see **Kelley**) says that given a closed subset $B \subseteq A$ and a continuous mapping $f: B \longrightarrow [0,1]$, f can be extended to a continuous mapping $g: A \longrightarrow [0,1]$ on the whole space. A well-known consequence of Urysohn's extension theorem is the so-called Urysohn lemma which says that given two disjoint closed subsets C, D of a normal space A, there exists a continuous mapping $f: A \longrightarrow [0,1]$ such that $f(B) = \{0\}$ and $f(C) = \{1\}$ (put $B = C \cup D$ in the extension theorem).

Let us recall also that a (strong) monomorphism $f: X \longrightarrow Y$ in Comp is a continuous injection (see 1.7.7.b). Since a continuous image of a compact subset is compact, $f(X) \subseteq Y$ is compact and $f: X \longrightarrow f(X)$ is a continuous bijection between two compact Hausdorff spaces; it is thus a homeomorphism. So up to a homeomorphism, we can identify the monomorphism $f: X \longrightarrow Y$ with a subspace $X \subseteq Y$, where X is compact, thus closed. The injectivity of $[0,1]$ in Comp is thus exactly attested by Urysohn's extension theorem.

Now choose two distinct morphisms $f, g: X \rightrightarrows Y$ in Comp. There exists $x \in X$ such that $f(x) \neq g(x)$. Since $\{f(x)\}$ and $\{g(x)\}$ are closed in Y, Urysohn's lemma implies the existence of a morphism $h: Y \longrightarrow [0,1]$ such that $h(f(x)) = 0$ and) $h(g(x)) = 1$. In particular $h \circ f \neq h \circ g$ and $[0,1]$ is a cogenerator. $\qquad\square$

4.8 Exercises

4.8.1 Consider a category $\mathscr{C}$, a family $(G_i)_{i \in I}$ of objects of $\mathscr{C}$ and the corresponding full subcategory $\mathscr{G} \subseteq \mathscr{C}$. Prove that $(G_i)_{i \in I}$ is a dense family of generators if and only if the left Kan extension of $\mathscr{G} \subseteq \mathscr{C}$ along $\mathscr{G} \subseteq \mathscr{C}$ is the identity on $\mathscr{C}$.

4.8.2 If R is a ring without a unit, prove that the free R-module on one generator is a generator for the category Mod_R of R-modules.

4.8.3 In the category Set of sets, an object is injective if and only if it is not empty.

4.8.4 In the category Comp of compact Hausdorff spaces, every monomorphism is regular.

4.8.5 In a category $\mathscr{C}$ with finite limits, a family $(f_i: A_i \longrightarrow B)_{i \in I}$ of morphisms is (collectively) strongly epimorphic when, given a monomorphism $z: X \rightarrowtail Y$ and morphisms $u_i: A_i \longrightarrow X$, $v: B \longrightarrow Y$ such that $z \circ u_i = v \circ f_i$ for every i, there exists a unique $t: B \longrightarrow X$ such that $z \circ t = v$ and $z \circ f_i = u_i$ for every i (see diagram 4.41). When $\mathscr{C}$ has

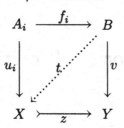

Diagram 4.41

coproducts, prove this is equivalent to the factorization

$$f: \coprod_{i \in I} A_i \longrightarrow B$$

being a strong epimorphism. Deduce an alternative definition of a strong family of generators.

4.8.6 Prove that in the category **Ab** of abelian groups, $\mathbb{Z}$ is a regular generator, but not a dense generator. [Hint: consider $\mathbb{Z} \times \mathbb{Z}$.]

5

Categories of fractions

5.1 Graphs and path categories

A graph is, roughly speaking, a "category without a composition law". Their interest in this book is limited to their use in constructing some new categories.

Definition 5.1.1 *A graph $\mathscr{G}$ consists of*

(1) a class $|\mathscr{G}|$ whose elements are called the objects (or vertices) of the graph,

(2) for each pair $(A, B) \in |\mathscr{G}| \times |\mathscr{G}|$, a set $\mathscr{G}(A, B)$ whose elements are called the morphisms (or arrows) from A to B.

The graph $\mathscr{G}$ is small when $|\mathscr{G}|$ itself is a set.

Definition 5.1.2 *A morphism of graphs $F \colon \mathscr{F} \longrightarrow \mathscr{G}$ between two graphs $\mathscr{F}, \mathscr{G}$ consists of*

(1) a mapping $F \colon |\mathscr{F}| \longrightarrow |\mathscr{G}|$,

(2) for each pair $(A, B) \in \mathscr{F} \times \mathscr{F}$ of objects, a mapping

$$\mathscr{F}(A, B) \longrightarrow \mathscr{G}(FA, FB).$$

Obviously every category is a graph (just forget the composition law) and every functor between categories is a morphism of graphs. So the category **Cat** of small categories is provided with a faithful (but not full) functor to the category **Graph** of small graphs. We intend to construct the left adjoint to this forgetful functor.

Definition 5.1.3 *Let $\mathscr{G}$ be a graph. A path in $\mathscr{G}$ is a non-empty finite sequence $(A_1, f_1, A_2, f_2, \ldots, A_n)$ alternating objects $A_i \in \mathscr{G}$ and arrows $f_i \in \mathscr{G}$; the first and the last term are required to be objects and each arrow f_i has domain A_i and codomain A_{i+1}.*

Proposition 5.1.4 *Given a small graph $\mathcal{G}$, one gets a small category $\mathcal{P}$ called the "path category of $\mathcal{G}$" by putting*

(1) $|\mathcal{P}| = |\mathcal{G}|$ as class of objects,

(2) $\mathcal{P}(A, B)$ as the set of paths in $\mathcal{G}$ starting at A and ending at B,

(3) $(A_n, f_n, \ldots, A_m) \circ (A_1, f_1, \ldots, A_n)$
$$= (A_1, f_1, \ldots, A_n, f_n, \ldots, A_m).$$

Together with the morphism of graphs

$$\Gamma \colon \mathcal{G} \longrightarrow \mathcal{P}, \quad \Gamma(A) = A, \quad \Gamma(A \xrightarrow{\;f\;} B) = (A, f, B),$$

this produces the reflection $(\mathcal{P}, \Gamma)$ of the graph $\mathcal{G}$ along the forgetful functor Cat $\longrightarrow$ Graph.

Proof Obviously $\mathcal{P}$ is a small category, with the path (A) as identity on A. Γ is by construction a morphism of graphs. Now given a category $\mathcal{D}$ and a morphism of graphs $F \colon \mathcal{G} \longrightarrow \mathcal{D}$, the unique functor $G \colon \mathcal{P} \longrightarrow \mathcal{D}$ such that $G \circ \Gamma = F$ is given by

$$G(A) = F(A) \;, \quad G(A_1, f_1, \ldots, A_{\dot{n}}) = F(1_{A_n}) \circ \ldots \circ F(f_1) \circ F(1_{A_1}). \quad \square$$

It should be observed that proposition 5.1.4 no longer holds if one removes the smallness assumptions. Of course we know already that between two large categories there is in general a proper class (not a set) of functors (see section 1.1) and an analogous observation can be made for large graphs. But moreover if $\mathcal{G}$ is a large graph, the construction of $\mathcal{P}$ as in 5.1.4 does not yield a category! Indeed $\mathcal{P}$ has now a class of objects, but there is no reason at all for each $\mathcal{P}(A, B)$ to be a set: when $\mathcal{G}$ has a proper class of objects, there is in general a proper class of paths connecting an object A with an object B.

Very often in the applications, the graphs we shall consider will be built up from categories and some additional data. In particular some composites of arrows will already exist in the original categories and certainly one wants to preserve them. Or some diagrams of arrows will appear in the graph which one wants to become commutative in the end. To treat these questions, we introduce the following definition.

Definition 5.1.5 *Let $\mathcal{G}$ be a graph. A commutativity condition on $\mathcal{G}$ is a pair of paths both defined from some given object A to some given object B.*

In the situation of 5.1.5, the problem is now to perform a quotient of the category $\mathcal{P}$ of paths of $\mathcal{G}$, in such a way that the two arrows $A \rightrightarrows B$ of $\mathcal{P}$ induced by the commutativity condition are identified in the quotient. In other words, if the two paths in the commutativity condition are

$$(A, f_1, A_2, \ldots, f_{n-1}, B), \quad (A, g_1, B_2, \ldots, g_{m-1}, B),$$

f_i and g_j can be identified with individual arrows of $\mathscr{P}$ (paths reduced to one single arrow, together with its domain and its codomain) and one forces the commutativity condition to give rise to an actual commutative diagram

$$f_{n-1} \circ \ldots \circ f_1 = g_{m-1} \circ \ldots \circ g_1$$

in the quotient.

Let us write CondGraph for the category whose objects are small graphs together with a set of commutativity conditions (= conditional graphs) and whose arrows are morphisms of graphs which map a commutativity condition to a commutativity condition. Viewing a small category $\mathscr{C}$ as a graph together with all the commutativity conditions given by the commutative diagrams in $\mathscr{C}$, we again get a faithful functor from the category Cat of small categories to the category CondGraph of small conditional graphs. As a matter of convention, if $C \in \mathscr{C}$ is an object of some category $\mathscr{C}$, we consider the pair $\big((C, 1_C, C), (C)\big)$ as a commutativity condition in $\mathscr{C}$.

Proposition 5.1.6 *The forgetful functor* Cat $\longrightarrow$ CondGraph *has a left adjoint.*

Proof   Consider a graph $\mathscr{G}$ and write $\mathscr{P}$ for the path category of $\mathscr{G}$. Given a set $\Sigma \subseteq \mathscr{G} \times \mathscr{G}$ of commutativity conditions on $\mathscr{G}$, consider the intersection $\mathscr{R}$ of all the subcategories $|\mathscr{S}| \subseteq \mathscr{P} \times \mathscr{P}$ which satisfy the conditions

(1) $\mathscr{S} = \big\{(A, A) \,\big|\, A \in \mathscr{G}\big\}$,
(2) $\mathrm{Ar}(\mathscr{S})$ is an equivalence relation on $\mathrm{Ar}(\mathscr{P})$ (where $\mathrm{Ar}(\mathscr{X})$ denotes the set of arrows of $\mathscr{X}$),
(3) $\Sigma \subseteq \mathrm{Ar}(\mathscr{S})$.

Clearly $\mathscr{R}$ still satisfies conditions (1), (2), (3).

Observe that given $\mathscr{S} \subseteq \mathscr{P} \times \mathscr{P}$ as above, the special form of the objects of $\mathscr{S}$ imposes the requirement that when a pair (φ, ψ) of morphims of $\mathscr{P}$ is in $\mathscr{S}$, then φ, ψ have the same domain and also the same codomain: $\varphi, \psi \colon A \rightrightarrows B$. This property is thus inherited by $\mathscr{R}$.

Since a pair (φ, ψ) in $\mathrm{Ar}(\mathscr{R})$ is such that φ, ψ have the same domain and the same codomain, we can define without any ambiguity a new category $\mathscr{Q}$:

- $|\mathcal{Q}| = |\mathcal{P}| = |\mathcal{G}|$,
- $|\mathcal{Q}|\,(A, B)$ is the set of equivalence classes of paths from A to B.

The category structure of $\mathcal{P}$ induces a category structure on $\mathcal{Q}$, just because $\mathcal{R}$ is a category. Indeed given morphisms

$$[\varphi]\colon A \longrightarrow B, \quad [\psi]\colon B \longrightarrow C$$

in $\mathcal{Q}$, we have corresponding morphisms $\varphi\colon A \longrightarrow B$, $\psi\colon B \longrightarrow \mathcal{C}$ in $\mathcal{P}$ and one can define $[\psi] \circ [\varphi] = [\psi \circ \varphi]$. Observe that when $[\varphi] = [\varphi']$ and $[\psi] = [\psi']$, then $(\varphi, \varphi') \in \mathcal{R}$ and $(\psi, \psi') \in \mathcal{R}$, thus also $(\psi \circ \varphi, \psi' \circ \varphi') \in \mathcal{R}$ since $\mathcal{R}$ is a category. From these observations it follows immediately that $\mathcal{Q}$ is a category and

$$\theta\colon \mathcal{P} \longrightarrow \mathcal{Q}, \quad \theta(A) = A, \quad \theta(\varphi) = [\varphi]$$

is a functor.

With the notation of 5.1.4, let us then consider the composite

$$\mathcal{G} \xrightarrow{\ \Gamma\ } \mathcal{P} \xrightarrow{\ \theta\ } \mathcal{Q},$$

which is a morphism of graphs. But $\mathcal{Q}$ is a category and moreover if $(\varphi, \psi) \in \Sigma$ is one of the given commutativity conditions, $\theta(\varphi) = \theta(\psi)$ since $\Sigma \subseteq \mathcal{R}$. Thus $\theta\Gamma$ is a morphism in $\mathsf{CondGraph}$, when $\mathcal{G}$ is provided with the set of commutativity conditions Σ.

Given another category $\mathcal{D}$ and a morphism of graphs $F\colon \mathcal{G} \longrightarrow \mathcal{D}$ preserving the commutativity conditions in Σ, we get a unique factorization functor $G\colon \mathcal{P} \longrightarrow \mathcal{D}$ such that $G \circ \Gamma = F$ (see 5.1.4). But if $(\varphi, \psi) \in \Sigma$, the assumption on F implies $G\varphi = G\psi$. Therefore we can define a functor $H\colon \mathcal{Q} \longrightarrow \mathcal{D}$ by

- $H(A) = G(A) = F(A)$,
- $H([\varphi]) = G(\varphi)$.

By construction, $H \circ \theta \circ \Gamma = F$ and clearly H is the unique functor with this property. $\qquad\square$

As an interesting corollary, we obtain

Proposition 5.1.7 *The category* Cat *of small categories is cocomplete.*

Proof We know already that Cat has coproducts, which are just disjoint unions (see 2.2.4.d). So it suffices to prove the existence of the coequalizer of two functors $F, G\colon \mathscr{A} \rightrightarrows \mathscr{B}$ (see 2.8.1). Let us consider the intersection $\mathcal{R}$ of all the subcategories $\mathscr{S} \subseteq \mathscr{B} \times \mathscr{B}$ which satisfy the following conditions:

(1) $|\mathscr{S}|$ is an equivalence relation on $|\mathscr{B}|$;
(2) $\forall A \in \mathscr{A}$, $(FA, GA) \in \mathscr{S}$;
(3) $\mathsf{Ar}(\mathscr{S})$ is an equivalence relation on $\mathsf{Ar}(\mathscr{B})$ (where $\mathsf{Ar}(\mathscr{X})$ denotes the set of arrows of $\mathscr{X}$);
(4) $\forall f \in \mathsf{Ar}(\mathscr{A})$ $(Ff, Gf) \in \mathsf{Ar}(\mathscr{S})$.

Clearly, $\mathscr{R}$ still satisfies properties (1) to (4).

Let us observe that given a pair $(g, h) \in \mathsf{Ar}(\mathscr{R})$, with $g\colon A \longrightarrow B$ and $h\colon C \longrightarrow D$, $(A, C) \in \mathscr{R}$ and $(B, D) \in \mathscr{R}$. Therefore there is no ambiguity in defining a graph $\mathscr{G}$ by

- $|\mathscr{G}|$ is the quotient of $|\mathscr{B}|$ by $|\mathscr{R}|$,
- $\mathsf{Ar}(\mathscr{G})$ is the quotient of $\mathsf{Ar}(\mathscr{B})$ by $\mathsf{Ar}(\mathscr{R})$,
- given an arrow $g\colon B \longrightarrow B'$ in $\mathscr{B}$, its equivalence class $[g]$ in $\mathscr{G}$ has domain $[B]$ and codomain $[B']$.

It should be observed that $\mathscr{G}$ is not, in general, a category. Indeed, given two non-composable arrows $g\colon A \longrightarrow B$ and $h\colon C \longrightarrow D$ in $\mathscr{B}$, the trouble occurs when $[B] = [C]$ in $\mathscr{G}$.

Let us now provide the graph $\mathscr{G}$ with the following set Σ of commutativity conditions

(1) $\forall B \in \mathscr{B}$ $\Big(([B], [1_B], [B]), ([B]) \Big) \in \Sigma$;
(2) $\forall g\colon A \longrightarrow B$ $\forall h\colon B \longrightarrow C$ in $\mathscr{B}$

$$\Big(([A], [g], [B], [h], [C]), ([A], [h \circ g], [C]) \Big) \in \Sigma;$$

(3) $\forall f\colon X \longrightarrow Y$ in $\mathscr{A}$

$$\Big(([FX], [Ff], [FY]), ([GX], [Gf], [GY]) \Big) \in \Sigma.$$

Observe that the last condition makes sense, since $[FX] = [GX]$, $[FY] = [GY]$ by definition of $\mathscr{R}$. It is the one which will force the coequalizing of F, G. The first two conditions are just there to recapture in $\mathscr{G}$ the identities and the composition law of $\mathscr{B}$.

Finally consider the category $\mathscr{D}$ universally associated with the pair $(\mathscr{G}, \Sigma)$ (see 5.1.6) and the corresponding morphism of graphs $\theta\colon \mathscr{G} \longrightarrow \mathscr{D}$. The composite $\mathscr{B} \xrightarrow{P} \mathscr{G} \xrightarrow{\theta} \mathscr{D}$, where P is the canonical projection, is a functor because of the first two conditions in the definition of Σ. This functor satisfies $\theta P F = \theta P G$ because of the third condition in the definition of Σ.

Now let us consider a category $\mathscr{D}$ and a functor $H\colon \mathscr{B} \longrightarrow \mathscr{D}$ such that $HF = HG$. Consider the kernel pair $\mathscr{K} \subseteq \mathscr{B} \times \mathscr{B}$ of H, i.e.

$$|\mathscr{K}| = \left\{(B_1, B_2) \in \mathscr{B} \times \mathscr{B} \,\middle|\, HB_1 = HB_2\right\}$$
$$\mathscr{K}\big((B_1, B_2), (B_1', B_2')\big) = \left\{(f, g) \in \mathscr{B}(B_1, B_1') \times \mathscr{B}(B_2, B_2') \,\middle|\, Hf = Hg\right\}.$$

The subcategory $\mathscr{K}$ satisfies in an obvious way the conditions (1) to (4) defining $\mathscr{R}$; thus $\mathscr{R} \subseteq \mathscr{K}$. Therefore H factors through the quotient graph $\mathscr{G}$ via a morphism of graphs $K\colon \mathscr{G} \longrightarrow \mathscr{D}$, with $K \circ P = H$. Since H is a functor, K respects the commutativity conditions (1), (2) in the definition of Σ; since $HF = HG$, K also respects the commutativity condition (3) in the definition of Σ. Therefore H factors uniquely through θ via a functor $L\colon \mathscr{Q} \longrightarrow \mathscr{D}$, with $L\theta = K$. Finally $L\theta P = KP = H$ and the uniqueness of L follows from that of K and L. $\qquad\square$

5.2 Calculus of fractions

This section introduces the main problem of this chapter: formally adding an inverse to some arrows of a given category. By analogy with the case of rings of fractions where one formally inverts (for the multiplication) a given (multiplicative) set of elements, we shall call our construction a "category of fractions".

Definition 5.2.1 *Consider a category $\mathscr{C}$ and a class Σ of arrows of $\mathscr{C}$. The category of fractions $\mathscr{C}[\Sigma^{-1}]$ is said to exist when a category $\mathscr{C}[\Sigma^{-1}]$ and a functor $\varphi\colon \mathscr{C} \longrightarrow \mathscr{C}[\Sigma^{-1}]$ can be found, with the following properties:*

(1) $\forall f \in \Sigma \;\; \varphi(f)$ is an isomorphism;

(2) if $\mathscr{D}$ is a category and $F\colon \mathscr{C} \longrightarrow \mathscr{D}$ is a functor such that for all morphisms $f \in \Sigma$, $F(f)$ is an isomorphism, there exists a unique functor $G\colon \mathscr{C}[\Sigma^{-1}] \longrightarrow \mathscr{D}$ such that $G \circ \varphi = F$.

The uniqueness condition on G implies immediately that when it exists, a category of fractions is defined uniquely up to isomorphism.

Proposition 5.2.2 *Consider a category $\mathscr{C}$ and a set Σ of arrows of $\mathscr{C}$. The category of fractions $\mathscr{C}[\Sigma^{-1}]$ exists. Moreover when $\mathscr{C}$ is small, $\mathscr{C}[\Sigma^{-1}]$ is small as well.*

Proof Let us first construct a graph $\mathscr{G}$:

- $|\mathscr{G}| = |\mathscr{C}|$;
- $\mathscr{G}(A, B) = \mathscr{C}(A, B) \amalg \left\{f \in \mathscr{C}(B, A) \,\middle|\, f \in \Sigma\right\}.$

Thus for each morphism $f: B \longrightarrow A$ in Σ ("to be inverted") we formally introduce a new arrow, which we shall write $f^{-1}: A \longrightarrow B$ and which will eventually produce the inverse of f.

On this graph $\mathcal{G}$, we introduce a class Θ of commutativity conditions given exactly by the following requirements:

(1) $\forall C \in \mathcal{C}$ $\big((C, 1_C, C), (C)\big) \in \Theta$;

(2) $\forall f: C \longrightarrow D$ and $\forall g: D \longrightarrow E$ in $\mathcal{C}$
$\big((C, f, D, g, E), (C, g \circ f, E)\big) \in \Theta$;

(3) $\forall f: C \longrightarrow D$ in Σ $\big((C, f, D, f^{-1}, C), (C, 1_C, C)\big) \in \Theta$
and $\big((D, f^{-1}, C, f, D), (D, 1_D, D)\big) \in \Theta$.

If we think of "class" as "element of a universe $\mathcal{V}$" and "set" as "element of a universe $\mathcal{U} \in \mathcal{V}$" (see section 1.1), the graph $\mathcal{G}$ is $\mathcal{V}$-small so that proposition 5.1.6 applies and we can consider the $\mathcal{V}$-small category $\mathcal{C}[\Sigma^{-1}]$ associated with the pair $(\mathcal{G}, \Theta)$ and the corresponding morphism $\tau: \mathcal{G} \longrightarrow \mathcal{C}[\Sigma^{-1}]$. We define $\varphi: \mathcal{C} \longrightarrow \mathcal{C}[\Sigma^{-1}]$ to be the composite $\tau \circ I$, where $I: \mathcal{C} \longrightarrow \mathcal{G}$ is the canonical inclusion (see 5.1.6).

When $\mathcal{C}$ is $\mathcal{U}$-small, $\mathcal{G}$ and $\mathcal{C}[\Sigma^{-1}]$ are small as well. But when $|\mathcal{C}| \in \mathcal{V}$ and $\mathcal{C}(A, B) \in \mathcal{U}$ for all $A, B \in |\mathcal{C}|$, it remains to prove that each $\mathcal{C}[\Sigma^{-1}](A, B)$ is still $\mathcal{U}$-small. By conditions (1), (2) in the definition of the class Θ, an arrow in $\mathcal{C}[\Sigma^{-1}]$ can always be presented as the equivalence class of a "reduced path" alternating arrows g_i of $\mathcal{C}$ and arrows of the type f_i^{-1}, for $f_i \in \Sigma$; we can even assume that the first and last arrows are some g_i's (replace two consecutive arrows of $\mathcal{C}$ by their composite and add identity arrows when necessary). Writing just the arrows and omitting the objects, a reduced path thus has the form

$$g_1 f_1^{-1} g_2 f_2^{-1} \cdots g_n f_n^{-1} g_{n+1}.$$

Since the class Σ is $\mathcal{U}$-small, the set of all possible finite sequences $f_1^{-1}, f_2^{-1}, \ldots, f_n^{-1}$ is $\mathcal{U}$-small. Once such a sequence of f_i^{-1}'s is fixed, the domain and the codomain of each g_i are fixed so that for each g_i, there is just a $\mathcal{U}$-small set of possibilities; $\mathcal{C}(X, Y)$ is $\mathcal{U}$-small for all X, Y. So between two objects A, B of $\mathcal{C}$, there is just a $\mathcal{U}$-small set of "reduced paths" as indicated, from which $\mathcal{C}[\Sigma^{-1}]$ is $\mathcal{U}$-small.

For each $f \in \Sigma$, $\varphi(f)$ is an isomorphism in $\mathcal{C}[\Sigma^{-1}]$ by condition (3) in the definition of Θ. Moreover given a functor $F: \mathcal{C} \longrightarrow \mathcal{D}$ such that $F(f)$ is an isomorphism for each $f \in \Sigma$, let us define a morphism $H: \mathcal{G} \longrightarrow \mathcal{D}$

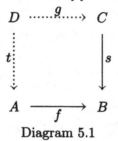

Diagram 5.1

by

$$H(A) = F(A) \quad \text{for} \quad A \in \mathscr{C},$$
$$H(f) = F(f) \quad \text{for} \quad f \in \mathscr{C}(A,B),$$
$$H(f^{-1}) = (Ff)^{-1} \quad \text{for} \quad f \in \mathscr{C}(B,A), \ f \in \Sigma.$$

Now H is a morphism of graphs. It preserves the commutativity conditions (1), (2) defining Θ because F is a functor; it preserves condition (3) just by definition. Therefore we find a unique functor $G: \mathscr{C}[\Sigma^{-1}] \longrightarrow \mathscr{D}$ such that $G \circ \tau = H$ and thus $G \circ \tau \circ I = F$. The uniqueness of G is obvious. $\qquad\square$

In a special case of interest, an easy description of the category of fractions $\mathscr{C}[\Sigma^{-1}]$ can be given.

Definition 5.2.3 *Consider a category $\mathscr{C}$ and a class Σ of morphisms of $\mathscr{C}$. The class Σ admits a right calculus of fractions when the following conditions hold:*

(1) $\forall C \in \mathscr{C} \ 1_C \in \Sigma$;

(2) *given $s: A \longrightarrow B$ and $t: B \longrightarrow C$, $(s \in \Sigma$ and $t \in \Sigma) \Rightarrow (t \circ s \in \Sigma)$;*

(3) *if $f: A \longrightarrow B$ is in $\mathscr{C}$ and $s: C \longrightarrow B$ is in Σ, there exist $g: D \longrightarrow C$ in $\mathscr{C}$ and $t: D \longrightarrow A$ in Σ such that $f \circ t = s \circ g$ (see diagram 5.1);*

(4) *if $f, g: A \rightrightarrows B$ are in $\mathscr{C}$ and $s: B \longrightarrow C$ is in Σ with the property $s \circ f = s \circ g$, there exists $t: D \longrightarrow A$ in Σ with the property $f \circ t = g \circ t$.*

Proposition 5.2.4 *Consider a category $\mathscr{C}$ and a class Σ of morphisms of $\mathscr{C}$ which admits a right calculus of fractions. When the category of fractions $\mathscr{C}[\Sigma^{-1}]$ exists, it can be described in the following way:*

(1) *the objects of $\mathscr{C}[\Sigma^{-1}]$ are those of $\mathscr{C}$;*

(2) *an arrow $f: A \longrightarrow B$ of $\mathscr{C}[\Sigma^{-1}]$ is an equivalence class of triples (s, I, f) where*

- *I is an object of $\mathscr{C}$,*
- *$s: I \longrightarrow A$ is a morphism of Σ,*
- *$f: I \longrightarrow B$ is an arbitrary morphism of $\mathscr{C}$,*

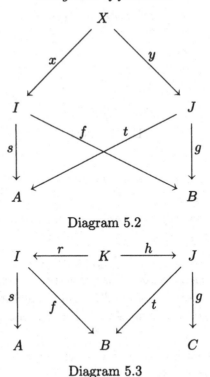

Diagram 5.2

$$I \xleftarrow{\quad r \quad} K \xrightarrow{\quad h \quad} J$$

Diagram 5.3

- the triple (s, I, f) is equivalent to the triple (t, J, g) when there exist x, y in $\mathscr{C}$ such that $s \circ x = t \circ y \in \Sigma$ and $f \circ x = g \circ y$ (see diagram 5.2);

(3) the composite of the equivalence classes
$$[(s, I, f)] \colon A \longrightarrow B, \quad [(t, J, g)] \colon B \longrightarrow C$$
in $\mathscr{C}[\Sigma^{-1}]$ is just $[(s \circ r, K, g \circ h)] \colon A \longrightarrow C$ where $r \in \Sigma$ and $h \in \mathscr{C}$ are any morphisms such that $f \circ r = t \circ h$ (see diagram 5.3).

Proof   In a first approach, let us assume that $\mathscr{C}$ is a small category. This implies that Σ itself is a set, so that 5.2.2 applies.

The relation of the statement used for defining the arrows is obviously reflexive and symmetric. To prove the transitivity, consider also an arrow $(u, K, h) \colon A \longrightarrow B$ with v, w such that $t \circ v = u \circ w \in \Sigma$ and $g \circ v = h \circ w$ (see diagram 5.4). Since $t \circ v \in \Sigma$, there exist $m \in \Sigma$ and $n \in \mathscr{C}$ such that $t \circ y \circ m = t \circ v \circ n$; see 5.2.3.(3). But since $t \in \Sigma$ and $t \circ y \circ m = t \circ v \circ n$, we get $r \in \Sigma$ such that $y \circ m \circ r = v \circ n \circ r$; see 5.2.3.(4). Finally one has

$$f \circ x \circ m \circ r = g \circ y \circ m \circ r = g \circ v \circ n \circ r = h \circ w \circ n \circ r$$

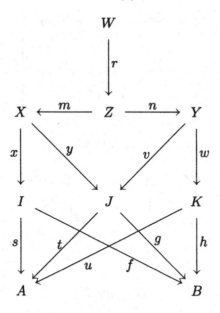

Diagram 5.4

and

$$s \circ x \circ m \circ r = t \circ y \circ m \circ r = t \circ v \circ n \circ r = u \circ w \circ n \circ r;$$

moreover $s \circ x \circ m \circ r \in \Sigma$ since $s \circ x \in \Sigma$, $m \in \Sigma$, $r \in \Sigma$; see 5.2.3.(2). This proves the equivalence of (s, I, f) and (u, K, h).

The definition of the composite $[(t, J, g)] \circ [(s, I, f)]$ given in the statement makes sense. Indeed by 5.2.3.(3) there always exist $r \in \Sigma$ and $h \in \mathscr{C}$ such that $f \circ r = t \circ h$ and $s \circ r \in \Sigma$ since $s, r \in \Sigma$. Moreover this definition is independent of the choices of f, s, g, t, h, r. This is lengthy but straightforward: the arguments are analogous to those for proving the transitivity of the equivalence relation defined on the arrows. We leave those details to the reader as well as the checking of the category axioms (the identity on A is just $[1_A, A, 1_A]$, which makes sense by 5.2.3.(1)). Let us for a while denote by $\mathscr{F}$ the category defined in the statement.

A functor $\varphi \colon \mathscr{C} \longrightarrow \mathscr{F}$ is easily defined:

- $\varphi(A) = A$ for $A \in \mathscr{C}$,
- $\varphi(f) = [(1_A, A, f)]$ for $f \colon A \longrightarrow B$ in $\mathscr{C}$;

observe that we again use 5.2.3.(1). By definition, $\varphi(1_A) = [1_A, A, 1_A]$, thus φ preserves identities. On the other hand, considering diagram 5.5, one concludes that $\varphi(g \circ f) = \varphi(g) \circ \varphi(f)$. Thus φ is a functor. Moreover,

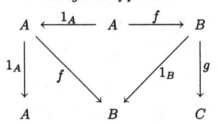

Diagram 5.5

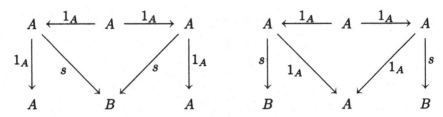

Diagram 5.6

if $s \in \Sigma$, consideration of diagram 5.6 indicates that

$$[s, A, 1_A] \circ [1_A, A, s] = [1_A, A, 1_A],$$
$$[1_A, A, s] \circ [s, A, 1_A] = [s, A, s] = [1_B, B, 1_B],$$

or in other words $[s, A, 1_A]$ is the inverse of $\varphi(s) = [1_A, A, s]$ in $\mathscr{C}[\Sigma^{-1}]$. So indeed $\varphi(s)$ is invertible as long as $s \in \Sigma$.

Now choose a functor $F: \mathscr{C} \longrightarrow \mathscr{D}$ such that Ff is invertible for every $f \in \Sigma$. A functor $G: \mathscr{F} \longrightarrow \mathscr{D}$ such that $G \circ \varphi = F$ satisfies

(1) $G(A) = G\varphi(A) = FA$ for each $A \in \mathscr{C}$,
(2) $G[(1_A, A, f)] = G\varphi(f) = Ff$ for each $f: A \longrightarrow B$ in $\mathscr{C}$,
(3) $G[(s, A, 1_A)] = G((\varphi s)^{-1}) = (Fs)^{-1}$ for each $s: A \longrightarrow B$ in Σ.

But an arbitrary morphism $[s, I, f]: A \longrightarrow B$ in $\mathscr{F}$ can be written as

$$[(s, I, f)] = [(1_I, I, f)] \circ [(s, I, 1_I)]$$

(see diagram 5.7) so that necessarily

(4) $G[s, I, f] = G[1_I, I, f] \circ G[s, I, 1_I] = (Ff) \circ (Fs)^{-1}$.

This last formula characterizes G on the arrows and, together with condition (1), proves the uniqueness of G.

To prove the existence of G, it suffices to take the relations (1) and (4) as a definition. Observe that the definition of $G[s, I, f]$ is independent of

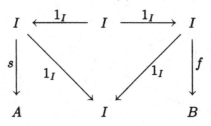

Diagram 5.7

the choices of s and f. Indeed going back to diagram 5.2, one observes that

$$
\begin{aligned}
Ff \circ (Fs)^{-1} &= Ff \circ (Fs)^{-1} \circ F(s \circ x) \circ \big(F(s \circ x)\big)^{-1} \\
&= Ff \circ (Fs)^{-1} \circ Fs \circ Fx \circ \big(F(s \circ x)\big)^{-1} \\
&= Ff \circ Fx \circ \big(F(s \circ x)\big)^{-1} \\
&= Fg \circ Fy \circ \big(F(t \circ y)\big)^{-1} \\
&= Fg \circ (Ft)^{-1} \circ Ft \circ Fy \circ \big(F(t \circ y)\big)^{-1} \\
&= Fg \circ (Ft)^{-1} \circ F(t \circ y) \circ \big(F(t \circ y)\big)^{-1} \\
&= Fg \circ (Ft)^{-1}.
\end{aligned}
$$

In an analogous way one checks that G is a functor. Moreover, given $f \colon A \longrightarrow B$ in $\mathscr{C}$,

$$
G\varphi(f) = G[1_A, A, f] = (Ff) \circ (F1_A)^{-1} = Ff,
$$

which concludes the proof that $\mathscr{F}$ is the category of fractions $\mathscr{C}[\Sigma^{-1}]$.

To conclude the proof of the proposition, consider an arbitrary category $\mathscr{C}$ and a class $\Sigma \subseteq \mathscr{C}$ of morphisms of $\mathscr{C}$ such that the category of fractions $\mathscr{C}[\Sigma^{-1}]$ exists. Using the axiom system of universes, $\mathscr{C}$ and Σ are $\mathscr{V}$-small with respect to some universe $\mathscr{V}$ and $\mathscr{C}(A, B)$ is $\mathscr{U}$-small for all $A, B \in |\mathscr{C}|$, for some universe $\mathscr{U} \in \mathscr{V}$. The proof we have just developed applies to $\mathscr{C}$ and Σ considered as $\mathscr{V}$-small; thus the category $\mathscr{F}$ described in this proof has the universal property of the category of fractions $\mathscr{C}[\Sigma^{-1}]$, with respect to all functors $F \colon \mathscr{C} \longrightarrow \mathscr{D}$ inverting the morphisms in Σ, with $\mathscr{D}$ a $\mathscr{V}$-small category. In particular this property holds for all categories $\mathscr{D}$ with $|\mathscr{D}| \in \mathscr{V}$ and each $\mathscr{D}(X, Y) \in \mathscr{U}$. Observe that the assumption on the existence of $\mathscr{C}[\Sigma^{-1}]$ is just there to ensure that each $\mathscr{C}[\Sigma^{-1}](A, B)$ is still $\mathscr{U}$-small. $\qquad\square$

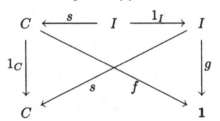

Diagram 5.8

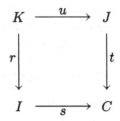

Diagram 5.9

Proposition 5.2.5 *Consider a category $\mathscr{C}$ and a class Σ of morphisms of $\mathscr{C}$ which admits a right calculus of fractions and such that the category of fractions $\mathscr{C}[\Sigma^{-1}]$ exists. When $\mathscr{C}$ is finitely complete, so is $\mathscr{C}[\Sigma^{-1}]$ and the canonical functor $\varphi\colon \mathscr{C} \longrightarrow \mathscr{C}[\Sigma^{-1}]$ preserves finite limits.*

Proof We use the notation of 5.2.4. If $\mathbf{1}$ is a terminal object in $\mathscr{C}$ and $f\colon C \longrightarrow \mathbf{1}$ is the unique arrow from a given object $C \in \mathscr{C}$, diagram 5.8 indicates that every morphism $[(s, I, g)]\colon C \longrightarrow \mathbf{1}$ in $\mathscr{C}[\Sigma^{-1}]$ is just $\varphi(f) = [(1_C, C, f)]$. Thus $\mathbf{1}$ is terminal in $\mathscr{C}[\Sigma^{-1}]$ as well.

Given two objects A, B in $\mathscr{C}$, consider their product $(A \times B, p_A, p_B)$ in $\mathscr{C}$. We shall prove that $A \times B$ together with the projections

$$[(1_{A \times B}, A \times B, p_A)], \quad [(1_{A \times B}, A \times B, p_B)]$$

is still the product of A, B in $\mathscr{C}[\Sigma^{-1}]$. Indeed consider an object $C \in \mathscr{C}$ and morphisms $[s, I, f]\colon C \longrightarrow A$, $[(t, J, g)]\colon C \longrightarrow B$ in $\mathscr{C}[\Sigma^{-1}]$. Applying 5.2.3.(3) we choose u, r such that $s \circ r = t \circ u$ and $r \in \Sigma$, as in diagram 5.9. From $r \in \Sigma$, $s \in \Sigma$ we deduce $t \circ u = s \circ r \in \Sigma$. But by definition of $\mathscr{C}[\Sigma^{-1}]$

$$[(s, I, f)] = [(sr, K, fr)], \quad [(t, J, g)] = [(tu, K, gu)].$$

But now, in $\mathscr{C}$, we have $f \circ r\colon K \longrightarrow A$, $g \circ u\colon K \longrightarrow B$ from which we get a unique $h\colon K \longrightarrow A \times B$ such that $p_A \circ h = f \circ r$, $p_B \circ h = g \circ u$.

This yields a morphism

$$[(sr, K, h)] : C \longrightarrow A \times B$$

in $\mathscr{C}[\Sigma^{-1}]$, with the properties

$$[(1_{A \times B}, A \times B, p_A)] \circ [(sr, K, h)] = [(s, I, f)],$$
$$[(1_{A \times B}, A \times B, p_B)] \circ [(sr, K, h)] = [(t, J, g)].$$

If $[(u, L, m)]$ is another factorization, one first uses axioms 5.2.3.(3,4) to replace L, M by a single object N (like in the proof of the transitivity, in 5.2.4) and then one can deduce the equality $[(u, L, m)] = [(s \circ r, K, h)]$ just using straightforward arguments.

Now consider two morphisms in $\mathscr{C}[\Sigma^{-1}]$

$$[(s, I, f)], [(t, J, g)] : A \Longrightarrow B;$$

we shall construct their equalizer. Applying 5.2.3.(3) we can find morphisms x, y such that $s \circ x \doteq t \circ y$, with $x \in \Sigma$ since $t \in \Sigma$ (see diagram 5.2). Clearly one has

$$[(s, I, f)] = [(s \circ x, X, f \circ x)], \quad [(t, J, g)] = [(t \circ y, X, g \circ y)],$$

with indeed $t \circ y = s \circ x \in \Sigma$ since $s \in \Sigma$ and $x \in \Sigma$. It suffices now to compute the equalizer

$$k = \mathsf{Ker} \, (f \circ x, g \circ y) : K \longrightarrow X$$

in $\mathscr{C}$. This yields a morphism

$$\varphi(s \circ x \circ k) = \varphi(t \circ y \circ k) : K \longrightarrow A$$

in $\mathscr{C}[\Sigma^{-1}]$ which is easily seen to be the equalizer of the original pair of morphisms.

Observe that when the two original morphisms have the form

$$\varphi(f) = [(1_A, A, f)], \quad \varphi(g) = [(1_A, A, g)]$$

we can choose $x = 1_A = y$ and conclude that the equalizer in $\mathscr{C}[\Sigma^{-1}]$ is just $[(1_K, K, k)] = \varphi(k)$ where $k = \mathsf{Ker} \, (f, g)$ in $\mathscr{C}$. This proves that φ preserves equalizers. $\qquad \qquad \square$

It should be observed that in the construction of a category of fractions $\varphi \colon \mathscr{C} \longrightarrow \mathscr{C}[\Sigma^{-1}]$, the morphisms $f \in \Sigma$ are inverted by φ, but in general there exist other morphisms $f \notin \Sigma$ which are inverted by φ. For example in the category Set of sets consider for Σ the class of all injections. So Σ admits a right calculus of fractions since

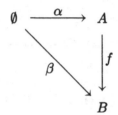

Diagram 5.10

(1) every identity is an injection,
(2) the composite of two injections is an injection,
(3) injections are stable under pullback,
(4) if $s \circ f = s \circ g$ with s injective and $f, g\colon A \rightrightarrows B$, then $f = g$ and thus $f \circ 1_A = g \circ 1_A$.

Observe that if a functor $F\colon \mathsf{Set} \longrightarrow \mathscr{D}$ to an arbitrary category $\mathscr{D}$ inverts all injections, it inverts all morphisms of Set. Indeed given an arbitrary morphism $f\colon A \longrightarrow B$ in Set, consider diagram 5.10 in Set where $\emptyset$ is the empty set and α, β are the obvious injections. From the equality $F\beta = Ff \circ F\alpha$, we deduce that Ff is an isomorphism since $F\alpha, F\beta$ are. Moreover given another morphism $g\colon A \longrightarrow B$ in Set, $g \circ \alpha = \beta$, thus $Fg \circ F\alpha = F\beta = Ff \circ F\alpha$; since $F\alpha$ is an isomorphism, $Ff = Fg$. So F maps all morphisms of $\mathsf{Set}(A, B)$ to a single isomorphism $FA \stackrel{\cong}{\longrightarrow} FB$. The category of fractions $\mathsf{Set}[\Sigma^{-1}]$ is thus obtained by taking one object for each set and one single isomorphism between any two objects.

Therefore the following definition is pertinent:

Definition 5.2.6 *Let $\mathscr{C}$ be a category and $\Sigma \subseteq \mathscr{C}$ a class of morphisms such that the category of fractions $\varphi\colon \mathscr{C} \longrightarrow \mathscr{C}[\Sigma^{-1}]$ exists. The class Σ is saturated when for every morphism $f \in \mathscr{C}$*

$$\varphi(f) \text{ is an isomorphism iff } f \in \Sigma.$$

5.3 Reflective subcategories as categories of fractions

Up to equivalence of categories, every reflection can be seen as a category of fractions.

Proposition 5.3.1 *Consider a category $\mathscr{B}$ and a reflective subcategory $i\colon \mathscr{A} \hookrightarrow \mathscr{B}$, with reflection $r \dashv i$. Write Σ for the class of all morphisms f of $\mathscr{B}$ such that $r(f)$ is an isomorphism. In this case the category* of fractions $\varphi\colon \mathscr{B} \longrightarrow \mathscr{B}[\Sigma^{-1}]$ exists and is equivalent to $r\colon \mathscr{B} \longrightarrow \mathscr{A}$. Moreover the class Σ admits a left calculus of fractions.

Proof   Let us first recall that the canonical morphisms $\varepsilon_A\colon riA \longrightarrow A$ of the adjunction are isomorphisms (see 3.4.1). Given $B \in \mathscr{B}$, the identity $\varepsilon_{rB} \circ r\eta_B = 1_{rB}$ implies that given the second canonical morphism $\eta_B\colon B \longrightarrow irB$, $r(\eta_B)$ is an isomorphism as well, with inverse the isomorphism ε_{rB}. Thus $\eta_B \in \Sigma$.

Let us write $\mathscr{X}$ for the following category:

- $|\mathscr{X}| = |\mathscr{B}|$;
- $\mathscr{X}(B, B') = \mathscr{A}(rB, rB')$;
- the composition is that of $\mathscr{A}$.

There is an obvious functor $\Gamma\colon \mathscr{A} \longrightarrow \mathscr{X}$:

- $\Gamma(A) = i(A)$ for $A \in \mathscr{A}$;
- $\Gamma(f) = ri(f)$ for $f \in \mathscr{A}(A, A')$.

This functor is full and faithful since ri is isomorphic to the identity on $\mathscr{A}$ via the isomorphism ε. Moreover, given $B \in \mathscr{X}$, we also have an isomorphism $r\eta_B \in \mathscr{X}(B, irB) = \mathscr{X}(B, \Gamma(rB))$. Therefore Γ is an equivalence of categories (see 3.4.3).

There is an obvious functor $\varphi\colon \mathscr{B} \longrightarrow \mathscr{X}$:

- $\varphi(B) = B$ for $B \in \mathscr{B}$;
- $\varphi(f) = r(f)$ for $f\colon B \longrightarrow B'$ in $\mathscr{B}$.

We shall prove that $(\mathscr{X}, \varphi\colon \mathscr{B} \longrightarrow \mathscr{X})$ is the category of fractions $\mathscr{B}[\Sigma^{-1}]$. Given a functor $F\colon \mathscr{B} \longrightarrow \mathscr{C}$ such that $F(f)$ is an isomorphism for each $f \in \Sigma$, a functor $G\colon \mathscr{X} \longrightarrow \mathscr{C}$ such that $G \circ \varphi = F$ must have the following characteristics, since each η_B is in Σ:

(1) $GB = G\varphi(B) = FB$;
(2) $Gr(\eta_B) = G\varphi(\eta_B) = F(\eta_B)$;
(3) $G(\varepsilon_{rB}) = G\big((r(\eta_B))^{-1}\big) = \big(G(r(\eta_B))\big)^{-1} = (F(\eta_B))^{-1}$;
(4) $\begin{aligned}[t] G(f) &= G\big(f \circ \varepsilon_{rB} \circ r(\eta_B)\big) \\ &= G\big(\varepsilon_{rB'} \circ r(if) \circ r(\eta_B)\big) \\ &= G(\varepsilon_{rB'}) \circ Gr(if) \circ Gr(\eta_B) \\ &= F(\eta_{B'})^{-1} \circ G\varphi i(f) \circ F(\eta_B) \\ &= F(\eta_{B'})^{-1} \circ Fi(f) \circ F(\eta_B) \end{aligned}$

for $B, B' \in \mathscr{X}$ and $f \in \mathscr{X}(B, B')$. This proves the uniqueness of G. On the other hand the naturalities of ε, η show that G, defined by

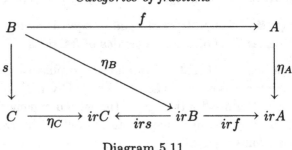

Diagram 5.11

- $GB = FB$ for $B \in \mathscr{X}$,
- $Gf = F(\eta_{B'})^{-1} \circ Fif \circ F(\eta_B)$ for $f \in \mathscr{X}(B, B')$,

is indeed a functor. It is such that $G \circ \varphi = F$ since

$$G\varphi(B) = GB = FB,$$
$$\begin{aligned}
G\varphi(f) &= Gr(f) \\
&= F(\eta_{B'})^{-1} \circ Fir(f) \circ F(\eta_B) \\
&= F(\eta_{B'})^{-1} \circ F(ir(f) \circ \eta_B) \\
&= F(\eta_{B'})^{-1} \circ F(\eta_{B'} \circ f) \\
&= F(\eta_{B'})^{-1} \circ F(\eta_{B'}) \circ Ff \\
&= Ff
\end{aligned}$$

for $B \in \mathscr{B}$ and $f \in \mathscr{B}(B, B')$.

It remains to prove that Σ satisfies the conditions dual to those of 5.2.3. Clearly $r(1_B)$ is an isomorphism for every $B \in \mathscr{B}$ and if $g \circ f$ exists in $\mathscr{B}$ with $r(g), r(f)$ isomorphisms, $r(g \circ f) = r(g) \circ r(f)$ is an isomorphism as well.

Let us check the dual of condition 5.2.3.(3). Diagram 5.11 commutes in $\mathscr{B}$ just by naturality of η. So if $s \in \Sigma$, $ir(s)$ is an isomorphism and

$$\eta_A \circ f = \left(ir(f) \circ \left(ir(s) \right)^{-1} \circ \eta_C \right) \circ s$$

with $\eta_A \in \Sigma$ by 3.4.1.

For the dual of condition 5.2.3.(4), consider the commutative diagram 5.12 in $\mathscr{B}$. If $f \circ s = g \circ s$ with $s \in \Sigma$, $ir(f) \circ ir(s) = ir(g) \circ ir(s)$ with $ir(s)$ an isomorphism, thus $ir(f) = ir(g)$. Therefore

$$\eta_A \circ f = ir(f) \circ \eta_B = ir(g) \circ \eta_B = \eta_A \circ g$$

with $\eta_A \in \Sigma$ by 3.4.1. $\qquad\qquad\qquad\qquad\qquad\qquad\qquad\square$

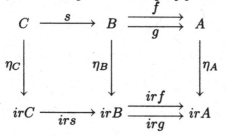

Diagram 5.12

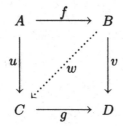

Diagram 5.13

5.4 The orthogonal subcategory problem

Considering a reflective subcategory as a category of fractions "up to equivalence" can lead finally to a clumsy presentation, as suggested by the proof of 5.3.1. There is an equivalent, but more elegant approach.

Definition 5.4.1 *Consider two arrows $f\colon A \longrightarrow B$, $g\colon C \longrightarrow D$ in a category $\mathscr{C}$. We say that "f is orthogonal to g" and we write $f \perp g$ when, given arbitrary morphisms u, v such that $v \circ f = g \circ u$ there exists a unique morphism w such that $w \circ f = u$, $g \circ w = v$ (see diagram 5.13).*

Let us make clear that this orthogonality relation is by no means symmetric.

Let us also observe that the definition of a strong epimorphism (see 4.3.5) can be rephrased in the following way:

An epimorphism f is strong when, for every monomorphism g, $f \perp g$.

Definition 5.4.2 *Given an arrow $f\colon A \longrightarrow B$ and objects X, Y of a category $\mathscr{C}$ (see diagram 5.14):*

(1) we say that f is orthogonal to X and write $f \perp X$ when for every morphism $a\colon A \longrightarrow X$, there exists a unique morphism $b\colon B \longrightarrow X$ such that $b \circ f = a$;

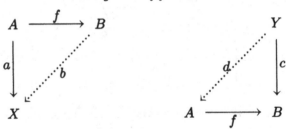

Diagram 5.14

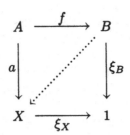

Diagram 5.15

(2) we say that Y is orthogonal to f and write $Y \perp f$ when for every morphism $c \colon Y \longrightarrow B$ there exists a unique morphism $d \colon Y \longrightarrow A$ such that $f \circ d = c$.

Clearly the two notions of 5.4.2 are dual to each other, while 5.4.1 is an autodual definition. There is generally no ambiguity between the notions of orthogonality in 5.4.1 and 5.4.2 since

Proposition 5.4.3 *Consider a category $\mathscr{C}$ with a terminal object **1**. Given an arrow $f \colon A \longrightarrow B$ of $\mathscr{C}$ and an object X,*

$$f \perp X \quad \text{iff} \quad f \perp \xi_X,$$

where $\xi_X \colon X \longrightarrow \mathbf{1}$ is the unique existing morphism.

Proof It suffices to consider diagram 5.15 where the outer square and the lower triangle are automatically commutative.   □

Here now is the way to use these orthogonality conditions to describe reflective subcategories.

Proposition 5.4.4 *Consider a reflective subcategory $i \colon \mathscr{A} \hookrightarrow \mathscr{B}$ with reflection $r \colon \mathscr{B} \longrightarrow \mathscr{A}$. Write Σ for the class of all morphisms $f \in \mathscr{B}$ inverted by r. Given $B \in \mathscr{B}$, write $\eta_B \colon B \longrightarrow irB$ for the canonical morphism of the adjunction.*

(1) For an object $X \in \mathscr{B}$, the following conditions are equivalent:

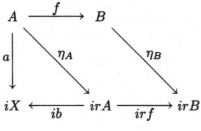

Diagram 5.16

(a) $X \in \mathscr{A}$;

(b) $\forall f \in \Sigma \quad f \perp X$;

(c) $\forall B \in \mathscr{B} \quad \eta_B \perp X$.

(2) For an arrow $f \in \mathscr{B}$, the following conditions are equivalent:

(a) $f \in \Sigma$;

(b) $\forall X \in \mathscr{A} \quad f \perp X$;

(c) $\forall g \in \mathscr{A} \quad f \perp g$.

Proof Choose $X \in \mathscr{A}$ and $f \colon A \longrightarrow B$ in Σ, so that η_X is an isomorphism (see 3.4.1) and consider diagram 5.16. Given $a \colon A \longrightarrow X = i(X)$, $\eta_X \circ a = ir(a) \circ \eta_A$ so that $a = i(\eta_X^{-1} \circ r(a)) \circ \eta_A$; we put $b \colon r(A) \longrightarrow X$, $b = \eta_X^{-1} \circ r(a)$, thus $a = i(b) \circ \eta_A$. Since $r(f)$ is an isomorphism, we obtain

$$i(b) \circ i(r(f))^{-1} \circ \eta_B \circ f = i(b) \circ \eta_A = a,$$

so that $i(b) \circ i(r(f))^{-1} \circ \eta_B$ is a factorization of a through f. Such a factorization is unique because given $h, k \colon B \rightrightarrows iX$ such that $h \circ f = a = k \circ f$, one has $r(h) \circ r(f) = r(k) \circ r(f)$ and thus $r(h) = r(k)$ since $r(f)$ is an isomorphism. Then

$$\eta_{iX} \circ h = ir(h) \circ \eta_B = ir(k) \circ \eta_B = \eta_{iX} \circ k$$

and finally $h = k$ because η_{iX} is an isomorphism (ε_X is an isomorphism by 3.4.1 and η_{iX} is the inverse of $i\varepsilon_X$ by 3.1.5).

That (b) implies (c) is obvious, since $\eta_B \in \Sigma$; notice that $\varepsilon_{r(B)}$ is an isomorphism by 3.4.1 and $r(\eta_B)$ is its inverse by 3.1.5.

Let us suppose now that $\eta_B \perp X$ for each $B \in \mathscr{B}$. Putting $B = X$, we find a unique morphism $x \colon ir(X) \longrightarrow X$ such that $x \circ \eta_X = 1_X$, as in diagram 5.17. But $\eta_X \circ x \circ \eta_X = \eta_X = 1_{ir(X)} \circ \eta_X$, so that by uniqueness of the factorization, $\eta_X \circ x = 1_{ir(X)}$. So η_X and x are inverse isomorphisms and $X \in \mathscr{A}$ because $r(X) \in \mathscr{A}$ and $\mathscr{A}$ is replete in $\mathscr{B}$; see 3.5.1.

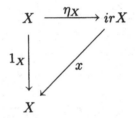

Diagram 5.17

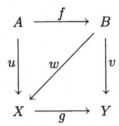

Diagram 5.18

Let us now prove the second set of equivalences. Given $f\colon A \longrightarrow B$ in Σ and $X \in \mathscr{A}$, we must prove that $f \perp X$. But this is precisely (a) $\Rightarrow$ (b) in the first part of the proof.

Choose now $f\colon A \longrightarrow B$ with $f \perp X$ for every $X \in \mathscr{A}$. Choose also an arbitrary morphism $g\colon X \longrightarrow Y$ in $\mathscr{A}$, as in diagram 5.18. Given u, v with $g \circ u = v \circ f$, the condition $f \perp X$ implies the existence of a unique w such that $w \circ f = u$. It remains to prove that $g \circ w = v$. But $g \circ w \circ f = g \circ u = v \circ f$ so that $g \circ w$ and v are two factorizations of $g \circ u$ through f; since $f \perp Y$, this implies $g \circ w = v$.

Finally choose $f\colon A \longrightarrow B$ in $\mathscr{B}$ orthogonal to every $g\colon X \longrightarrow Y$ in $\mathscr{A}$. Considering diagram 5.19, we have $ir(f) \circ \eta_A = \eta_B \circ f$; therefore we obtain g such that $g \circ f = \eta_A$, $ir(f) \circ g = \eta_B$. Let us prove that $\varepsilon_{r(A)} \circ r(g)$ is the inverse of $r(f)$. Applying 3.1.5,

$$\varepsilon_{r(A)} \circ r(g) \circ r(f) = \varepsilon_{r(A)} \circ r(\eta_A) = 1_{r(A)},$$
$$r(f) \circ \varepsilon_{r(A)} \circ r(g) = \varepsilon_{r(B)} \circ rir(f) \circ r(g)$$
$$= \varepsilon_{r(B)} \circ r(\eta_B) = 1_{r(B)}. \qquad \square$$

The equivalence (a)$\Leftrightarrow$(b) in the first part of 5.4.4 suggests the following definition:

Definition 5.4.5 *Let $\mathscr{C}$ be a category and Σ a class of morphisms of $\mathscr{C}$. By the orthogonal subcategory of $\mathscr{C}$ determined by Σ, we mean the full*

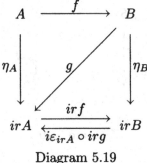

Diagram 5.19

subcategory $\mathscr{C}_\Sigma$ of $\mathscr{C}$ whose objects are those $X \in \mathscr{C}$ such that $f \perp X$ for every $f \in \Sigma$.

5.4.4 shows that given a reflective subcategory $i\colon \mathscr{A} \hookrightarrow \mathscr{B}$ with reflection $r\colon \mathscr{B} \longrightarrow \mathscr{A}$, $\mathscr{A}$ is precisely the orthogonal subcategory $\mathscr{C}_\Sigma$ determined by the class Σ of those morphisms inverted by r. The orthogonal subcategory problem consists in finding conditions such that, given a category $\mathscr{C}$ and a class Σ of morphisms of $\mathscr{C}$, the orthogonal subcategory $\mathscr{C}_\Sigma$ is reflective in $\mathscr{C}$.

Observe first an obvious fact (see 3.5.1).

Proposition 5.4.6 *Let $\mathscr{C}$ be a category and Σ a class of morphisms of $\mathscr{C}$. The full subcategory $\mathscr{C}_\Sigma$ of $\mathscr{C}$ is replete.*   □

The solution of the orthogonal subcategory problem uses in an essential way a notion which will be studied more systematically in chapter 5 of volume 2. More precisely,

> an object C of a category $\mathscr{C}$ is α-presentable, for some regular cardinal α, when the representable functor $\mathscr{C}(C, -)\colon \mathscr{C} \longrightarrow \mathrm{Set}$ preserves α-filtered colimits.

More generally

> An object C of a category $\mathscr{C}$ is presentable when it is α-presentable for some regular cardinal α.

(See section 6.4 for the notion of regular cardinal.) In chapter 5 of volume 2, we shall study a wide class of categories which satisfy the assumptions of the following theorem, namely the locally presentable categories (see 5.2.10, volume 2).

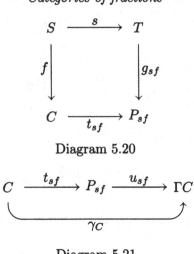

Diagram 5.20

$$C \xrightarrow{\ t_{sf}\ } P_{sf} \xrightarrow{\ u_{sf}\ } \Gamma C$$

$$\gamma_C$$

Diagram 5.21

Theorem 5.4.7 *Let $\mathscr{C}$ be a cocomplete category in which every object is presentable. Given a set Σ of morphisms of $\mathscr{C}$, the corresponding orthogonal subcategory $\mathscr{C}_\Sigma$ is reflective in $\mathscr{C}$.*

Proof Let us first indicate a general construction, starting from an arbitrary object $C \in \mathscr{C}$. For every pair (s, f) where $s\colon S \longrightarrow T$ is in Σ and $f\colon S \longrightarrow C$ is arbitrary, consider the pushout in diagram 5.20. Consider now the diagram constituted of all the arrows t_{sf}, for all possible pairs (s, f), and compute its colimit $(\Gamma C, (u_{sf})_{sf})$ (a sort of "infinite pushout" of arrows with a common codomain C). The diagram is small since Σ is a set, thus the colimit exists by assumption. We write $\gamma_C = u_{sf} \circ t_{sf}$ for the unique composite obtained in this way (see diagram 5.21).

So given (s, f) as above, we do not have in general a (unique) factorization $T \longrightarrow C$, but we certainly have a factorization $T \longrightarrow \Gamma C$ (see diagram 5.22), namely $u_{sf} \circ g_{sf}$; indeed

$$u_{sf} \circ g_{sf} \circ s = u_{sf} \circ t_{sf} \circ f = \gamma_C \circ f.$$

The factorization $T \longrightarrow \Gamma C$ has no reason to be unique. Therefore we consider all the morphisms $h\colon T \longrightarrow \Gamma C$ such that $h \circ s = \gamma_C \circ f$, for a fixed pair (s, f). We consider the colimit $q_{sf}\colon \Gamma C \longrightarrow Q_{sf}$ of all those factorizations h (a sort of "infinite coequalizer"); this makes sense since $\mathscr{C}(T, \Gamma C)$ is a set and $\mathscr{C}$ is cocomplete. Then we consider the diagram constituted of all the arrows q_{sf}, for all possible pairs (s, f), and we compute its colimit $(\Delta C, (v_{sf})_{sf})$ (a sort of "infinite pushout"); again this colimit does exist because Σ is small and $\mathscr{C}$ is cocomplete. We write

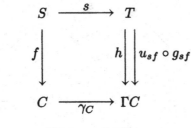

Diagram 5.22

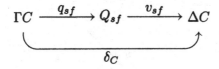

Diagram 5.23

$\delta_C = v_{sf} \circ q_{sf}$ for the unique composite obtained in this way (see diagram 5.23).

To construct the reflection of $C \in \mathscr{C}$ in $\mathscr{C}_\Sigma$, it will now suffice to iterate the previous construction. By transfinite induction, let us define a functor

$$F \colon \mathrm{Ord} \longrightarrow \mathscr{C}$$

where Ord is the preordered class of ordinals. Defining F is just giving a transfinite sequence

$$C_0 \xrightarrow{\theta_0} C_1 \xrightarrow{\theta_1} C_2 \xrightarrow{\theta_2} \ldots$$

in the category $\mathscr{C}$; we put

(1) $C_0 = C$,
(2) if C_β is defined, $C_{\beta+1} = \Delta C_\beta$ with connecting morphism $\theta_\beta = \delta_{C_\beta} \circ \gamma_{C_\beta}$

$$C_\beta \xrightarrow{\gamma_{C_\beta}} \Gamma C_\beta \xrightarrow{\delta_{C_\beta}} \Delta C_\beta,$$

(3) if β is a limit ordinal, $C_\beta = \mathrm{colim}_{\varepsilon < \beta} C_\varepsilon$, with the canonical morphisms of the colimit as connecting morphisms.

Such a construction can be performed in every cocomplete category, but has no reason to become stationary at some stage. But using the assumption that every object is presentable, we shall be able to reach our conclusion.

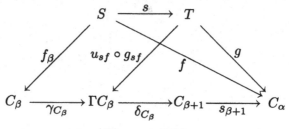

Diagram 5.24

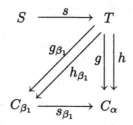

Diagram 5.25

Each object of $\mathscr{C}$ is presentable. Since Σ is a set, we can choose a sufficiently big regular cardinal α such that for each $s\colon S \longrightarrow T$ in Σ, both S and T are α-presentable. We shall prove that C_α, together with the connecting morphism $C \longrightarrow C_\alpha$, is the reflection of C in $\mathscr{C}_\Sigma$.

First of all let us consider $s\colon S \longrightarrow T$ in Σ and $f\colon S \longrightarrow C_\alpha$ as in diagram 5.24. Since S is α-presentable and the colimit defining C_α is α-filtered, f factors through some C_β, with $\beta < \alpha$ (see 5.1.3 of volume 2). Writing $s_\delta\colon C_\delta \longrightarrow C_\alpha$ for the canonical morphisms of the colimit, we have thus some f_β with $s_\beta \circ f_\beta = f$. By definition of ΓC_β we get

$$f = s_\beta \circ f_\beta = s_{\beta+1} \circ \theta_\beta \circ f_\beta = s_{\beta+1} \circ \delta_{C_\beta} \circ \gamma_{C_\beta} \circ f_\beta$$
$$= s_{\beta+1} \circ \delta_{C_\beta} \circ u_{sf} \circ g_{sf} \circ s,$$

which shows that $g = s_{\beta+1} \circ \delta_{C_\beta} \circ u_{sf} \circ g_{sf}$ is a factorization satisfying $g \circ s = f$.

Now let us suppose we have two factorizations $g, h\colon T \rightrightarrows C_\alpha$ with the property $g \circ s = f = h \circ s$. We must prove the equality $g = h$. Since T is α-presentable and the colimit defining C_α is α-filtered, g and h factor through some terms C_{β_1}, C_{β_2} of the colimit. Since the colimit is filtered, there is no restriction in supposing $\beta_1 = \beta_2$; this yields diagram 5.25. Writing g_{β_1}, h_{β_1} for the factorizations and again s_{β_1} for the canonical

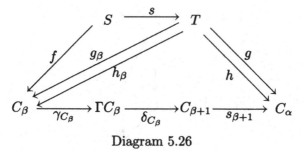

Diagram 5.26

morphism of the colimit, we have

$$s_{\beta_1} \circ g_{\beta_1} \circ s = g \circ s = f = h \circ s = s_{\beta_1} \circ h_{\beta_1} \circ s.$$

This shows that the two morphisms $(g_{\beta_1} \circ s, h_{\beta_1} \circ s)$ are identified in the colimit C_α; since this colimit is filtered, there exists $\beta_1 < \beta < \alpha$ such that those two morphisms are already coequalized at the level C_β (see 2.13.3). Finally we have got the situation of diagram 5.26 with $g_\beta \circ s = h_\beta \circ s$ and $g = s_\beta \circ g_\beta$, $h = s_\beta \circ h_\beta$. Putting $f = g_\beta \circ s$ in the first part of proof, we get

$$(\gamma_{C_\beta} \circ g_\beta) \circ s = \gamma_{C_\beta} \circ f = (\gamma_{C_\beta} \circ h_\beta) \circ s$$

thus $\gamma_{C_\beta} \circ g_\beta$ and $\gamma_{C_\beta} \circ h_\beta$ are two factorizations of $\gamma_{C_\beta} \circ f$ through s. This implies $q_{sf} \circ \gamma_{C_\beta} \circ g_\beta = q_{sf} \circ \gamma_{C_\beta} \circ h_\beta$ by definition of q_{sf}. Finally

$$
\begin{aligned}
g &= s_\beta \circ g_\beta \\
&= s_{\beta+1} \circ \theta_\beta \circ g_\beta \\
&= s_{\beta+1} \circ \delta_{C_\beta} \circ \gamma_{C_\beta} \circ g_\beta \\
&= s_{\beta+1} \circ v_{sf} \circ q_{sf} \circ \gamma_{C_\beta} \circ g_\beta \\
&= s_{\beta+1} \circ v_{sf} \circ q_{sf} \circ \gamma_{C_\beta} \circ h_\beta \\
&= s_{\beta+1} \circ \delta_{C_\beta} \circ \gamma_{C_\beta} \circ h_\beta \\
&= s_{\beta+1} \circ \theta_\beta \circ h_\beta = s_\beta \circ h_\beta \\
&= h.
\end{aligned}
$$

This ends the proof that C_α is an object of $\mathscr{C}_\Sigma$.

Now let us consider the canonical morphism $s_0 \colon C = C_0 \longrightarrow C_\alpha$ of the colimit. Given $D \in \mathscr{C}_\Sigma$ and $m \colon C_0 \longrightarrow D$, we must find a unique $n \colon C_\alpha \longrightarrow D$ such that $n \circ s_0 = m$. By transfinite induction, we shall construct a cocone of morphisms $n_\beta \colon C_\beta \longrightarrow D$ on the diagram constituted of the C_β, θ_β and we shall prove that n_α is the expected factorization. Clearly we put $n_0 = m$ and when β is a limit ordinal, we define

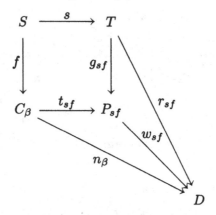

Diagram 5.27

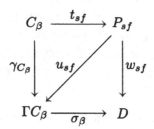

Diagram 5.28

n_β as the unique factorization of the cocone $(n_\varepsilon)_{\varepsilon < \beta}$ through the colimit $C_\beta = \text{colim}_{\varepsilon < \beta} C_\varepsilon$. There remains the case of a successor ordinal.

So we suppose that a cone $n_\varepsilon \colon C_\varepsilon \longrightarrow D$ has already been defined for all $\varepsilon \leq \beta$; we must define $n_{\beta+1} \colon C_{\beta+1} \longrightarrow D$ such that $n_{\beta+1} \circ \theta_\beta = n_\beta$. Consider $s \colon S \longrightarrow T$ in Σ and $f \colon S \longrightarrow C_\beta$. In diagram 5.27, we get a unique factorization r_{sf} such that $r_{sf} \circ s = n_\beta \circ f$, just because $D \in \mathscr{C}_\Sigma$ and $s \in \Sigma$. Since the square is by definition a pushout, this yields a new factorization w_{sf} making the whole diagram 5.27 commutative. The relations $w_{sf} \circ t_{sf} = n_\beta$ indicate that the morphisms w_{sf} constitute a cocone on the generalized pushout diagram constituted of the morphisms t_{sf}. Since $\bigl(\Gamma C_\beta, (u_{sf})_{sf}\bigr)$ is the colimit of this last diagram, we find a unique morphism σ_β as in diagram 5.28 such that $\sigma_\beta \circ u_{sf} = w_{sf}$. Composing with t_{sf} we obtain

$$\sigma_\beta \circ \gamma_{C_\beta} = \sigma_\beta \circ u_{sf} \circ t_{sf} = w_{sf} \circ t_{sf} = n_\beta.$$

We must now extend the factorization at the level $\Delta \Gamma C_\beta$. Consider diagram 5.29. If g, h are such that $h \circ s = \gamma_{C_\beta} \circ f = g \circ s$, then $\sigma_\beta \circ g$

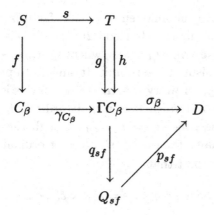

Diagram 5.29

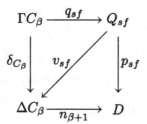

Diagram 5.30

and $\sigma_\beta \circ h$ are two factorizations of $\sigma_\beta \circ \gamma_{C_\beta} \circ f$ through s; since $s \in \Sigma$ and $D \in \mathscr{C}_\Sigma$, this implies $\sigma_\beta \circ g = \sigma_\beta \circ h$. This implies that σ_β factors through the coequalizer q_{sf} of all those possible factorizations $g, h, \ldots$, yielding a morphism p_{sf} such that $p_{sf} \circ q_{sf} = \sigma_\beta$. This last equality indicates precisely that the morphisms p_{sf} constitute a cocone on the generalized pushout diagram constituted of the morphisms q_{sf}. Since $\left(\Delta C_\beta, (v_{sf})_{sf}\right)$ is the colimit of this last diagram, we find in diagram 5.30 a unique morphism $n_{\beta+1} \colon C_{\beta+1} = \Delta C_\beta \longrightarrow D$ such that $n_{\beta+1} \circ v_{sf} = p_{sf}$. Composing with q_{sf} we obtain

$$n_{\beta+1} \circ \delta_{C_\beta} = n_{\beta+1} \circ v_{sf} \circ q_{sf} = p_{sf} \circ q_{sf} = \sigma_\beta$$

and finally

$$n_{\beta+1} \circ \theta_\beta = n_{\beta+1} \circ \delta_{C_\beta} \circ \gamma_{C_\beta} = \sigma_\beta \circ \gamma_{C_\beta} = n_\beta.$$

This constructs the cocone $(n_\beta)_{\beta \in \mathrm{Ord}}$ and in particular we have $n_\alpha \circ s_0 = n_0 = m$, which yields the expected factorization $n = n_\alpha$.

It remains to prove the uniqueness of that factorization n_α, but it is now a straightforward matter. Indeed if $l \colon C_\alpha \longrightarrow D$ is such that $l \circ s_0 = m$, for every $\beta \le \alpha$ we define $l_\beta = l \circ s_\beta$ where $s_\beta \colon C_\beta \longrightarrow C_\alpha$ is once more the canonical morphism of the colimit. It suffices to prove by induction the equality $l_\beta = n_\beta$, for every $\beta \le \alpha$. For $\alpha = 0$ this is just the relation $l \circ s_0 = m = n_0$. And when β is a limit ordinal, l_β and n_β are two factorizations of the cone $(l_\varepsilon = n_\varepsilon)_{\varepsilon < \beta}$ through the colimit C_β, thus $l_\beta = n_\beta$. There remains the case of a successor ordinal.

If $l_\beta = n_\beta$ observe first that

$$
\begin{aligned}
l_{\beta+1} \circ \delta_{C_\beta} \circ u_{sf} \circ g_{sf} \circ s &= l_{\beta+1} \circ \delta_{C_\beta} \circ u_{sf} \circ t_{sf} \circ f \\
&= l_{\beta+1} \circ \delta_{C_\beta} \circ \gamma_{C_\beta} \circ f \\
&= l_{\beta+1} \circ \theta_\beta \circ f \\
&= l_\beta \circ f \\
&= n_\beta \circ f \\
&= w_{sf} \circ g_{sf} \circ s \\
&= \sigma_\beta \circ u_{sf} \circ g_{sf} \circ s.
\end{aligned}
$$

Since $D \in \mathscr{C}_\Sigma$ and $s \in \Sigma$, this implies

$$
l_{\beta+1} \circ \delta_{C_\beta} \circ u_{sf} \circ g_{sf} = \sigma_\beta \circ u_{sf} \circ g_{sf}.
$$

On the other hand one has directly

$$
\begin{aligned}
l_{\beta+1} \circ \delta_{C_\beta} \circ u_{sf} \circ t_{sf} &= l_{\beta+1} \circ \delta_{C_\beta} \circ \gamma_{C_\beta} \\
&= l_{\beta+1} \circ \theta_\beta \\
&= l_\beta \\
&= n_\beta \\
&= w_{sf} \circ t_{sf} \\
&= \sigma_\beta \circ u_{sf} \circ t_{sf}.
\end{aligned}
$$

Thus the morphisms $l_{\beta+1} \circ \delta_{C_\beta} \circ u_{sf}$ and $\sigma_\beta \circ u_{sf}$ are equal when composed with the two canonical morphisms t_{sf}, g_{sf} of the pushout of f, s. Therefore

$$
l_{\beta+1} \circ \delta_{C_\beta} \circ u_{sf} = \sigma_\beta \circ u_{sf}
$$

and since the u_{sf} are the canonical morphisms of a colimit,

$$
l_{\beta+1} \circ \delta_{C_\beta} = \sigma_\beta.
$$

This last relation can be rewritten

$$l_{\beta+1} \circ v_{sf} \circ q_{sf} = p_{sf} \circ q_{sf}$$

and since q_{sf} is an epimorphism (coequalizer of a family of morphisms),

$$l_{\beta+1} \circ v_{sf} = p_{sf}.$$

But $n_{\beta+1}$ is by definition the unique arrow with that property, thus $l_{\beta+1} = n_{\beta+1}$. ☐

Corollary 5.4.8 *Let $\mathscr{C}$ be a locally presentable category and Σ a set of morphisms of $\mathscr{C}$. The corresponding orthogonal subcategory $\mathscr{C}_\Sigma$ is reflective in $\mathscr{C}$.*

Proof By 5.4.7, this volume, and 5.2.10, volume 2. ☐

Since a reflection is completely characterized by the class Σ of inverted morphisms (see 5.3.1), the properties of the reflection will depend heavily on the structure of the class Σ. Therefore it is very important to know when a class Σ is the class of inverted morphisms for a reflection.

Definition 5.4.9 *Let $\mathscr{C}$ be a cocomplete category and $\mathcal{E}$ a class of morphisms of $\mathscr{C}$. The class $\mathcal{E}$ is closed under colimits when given a small category $\mathscr{D}$, two functors $F, G: \mathscr{D} \rightrightarrows \mathscr{C}$ and a natural transformation $\alpha: F \Rightarrow G$, if all the morphisms $\alpha_D: FD \longrightarrow GD$ are in $\mathcal{E}$, then the corresponding factorization $\operatorname{colim} \alpha_D: \operatorname{colim} FD \longrightarrow \operatorname{colim} GD$ is in $\mathcal{E}$ as well.*

Proposition 5.4.10 *Let $\mathscr{C}$ be a cocomplete category in which every object is presentable. Consider a set Σ of morphisms of $\mathscr{C}$ and the corresponding reflective subcategory $r \dashv i: \mathscr{C}_\Sigma \leftrightarrows \mathscr{C}$ of those objects orthogonal to the morphisms of Σ. The class $\mathcal{E}$ of those morphisms inverted by r is the smallest class $\mathcal{E}$ with the following properties:*

(1) $\Sigma \subseteq \mathcal{E}$;
(2) every isomorphism is in $\mathcal{E}$;
(3) if two sides of a commutative triangle are in $\mathcal{E}$, so is the third side;
(4) $\mathcal{E}$ is closed under colimits.

Proof If $s: S \longrightarrow T$ is in Σ, consider diagram 5.31. Since $s \perp ir(S)$, we get a unique morphism $f: T \longrightarrow ir(S)$ such that $f \circ s = \eta_S$; see 5.4.4. Since $(r(T), \eta_T)$ is the reflection of T and $r(S) \in \mathscr{C}_\Sigma$, we get a unique morphism g such that $g \circ \eta_T = f$. From

$$g \circ ir(s) \circ \eta_S = g \circ \eta_T \circ s = f \circ s = \eta_S$$

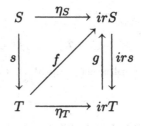

Diagram 5.31

and the universality of η_S, we get $g \circ ir(s) = 1_{ir(S)}$. From

$$ir(s) \circ g \circ \eta_T \circ s = ir(s) \circ f \circ s = ir(s) \circ \eta_S = \eta_T \circ s$$

we get $ir(s) \circ g \circ \eta_T = \eta_T$ since $s \perp ir(T)$; consequently $ir(s) \circ g = 1_{ir(T)}$ by universality of η_T. Thus $ir(s)$ is an isomorphism and $\Sigma \subseteq \mathcal{E}$.

Clearly if f is an isomorphism, so is $r(f)$ and thus $f \in \mathcal{E}$. In the same way if $f = g \circ h$, then $r(f) = r(g) \circ r(h)$ and if two of the morphisms $r(f)$, $r(g)$, $r(h)$ are isomorphisms, so is the third one.

Finally, with the notation of 5.4.9, $r(\operatorname{colim} \alpha_D) = \operatorname{colim} r(\alpha_D)$ since r preserves colimits (see 3.2.2). Thus when each $r(\alpha_D)$ is an isomorphism, $r(\operatorname{colim} \alpha_D)$ is an isomorphism as well as a colimit of isomorphisms.

Thus the class $\mathcal{E}$ of inverted morphisms certainly satisfies conditions (1), (2), (3), (4). From now on, let us write $\mathcal{E}$ for an arbitrary class which satisfies conditions (1), (2), (3), (4). Let us prove first that given $C \in \mathcal{C}$ the canonical morphism $\eta_C \colon C \longrightarrow ir(C)$ is in $\mathcal{E}$. We refer without further notice to the construction of η_C given in the proof of 5.4.7.

For every $s \colon S \longrightarrow T$ in Σ and $f \colon S \longrightarrow C$, consider diagram 5.32 where the front face is a pushout by definition; the back face is obviously a pushout as well. Since $s, 1_S$ and 1_C are in $\mathcal{E}$, $t_{sf} \in \mathcal{E}$. Now consider diagram 5.33, where (s, f) runs through all the possible pairs described before. We compute the generalized pushout of the morphisms t_{sf} on the bottom line and the corresponding generalized pushout of the morphisms 1_C on the top line. The vertical morphisms 1_C, t_{sf} connecting the two diagrams are in $\mathcal{E}$, thus $\gamma_C \in \mathcal{E}$.

The pair (s, f) being fixed again, consider diagram 5.34 where h_i runs through all the morphisms h such that $h_i \circ s = \gamma_C \circ f$; we take as many copies f_i of f as there are such morphisms h_i. Since all the f_i's are equal to f, the top line is a colimit. The bottom line is a colimit by definition. Since s and γ_C are in $\mathcal{E}$, $q_{sf} \circ \gamma_C \in \mathcal{E}$. But since $\gamma_C \in \mathcal{E}$, $q_{sf} \in \mathcal{E}$ as well. Now considering diagram 5.35 where (s, f) runs through all the possible

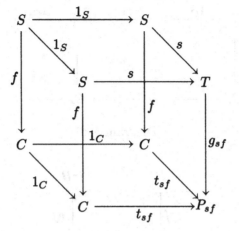

Diagram 5.32

$$C \xrightarrow{1_C} C \xrightarrow{1_C} C$$

$$1_C \downarrow \qquad t_{sf} \downarrow \qquad \downarrow \gamma_C$$

$$C \xrightarrow{t_{sf}} P_{sf} \xrightarrow{u_{sf}} \Gamma C$$

Diagram 5.33

$$S \xrightarrow{f_i} C \xrightarrow{1_C} C$$

$$s \downarrow \qquad \gamma_C \downarrow \qquad \downarrow q_{sf} \circ \gamma_C$$

$$T \xrightarrow{h_i} \Gamma C \xrightarrow{q_{sf}} Q_{sf}$$

Diagram 5.34

$$\Gamma C \xrightarrow{1_{\Gamma C}} \Gamma C \xrightarrow{1_{\Gamma C}} \Gamma C$$

$$1_{\Gamma C} \downarrow \qquad q_{sf} \downarrow \qquad \downarrow \delta_C$$

$$\Gamma C \xrightarrow{q_{sf}} Q_{sf} \xrightarrow{v_{sf}} \Delta C$$

Diagram 5.35

$$C_0 \xrightarrow{1_{C_0}} C_0 \xrightarrow{1_{C_0}} \cdots \xrightarrow{1_{C_0}} C_0 \xrightarrow{1_{C_0}} \cdots \longrightarrow C_0$$

$$\tau_0 \downarrow \qquad \tau_1 \downarrow \qquad\qquad \tau_\varepsilon \downarrow \qquad\qquad\qquad \tau_\beta \downarrow$$

$$C_0 \longrightarrow C_1 \longrightarrow \cdots \longrightarrow C_\varepsilon \longrightarrow \cdots \longrightarrow C_\beta$$

Diagram 5.36

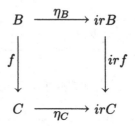

Diagram 5.37

pairs (s, f), the bottom part is a colimit by definition and the top part is a colimit since the indexing diagram is connected (see 2.6.7.e). Since $1_{\Gamma C}$ and q_{sf} are in $\mathcal{E}$, $\delta_C \in \mathcal{E}$.

Let us finally consider the various objects C_β and the corresponding connecting morphisms $C_0 \longrightarrow C_\beta$. We prove by induction on α that all those morphisms are in $\mathcal{E}$:

(1) if $\beta = 0$, the morphism is the identity on C_0, which is in $\mathcal{E}$;

(2) if $\tau_\beta \colon C_0 \longrightarrow C_\beta$ is in $\mathcal{E}$, then $\tau_{\beta+1} \colon C_0 \longrightarrow C_{\beta+1}$ is just $\delta_{C_\beta} \circ \gamma_{C_\beta} \circ \tau_\beta$, which is in $\mathcal{E}$ as composite of morphisms in $\mathcal{E}$;

(3) if β is a limit ordinal, it suffices to consider diagram 5.36; the top line is a colimit diagram, just because the indexing diagram is connected (see 2.6.7.e); the bottom line is a colimit diagram by definition. Since $\tau_\varepsilon \in \mathcal{E}$ for $\varepsilon < \beta$, $\tau_\beta \in \mathcal{E}$.

In particular $\tau_\alpha = \eta_C \in \mathcal{E}$.

Finally let us consider a morphism $f \colon B \longrightarrow C$ in $\mathscr{C}$ such that $r(f)$ is an isomorphism. Considering the commutative diagram 5.37, we have η_B, η_C and $ir(f)$ in $\mathcal{E}$, thus f is in $\mathcal{E}$ as well. $\square$

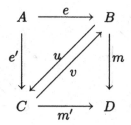

Diagram 5.38

5.5 Factorization systems

5.3.1 shows that a reflection $r \dashv i \colon \mathscr{A} \longrightarrow \mathscr{B}$ of a category $\mathscr{B}$ is completely characterized by the class Σ of those morphisms of $\mathscr{B}$ inverted by r. Here we shall prove that in most cases, Σ is part of a factorization system for the arrows of $\mathscr{B}$.

Definition 5.5.1 *By a factorization system on a category $\mathscr{B}$ we mean a pair $(\mathcal{E}, \mathcal{M})$ where both $\mathcal{E}$ and $\mathcal{M}$ are classes of morphisms of $\mathscr{B}$ and*

(1) every isomorphism belongs to both $\mathcal{E}$ and $\mathcal{M}$,
(2) both $\mathcal{E}$ and $\mathcal{M}$ are closed under composition,
(3) $\forall e \in \mathcal{E} \quad \forall m \in \mathcal{M} \quad e \perp m$,
(4) every morphism $f \in \mathscr{B}$ can be factored as $f = m \circ e$, with $e \in \mathcal{E}$ and $m \in \mathcal{M}$.

Let us make clear that in 5.5.1, nothing is required about some morphisms being monomorphisms or epimorphisms.

Proposition 5.5.2 *Under the conditions of 5.5.1, the factorization $f = m \circ e$ referred to in 5.5.1.(4) is unique up to an isomorphism.*

Proof Suppose $f = m' \circ e'$ with $e' \in \mathcal{E}$ and $m' \in \mathcal{M}$. Consider diagram 5.38. It suffices to apply 5.5.1.(3) to get morphisms u, v such that $u \circ e = e'$, $m' \circ u = m$, $v \circ e' = e$, $m \circ v = m'$. In particular considering the situations of diagram 5.39, one has $v \circ u \circ e = v \circ e' = e$, $m \circ v \circ u = m' \circ u = m$, $u \circ v \circ e' = u \circ e = e'$, $m' \circ u \circ v = m \circ v = m'$. By the uniqueness condition in the definition of orthogonality (see 5.4.1) one deduces $v \circ u = 1_B$ and $u \circ v = 1_C$. $\qquad\square$

Proposition 5.5.3 *Under the conditions of 5.5.1 and given a morphism $f \in \mathscr{B}$,*

$$f \in \mathcal{E} \Leftrightarrow \forall m \in \mathcal{M} \quad f \perp m,$$
$$f \in \mathcal{M} \Leftrightarrow \forall e \in \mathcal{E} \quad e \perp f.$$

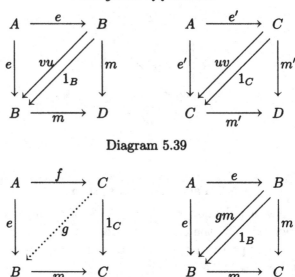

Diagram 5.39

Diagram 5.40

Proof By duality, it suffices to prove one of the equivalences. If $f \in \mathcal{E}$, then $f \perp m$ for every $m \in \mathcal{M}$, just by 5.5.1.(3). Conversely suppose $f \perp m$ for all $m \in \mathcal{M}$ and write $f = m \circ e$ with $e \in \mathcal{E}$, $m \in \mathcal{M}$ (see 5.5.1.(4)). Considering the situations of diagram 5.40, we first get g such that $g \circ f = e$, $m \circ g = 1_C$, by assumption on f. Considering the second square, $g \circ m \circ e = g \circ f = e$ and $m \circ g \circ m = m$, thus by the uniqueness condition in 5.4.1 we have $g \circ m = 1_B$. So m is an isomorphism and $m \in \mathcal{E}$ (5.5.1.(1)). Finally $f = m \circ e \in \mathcal{E}$ (5.5.1.(2)).   □

Proposition 5.5.3 tells us in some sense that definition 5.5.1 is redundant, since each one of the classes $\mathcal{E}$, $\mathcal{M}$ can be completely described in terms of the other one. Nevertheless the fact of using both classes allows in general more elegance and simplicity in the treatment of the problems.

The classes $\mathcal{E}, \mathcal{M}$ involved in a factorization system have quite a lot of stability properties, for example under some types of limits or colimits (see 5.9.1). Let us just emphasize the following facts.

Proposition 5.5.4 *Under the conditions of 5.5.1, consider a composite* $f \circ g$ *of two morphisms of $\mathcal{B}$:*

(1) $(f \circ g \in \mathcal{E}$ *and* $g \in \mathcal{E}) \Rightarrow (f \in \mathcal{E})$;

(2) $(f \circ g \in \mathcal{M}$ *and* $f \in \mathcal{M}) \Rightarrow (g \in \mathcal{M})$;

(3) $f \in \mathcal{E} \cap \mathcal{M} \Rightarrow f$ *is an isomorphism.*

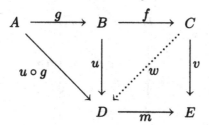

Diagram 5.41

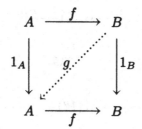

Diagram 5.42

Proof Let us prove the first assertion. Consider diagram 5.41 where $m \circ u = v \circ f$ and $m \in \mathcal{M}$. Since $f \circ g \in \mathcal{E}$, we get a unique w such that $w \circ f \circ g = u \circ g$ and $m \circ w = v$. But since $g \in \mathcal{E}$, the uniqueness condition in 5.4.1 implies $w \circ f = u$. Thus w is the expected factorization. Now w is unique with that property since $m \circ w' = v$ and $w' \circ f = u$ imply $w' \circ f \circ g = u \circ g$ and thus $w' = w$.

The second assertion follows by duality. Finally if $f \in \mathcal{E} \cap \mathcal{M}$ it suffices to consider the square of diagram 5.42 to get a unique g such that $g \circ f = 1_A$, $f \circ g = 1_B$. □

We shall now indicate the relations between reflective subcategories and factorization systems.

Proposition 5.5.5 *Let $\mathcal{B}$ be a category with a terminal object. Every factorization system $(\mathcal{E}, \mathcal{M})$ on the category $\mathcal{B}$ induces a reflective subcategory $r \dashv i \colon \mathcal{A} \xrightarrow{\longleftarrow} \mathcal{B}$, with the following properties:*

(1) *given $B \in \mathcal{B}$ and the $(\mathcal{E}, \mathcal{M})$-factorization*

$$B \xrightarrow{\ e_B\ } r(B) \xrightarrow{\ m_B\ } 1$$

of the unique morphisms $t_B \colon B \longrightarrow 1$, $(r(B), e_B)$ is the reflection of B in $\mathcal{A}$;

(2) *every morphism $f \in \mathcal{E}$ is inverted by r.*

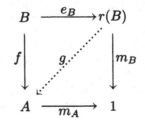

Diagram 5.43

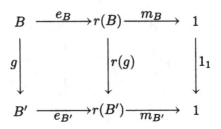

Diagram 5.44

Proof Given $B \in \mathcal{B}$, consider the unique morphism $\xi_B \colon B \longrightarrow 1$ and its $(\mathcal{E}, \mathcal{M})$-factorization $\xi_B = m_B \circ e_B$, $m_B \in \mathcal{M}$, $e_B \in \mathcal{E}$; write $r(B)$ for the corresponding object

$$B \xrightarrow{\;e_B\;} r(B) \xrightarrow{\;m_B\;} 1$$

We define $\mathcal{A}$ to be the full subcategory of $\mathcal{B}$ whose objects are those B's for which e_B is an isomorphism.

By uniqueness of the $(\mathcal{E}, \mathcal{M})$-factorization (see 5.5.2), the $(\mathcal{E}, \mathcal{M})$-factorization of m_B is just $m_B \circ 1_{rB}$. In other words, $r(B) \in \mathcal{A}$. Considering diagram 5.43, let us prove that (rB, e_B) is the reflection of B in $\mathcal{A}$. Given $A \in \mathcal{A}$, the unique arrow $m_A \colon A \longrightarrow 1$ is in $\mathcal{M}$ by definition. Then given $f \colon B \longrightarrow A$, the square shown is commutative since 1 is terminal. From $e_B \in \mathcal{E}$ and $m_A \in \mathcal{M}$, we get a unique g such that $g \circ e_B = f$ and, of course, $m_A \circ g = m_B$.

Choose now $g \colon B \longrightarrow B'$ in $\mathcal{E}$. In diagram 5.44, the outer rectangle is commutative since 1 is terminal. From $e_B \in \mathcal{E}$ and $m_{B'} \in \mathcal{M}$, we get a unique $r(g)$ making the whole diagram 5.44 commutative. This is precisely the definition of the functor r on the morphisms (see 3.1.3). But from $g \in \mathcal{E}$ and $e_{B'} \in \mathcal{E}$, we get $e_{B'} \circ g \in \mathcal{E}$; since $e_B \in \mathcal{E}$, this implies $r(g) \in \mathcal{E}$; see 5.5.4. In the same way from $m_B \in \mathcal{M}$ and $m_{B'} \in \mathcal{M}$ we deduce $r(g) \in \mathcal{M}$. Then $r(g)$ is both in $\mathcal{E}$ and in $\mathcal{M}$: it is an isomorphism (see 5.5.4). $\square$

It should be observed that in 5.5.5, $\mathcal{E}$ is not in general the class of morphisms of $\mathcal{B}$ inverted by the reflection r. For example in the category Set of sets and mappings every monomorphism is strong and every epimorphism is strong (see 4.3.10.a). Therefore we immediately get a system of factorization by defining

$$f \in \mathcal{E} \quad \text{iff} \quad f \text{ is a surjection,}$$

$$f \in \mathcal{M} \quad \text{iff} \quad f \text{ is an injection.}$$

Given a set B, the corresponding reflection $r(B)$ of $\mathcal{B}$ is the image of the unique mapping $B \longrightarrow 1$. Thus $r(B) = 1$ if $B \neq \emptyset$ and $r(\emptyset) = \emptyset$. The corresponding reflective subcategory is just $\{\emptyset \hookrightarrow 1\}$. Observe that every morphism $f \colon A \longrightarrow B$ between non-empty sets A, B is mapped to the identity on the singleton, thus is inverted by the reflection.

The reader should compare the previous example with the statement of our next proposition. We refer to 4.4.1 for the notion of a "finitely well-complete" category.

Proposition 5.5.6 *Let $\mathcal{B}$ be a finitely well-complete category. There exists a bijection between*

(1) the reflective subcategories $r \dashv i \colon \mathcal{A} \overset{\longleftarrow}{\underset{\longrightarrow}{}} \mathcal{B}$ of $\mathcal{B}$,

(2) the $(\mathcal{E}, \mathcal{M})$ factorization systems on $\mathcal{B}$ which satisfy the additional condition

$$f \circ g \in \mathcal{E} \text{ and } f \in \mathcal{E} \Rightarrow g \in \mathcal{E}.$$

Moreover, under this bijection, $\mathcal{E}$ is the class of those morphisms of $\mathcal{B}$ inverted by the reflection r and $\mathcal{M}$ contains all the arrows of $\mathcal{A}$.

Proof Let us start with a reflective subcategory $r \dashv i \colon \mathcal{A} \overset{\longleftarrow}{\underset{\longrightarrow}{}} \mathcal{B}$ and define the two classes $\mathcal{E}, \mathcal{M}$ of morphisms by

$$f \in \mathcal{E} \quad \text{iff} \quad r(f) \text{ is an isomorphism,}$$

$$f \in \mathcal{M} \quad \text{iff} \quad \forall e \in \mathcal{E} \ e \perp f.$$

We shall prove that $(\mathcal{E}, \mathcal{M})$ is a factorization system with $\mathcal{M}$ containing all the arrows of $\mathcal{A}$.

Clearly every isomorphism is in $\mathcal{E}$ and $\mathcal{E}$ is stable under composition. Moreover an isomorphism is always orthogonal (on the left or on the right) to any morphism, thus certainly all isomorphisms are in $\mathcal{M}$.

Now consider a composite $m_2 \circ m_1$, with $m_1, m_2 \in \mathcal{M}$. Considering diagram 5.45 where $e \in \mathcal{E}$ and $m_2 \circ m_1 \circ u = v \circ e$ one gets a unique x such that $m_2 \circ x = v$, $x \circ e = m_1 \circ u$ ($e \in \mathcal{E}, m_2 \in \mathcal{M}$) and then a unique y such that $m_1 \circ y = x$, $y \circ e = u$ ($e \in \mathcal{E}, m_1 \in \mathcal{M}$). This implies

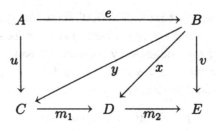

Diagram 5.45

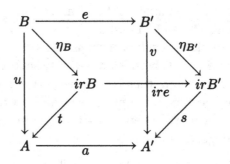

Diagram 5.46

$m_2 \circ m_1 \circ y = m_2 \circ x = v$, so that y is a factorization. If y' is another factorization such that $y' \circ e = u$, $m_2 \circ m_1 \circ y' = v$, one has

$$m_2 \circ (m_1 \circ y') = v = m_2 \circ x, \quad (m_1 \circ y') \circ e = (m_1 \circ u) = x \circ e,$$

so that $m_1 \circ y' = x$ by definition of x. Since moreover $y' \circ e = u$, $y = y'$ by definition of y. This proves that $\mathcal{M}$ is stable under composition.

By definition of $\mathcal{M}$, $e \in \mathcal{E}$ and $m \in \mathcal{M}$ imply $e \perp m$. Moreover if $a: A \longrightarrow A'$ is an arrow of $\mathcal{A}$ and $e: B \longrightarrow B'$ is in $\mathcal{E}$, let us prove the relation $e \perp a$. Consider diagram 5.46 where $a \circ u = v \circ e$. If $\eta_B, \eta_{B'}$ are the canonical morphisms of the reflection, we get factorizations t, s making the whole diagram 5.46 commute. This yields a morphism $w = t \circ \left(ir(e)\right)^{-1} \circ \eta_{B'}$ from B' to A, with the properties

$$w \circ e = t \circ \left(ir(e)\right)^{-1} \circ \eta_{B'} \circ e = t \circ \left(ir(e)\right)^{-1} \circ ir(e) \circ \eta_B = r\eta_B = u,$$

$$a \circ w = a \circ t \circ \left(ir(e)\right)^{-1} \circ \eta_{B'} = s \circ ir(e) \circ \left(ir(e)\right)^{-1} \circ \eta_{B'} = s \circ \eta_{B'} = v.$$

If $w': B' \longrightarrow A$ is another morphism with the properties $w' \circ e = u$, $a \circ w' = v$, the universal property of $\left(ir(B'), \eta_{B'}\right)$ implies the existence of $w'': ir(B') \longrightarrow A$ such that $w'' \circ \eta_{B'} = w'$. Therefore

$$w'' \circ ir(e) \circ \eta_B = w'' \circ \eta_{B'} \circ e = w' \circ e = u = t \circ \eta_B,$$

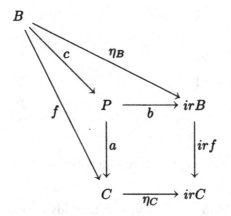

Diagram 5.47

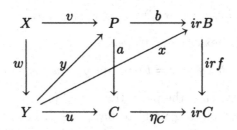

Diagram 5.48

from which $w'' \circ ir(e) = t$ by uniqueness of t. Finally,

$$w' = w'' \circ \eta_{B'} = w'' \circ ir(e) \circ \big(ir(e)\big)^{-1} \circ \eta_{B'} = t \circ \big(ir(e)\big)^{-1} \circ \eta_{B'} = w,$$

which proves $e \perp a$. Thus every morphism of $\mathscr{A}$ is in $\mathcal{M}$.

Let us now consider a composite $f \circ g$ in $\mathscr{B}$ with $f \circ g \in \mathcal{E}$ and $f \in \mathcal{E}$. Since $r(f \circ g) = r(f) \circ r(g)$ and $r(f \circ g)$, $r(f)$ are isomorphisms, $r(g) = \big(r(f)\big)^{-1} \circ r(f \circ g)$ is an isomorphism as well. Thus $g \in \mathcal{E}$.

It remains to prove that every morphism $f \colon B \longrightarrow C$ in $\mathscr{B}$ can be factored as $f = m \circ e$, with $e \in \mathcal{E}$ and $m \in \mathcal{M}$. We consider diagram 5.47 where the square is a pullback and c is the unique factorization making the whole diagram 5.47 commute.

Let us observe first that $a \in \mathcal{M}$. Indeed consider diagram 5.48 with $u \circ w = a \circ v$ and $w \in \mathcal{E}$. Since $w \in \mathcal{E}$ and $ir(f) \in \mathscr{A} \subseteq \mathcal{M}$, we get a unique x such that $x \circ w = b \circ v$ and $ir(f) \circ x = \eta_C \circ u$. Since (a, b) is a pullback and $ir(f) \circ x = \eta_C \circ u$, we obtain a unique y such that $a \circ y = u$, $b \circ y = x$. Observe that $b \circ y \circ w = x \circ w = b \circ v$ and $a \circ y \circ w = u \circ w = a \circ v$, from which $y \circ w = v$ since (a, b) is a pullback.

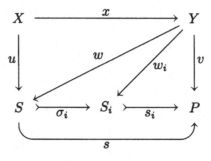

Diagram 5.49

Thus $a \circ y = u$ and $y \circ w = v$, which yields a factorization y. If y' is another such factorization, the relations

$$b \circ y' \circ w = b \circ v = x \circ w, \quad ir(f) \circ b \circ y' = \eta_C \circ a \circ y' = \eta_C \circ u = ir(f) \circ x$$

imply $b \circ y' = x$. Next, the relations

$$b \circ y' = x = b \circ y, \quad a \circ y' = u = a \circ y$$

imply $y' = y$. This ends the proof that $a \in \mathcal{M}$.

There is no reason in general to have $c \in \mathcal{E}$. Let us consider all the subobjects $s_i \colon S_i \rightarrowtail P$ with $s_i \in \mathcal{M}$ and through which c factors, let us say as $c = s_i \circ t_i$. We can compute the intersection $s \colon S \rightarrowtail P$ of all these subobjects ($\mathscr{B}$ is finitely well-complete) and get a factorization $c = s \circ t$ through that intersection. Observe that 1_P is one of the s_i's. Let us prove that $s \in \mathcal{M}$; see diagram 5.49. Given $x \in \mathcal{E}$ and $s \circ u = v \circ x$, we get a unique w_i such that $s_i \circ w_i = v$, $w_i \circ x = \sigma_i \circ u$ since $s_i \in \mathcal{M}$. Since all composites $s_i \circ w_i$ are just v, we get a unique factorization w through the intersection S, with $\sigma_i \circ w = w_i$. Therefore

$$s \circ w = s_i \sigma_i \circ w = s_i \circ w_i = v.$$

Moreover from $s \circ w \circ x = v \circ x = s \circ u$ we deduce $w \circ x = u$ since s is a monomorphism. So w is the expected factorization and it is necessarily unique because s is a monomorphism. This proves $s \in \mathcal{M}$.

We already have $f = a \circ c = a \circ s \circ t$ with a and s in $\mathcal{M}$, thus $a \circ s \in \mathcal{M}$. It remains to prove that $t \in \mathcal{E}$. Let us consider diagram 5.50 where the square is a pullback. From

$$r(b \circ s) \circ r(t) = r(b \circ c) = r(\eta_B)$$

and the fact that $r(\eta_B)$ is an isomorphism (with inverse $\varepsilon_{r(B)}$; see 3.1.5 and 3.4.1), we deduce that $r(t) = ir(t)$ is a monomorphism and thus

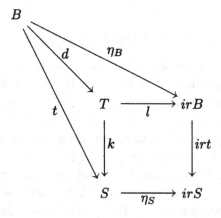

Diagram 5.50

k is a monomorphism as well (see 2.5.3). But since $ir(t) \in \mathcal{A} \subseteq \mathcal{M}$, the same argument developed for diagram 5.47 shows that $k \in \mathcal{M}$. We have then $c = s \circ t = s \circ k \circ d$ with $s \circ k$ a monomorphism in $\mathcal{M}$, since s and k are. By definition of s, k must be an isomorphism. Finally, recalling once more that $r(\eta_X)$ is always invertible, with inverse $\varepsilon_{r(X)}$ (see 3.1.5 and 3.4.1) we observe that $r(l) \circ r(d) = r(\eta_B)$ is an isomorphism, thus $r(d)$ is a strong monomorphism (see 4.3.6). Since k is an isomorphism, $r(t)$ is thus a strong monomorphism as well. But from $r(\eta_S) \circ r(k) = rir(t) \circ r(l) \cong r(t) \circ r(l)$ we deduce that $r(t)$ is a strong epimorphism, since both $r(\eta_S)$ and $r(k)$ are isomorphisms. Thus $r(t)$ is an isomorphism (see 4.3.6) and $f = (a \circ s) \circ t$, with $a \circ s \in \mathcal{M}$ and $t \in \mathcal{E}$, is the required factorization.

Considering the construction we have just developed and that described in 5.5.5, it remains to prove that they induce the bijection announced in the statement.

Let us start with a reflective subcategory $r \dashv i \colon \mathcal{A} \overset{\longleftarrow}{\underset{\longrightarrow}{}} \mathcal{B}$ and the corresponding factorization system $(\mathcal{E}, \mathcal{M})$ described in this proof. For every $B \in \mathcal{B}$ let us consider the composite

$$B \xrightarrow{\ \eta_B\ } ir(B) \xrightarrow{\ m_B\ } 1.$$

Since $r(\eta_B)$ is an isomorphism, $\eta_B \in \mathcal{E}$. Since $ir(B) \in \mathcal{A}$ and $1 \in \mathcal{A}$, $m_B \in \mathcal{A}$ and thus $m_B \in \mathcal{M}$, as proved previously. Thus $r(B)$ coincides with the reflection of B constructed in 5.5.5.

Conversely consider a factorization system $(\mathcal{E}, \mathcal{M})$ satisfying the additional property

$$f \circ g \in \mathcal{E} \text{ and } f \in \mathcal{E} \Rightarrow g \in \mathcal{E}.$$

We consider the corresponding reflection as constructed in 5.5.5. Let us write Σ for the class of those morphisms of $\mathscr{B}$ inverted by r. The factorization system associated with the reflection has, by construction, the form $(\Sigma, \mathcal{M}')$. To prove the equality $(\mathcal{E}, \mathcal{M}) = (\Sigma, \mathcal{M}')$, it suffices to prove $\mathcal{E} = \Sigma$ (see 5.5.3) since the class $\mathcal{M}$ (or $\mathcal{M}'$) is completely characterized by the class $\mathcal{E}$ (or Σ). We know already that $\mathcal{E} \subseteq \Sigma$ (see 5.5.5), so it remains to choose $g \in \Sigma$ and prove it is in $\mathcal{E}$. Going back to the defininition of $r(g)$ in the proof of 5.5.5, we observe that $e_{B'} \circ g = r(g) \circ e_B$. The morphisms e_B and $e_{B'}$ are in $\mathcal{E}$ by definition and $r(g) \in \mathcal{E}$ since it is an isomorphism. Finally $e_{B'}$ and $r(g) \circ e_B$ are in $\mathcal{E}$, which implies $g \in \mathcal{E}$ by the additional assumption on the factorization system $(\mathcal{E}, \mathcal{M})$. $\square$

5.6 The case of localizations

We recall that a reflective subcategory $r \dashv i\colon \mathscr{A} \overset{\longleftarrow}{\underset{\longrightarrow}{}} \mathscr{B}$ of a finitely complete category $\mathscr{B}$ is called a localization when the reflection r preserves finite limits (see 3.5.5).

We shall now particularize the results of sections 5.3, 5.5 to the case of localizations.

Proposition 5.6.1 *Consider a finitely complete category $\mathscr{B}$ and a reflective subcategory $i\colon \mathscr{A} \hookrightarrow \mathscr{B}$, with reflection $r \dashv i$. Write Σ for the class of all morphisms $f \in \mathscr{B}$ such that $r(f)$ is an isomorphism. The following conditions are equivalent:*

(1) Σ admits a right calculus of fractions;

(2) the reflective subcategory $r \dashv i\colon \mathscr{A} \overset{\longleftarrow}{\underset{\longrightarrow}{}} \mathscr{B}$ is a localization;

(3) Σ is stable under pullbacks, i.e. given a pullback square in $\mathscr{B}$ as in diagram 5.51, if $s \in \Sigma$, then $t \in \Sigma$.

Proof   (1) $\Rightarrow$ (2). The reflection $r\colon \mathscr{B} \longrightarrow \mathscr{A}$ is equivalent to the category of fractions $\varphi\colon \mathscr{B} \longrightarrow \mathscr{B}[\Sigma^{-1}]$; see 5.3.1. If Σ admits a right calculus of fractions, $\mathscr{B}[\Sigma^{-1}]$ is finitely complete and φ preserves finite limits. Applying 3.4.5 (finite case), we conclude that $\mathscr{A}$ is finitely complete and r preserves finite limits. This is precisely the definition of being a localization (see 3.5.5).

(2) $\Rightarrow$ (1). Given $B \in \mathscr{B}$, $r(1_B) = 1_{rB}$ is an isomorphism, thus $1_B \in \Sigma$. Moreover given a composite $f \circ g$ in $\mathscr{B}$, if $r(f)$ and $r(g)$ are isomorphisms,

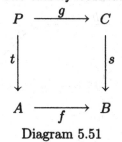

Diagram 5.51

$r(f \circ g) = r(f) \circ r(g)$ is an isomorphism and $f \circ g \in \Sigma$. Given (f, s) as in 5.2.3.(3), define (t, g) as the pullback of (f, s). Applying r to this pullback we get another pullback with $r(s)$ an isomorphism, thus $r(t)$ is an isomorphism as well (see 2.5.3) and $t \in \Sigma$. Finally given $f, g \colon A \rightrightarrows B$ and $s \in \Sigma$ such that $s \circ f = s \circ g$, one has $r(s) \circ r(f) = r(s) \circ r(g)$ and thus $r(f) = r(g)$ since $r(s)$ is an isomorphism. Putting $k = \mathsf{Ker}\,(f, g)$, one clearly has $f \circ k = g \circ k$. But $r(k) = \mathsf{Ker}\,\big(r(f), r(g)\big)$ and since $r(f) = r(g)$, $r(k) \cong 1_{r(A)}$. Thus $r(k)$ is an isomorphism and $k \in \Sigma$.

(2) $\Rightarrow$ (3). Considering the pullback in diagram 5.51, with $s \in \Sigma$, the image of this pullback under r is again a pullback with $r(s)$ an isomorphism; therefore $r(t)$ is an isomorphism as well (see 2.5.3) and $t \in \Sigma$.

(3) $\Rightarrow$ (2). Since $\mathscr{B}$ is finitely complete, $\mathscr{A}$ is finitely complete and finite limits in $\mathscr{A}$ are computed as in $\mathscr{B}$; see 3.5.3. That implies in particular that the terminal object $1 \in \mathscr{B}$ belongs to $\mathscr{A}$. But since $1 \in \mathscr{A}$, one has $r(1) \cong 1$ (see 3.4.1) and r preserves the terminal object. By 2.8.2, it remains to prove that r preserves pullbacks. Consider diagram 5.52 where (k, h) is the pullback of (f, g) in $\mathscr{B}$ and (u, v) is the pullback of $\big(ir(f), ir(g)\big)$ in $\mathscr{A}$. The morphism w is the unique factorization making diagram 5.52 commutative. It suffices to prove that $w \in \Sigma$, which will imply $P \cong ir(P) \cong ir(A)$ since $P \in \mathscr{A}$ and $w \in \Sigma$, and thus the preservation by r of the original pullback of (f, g).

To prove that $w \in \Sigma$, let us first consider diagram 5.53 where all the squares are pullbacks. We know at once that $\eta_B, \eta_C \in \Sigma$ since $r(\eta_B) \circ \varepsilon_{r(B)} = 1_{r(B)}$ and $\varepsilon_{r(B)}$ is an isomorphism (see 3.1.5 and 3.4.1). Therefore x, y, m, n are in Σ as well. Since w is the unique arrow such that $u \circ w = \eta_C \circ k$, $v \circ w = \eta_B \circ h$, one has $w = x \circ m \circ l = y \circ n \circ l$. Since Σ is obviously closed under composition, it remains to show that $l \in \Sigma$.

Now let us consider diagram 5.54 where the square is a pullback. Since $\eta_D \in \Sigma$, we get $p, q \in \Sigma$. Considering the unique factorization d such that

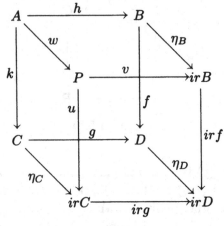

Diagram 5.52

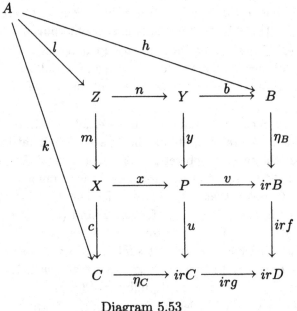

Diagram 5.53

$p \circ d = 1_D = q \circ d$, we get $r(p) \circ r(d) = 1_{r(D)}$ with $r(p)$ an isomorphism. Thus $r(d)$ is an isomorphism and $d \in \Sigma$. Now the relations

$$\eta_D \circ g \circ c \circ m = ir(g) \circ \eta_C \circ c \circ m$$
$$= ir(f) \circ \eta_B \circ b \circ n$$
$$= \eta_D \circ f \circ b \circ n$$

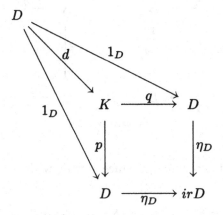

Diagram 5.54

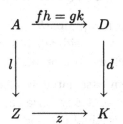

Diagram 5.55

imply the existence of a unique $z\colon Z \longrightarrow K$ such that $g \circ c \circ m = p \circ z$, $f \circ b \circ n = q \circ z$.

Now consider diagram 5.55; we shall prove it is a pullback. Composing with p and q we find

$$p \circ z \circ l = g \circ c \circ m \circ l = g \circ k = f \circ h = p \circ d \circ f \circ h,$$
$$q \circ z \circ l = f \circ b \circ n \circ l = f \circ h = q \circ d \circ f \circ h.$$

Since (p, q) is a pullback, this implies $z \circ l = d \circ f \circ h$, thus the commutativity of the square. Moreover, given α, β such that $z \circ \alpha = d \circ \beta$, one has $f \circ b \circ n \circ \alpha = q \circ z \circ \alpha = q \circ d \circ \beta = \beta = p \circ d \circ \beta = p \circ z \circ \alpha = g \circ c \circ m \circ \alpha$. This implies the existence of a unique γ such that $b \circ n \circ \alpha = h \circ \gamma$, $c \circ m \circ \alpha = k \circ \gamma$. This γ is the expected factorization. Indeed

$$c \circ m \circ l \circ \gamma = k \circ \gamma = c \circ m \circ \alpha, \quad b \circ n \circ l \circ \gamma = h \circ \gamma = b \circ n \circ \alpha$$

imply $l \circ \gamma = \alpha$ since $(c \circ m, b \circ n)$ is a pullback; on the other hand

$$f \circ h \circ \gamma = f \circ b \circ n \circ \alpha = q \circ z \circ \alpha = q \circ d \circ \beta = \beta.$$

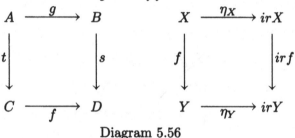

Diagram 5.56

Such a factorization γ is unique since given δ with $l \circ \delta = \alpha$, $f \circ h \circ \delta = \beta$ one gets

$$k \circ \delta = c \circ m \circ l \circ \delta = c \circ m \circ \alpha = k \circ \gamma,$$
$$h \circ \delta = b \circ n \circ l \circ \delta = b \circ n \circ \alpha = h \circ \gamma,$$

and thus $\delta = \gamma$ since (k, h) is a pullback.

Finally the last square we considered is a pullback and since $d \in \Sigma$, one gets $l \in \Sigma$, which concludes the proof.   ☐

The factorization system associated with a localization can be described completely and without any "size condition" (like finitely well-complete, in 5.5.6) on the original category.

Proposition 5.6.2 *Consider a finitely complete category $\mathscr{B}$. There exists a bijection between*

(1) *the localizations $r \dashv i$: $\mathscr{A} \overset{\longleftarrow}{\longrightarrow} \mathscr{B}$ of $\mathscr{B}$,*
(2) *the factorization systems $(\mathcal{E}, \mathcal{M})$ on $\mathscr{B}$ which satisfy the two following additional conditions:*

    (a) *$f \circ g \in \mathcal{E}$ and $f \in \mathcal{E} \implies g \in \mathcal{E}$;*
    (b) *$\mathcal{E}$ is stable under pullbacks, i.e. given a pullback square in $\mathscr{B}$ as in the left part of diagram 5.56, if $s \in \mathcal{E}$, then $t \in \mathcal{E}$.*

Moreover, under this correspondence,

(1) *$f \in \mathcal{E}$ iff $r(f)$ is an isomorphism,*
(2) *$f \in \mathcal{M}$ iff the right-hand square of diagram 5.56 is a pullback, where η_X, η_Y are the canonical morphisms of the adjunction.*

Proof For a finitely well-complete category $\mathscr{B}$, the result is an immediate consequence of 5.5.6 and 5.6.1, with the exception of the description of the class $\mathcal{M}$. In the more general case where $\mathscr{B}$ is just finitely complete, the proof is an easy modification of that of 5.5.6.

Let us start with a localization $r \dashv i$: $\mathscr{A} \overset{\longleftarrow}{\longrightarrow} \mathscr{B}$ and let us consider the two classes $\mathcal{E}, \mathcal{M}$ of morphisms as defined in the statement. Obviously

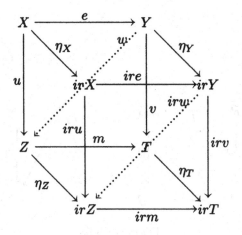

Diagram 5.57

every isomorphism is in $\mathcal{E}$ and $\mathcal{E}$ is stable under composition. In the same way every isomorphism is in $\mathcal{M}$ (see 2.5.3) and $\mathcal{M}$ is closed under composition (see 2.5.9).

Now choose $e \in \mathcal{E}$, $m \in \mathcal{M}$ and consider diagram 5.57 where $v \circ e = m \circ u$. Since $ir(e)$ is an isomorphism we have

$$ir(m) \circ ir(u) \circ \big(ir(e)\big)^{-1} \circ \eta_Y = ir(v) \circ \eta_Y = \eta_T \circ v.$$

Since the bottom face is a pullback we get a unique factorization w such that

$$\eta_Z \circ w = ir(u) \circ \big(ir(e)\big)^{-1} \circ \eta_Y, \quad m \circ w = v.$$

From the relations

$$m \circ w \circ e = v \circ e = m \circ u,$$

we deduce

$$\eta_Z \circ w \circ e = ir(u) \circ \big(ir(e)\big)^{-1} \circ \eta_Y \circ e$$
$$= ir(u) \circ \big(ir(e)\big)^{-1} \circ ir(e) \circ \eta_X = ir(u) \circ \eta_X = \eta_Z \circ u,$$

and from the fact that the bottom square is a pullback, we deduce that $w \circ e = u$. Thus w is an acceptable factorization. If w' is another morphism such that $w' \circ e = u$, $m \circ w' = v$ one has

$$\eta_Z \circ w' = i(rw') \circ \eta_Y = ir(u) \circ \big(ir(e)\big)^{-1} \circ \eta_Y = \eta_Z \circ w,$$

$$m \circ w' = v = m \circ w,$$

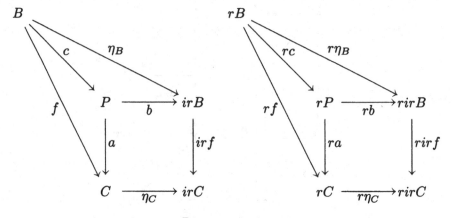

Diagram 5.58

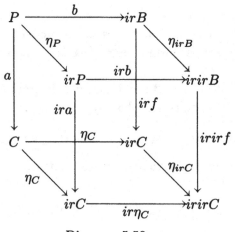

Diagram 5.59

from which $w = w'$ since the bottom square is a pullback. Thus we have proved that $e \perp m$.

Let us now consider a morphism $f\colon B \longrightarrow C$ and let us factor it as $f = m \circ e$, with $e \in \mathcal{E}$ and $m \in \mathcal{M}$. We consider diagram 5.58 where (a, b) is the pullback of $(\eta_C, ir(f))$. The second diagram is the image of the first one under the reflection r. Since r preserves finite limits, the square is a pullback. But since $r\eta_B$ and $r\eta_C$ are isomorphisms with inverse $\varepsilon_{r(C)}$ (see 3.4.1 and 3.1.5), the outer diagram is a pullback as well (see 2.5.3). By uniqueness of a pullback, rc is an isomorphism and thus $c \in \mathcal{E}$. On the other hand, considering diagram 5.59,

- η_{irB} and $\eta_{ir(C)}$ are isomorphisms with inverse $i\varepsilon_{r(B)}$ (see 3.4.1 and

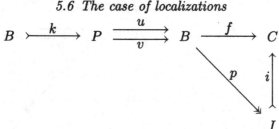

Diagram 5.60

3.1.5),
- the back face is a pullback by definition,
- the right lateral face is a pullback since $\eta_{ir(C)}$ and $\eta_{ir(B)}$ are isomorphisms (see 2.5.3),
- the front face is a pullback as image of the back face under the functor ir, which preserves pullbacks;

therefore the left lateral face is a pullback as well by 2.5.3 and thus $a \in \mathcal{M}$.

So $(\mathcal{E}, \mathcal{M})$ is a factorization system and, by 5.6.1, it satisfies the second additional condition. Since a factorization system is completely characterized by its class $\mathcal{E}$, this factorization system is the same as that constructed in 5.5.6. In particular all the arrows of $\mathscr{A}$ are in $\mathcal{M}$ and the factorization system satisfies the first additional condition.

By 5.6.1, the bijection described in 5.5.6 restricts to the bijection announced in 5.6.2. $\qquad\qquad\qquad\qquad\qquad\qquad\qquad\qquad\square$

Finally localizations have the particular property of admitting a description in terms of "inverted monomorphisms". Given a finitely complete category $\mathscr{B}$ with strong-epi–mono factorizations (e.g. a finitely well-complete $\mathscr{B}$, see 4.4.1) consider a morphism f factored as $f = i \circ p$, with i a monomorphism and p a strong epimorphism; see 4.4.3. Often i is called the "image of f". Let us consider the kernel pair (u, v) of p (see 2.5.4) and its equalizer $k = \mathsf{Ker}\,(u, v)$, as in diagram 5.60.

Lemma 5.6.3 *Let* $f\colon B \longrightarrow C$ *be a morphism in a finitely complete category and* (P, u, v) *its kernel pair. The equalizer* $k = \mathsf{Ker}\,(u, v)$ *of this kernel pair is the unique morphism* $k\colon B \longrightarrow P$ *such that* $u \circ k = 1_B = v \circ k$.

Proof From $f \circ 1_B = f \circ 1_B$ we get this unique morphism $k\colon B \longrightarrow P$ such that $u \circ k = 1_B = v \circ k$. In particular, k is a monomorphism. Now given $x\colon X \longrightarrow P$ such that $u \circ x = v \circ x$, one has $u \circ k \circ u \circ x = u \circ x$

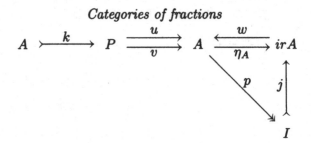

Diagram 5.61

and $v \circ k \circ u \circ x = u \circ x = v \circ x$. Therefore $k \circ u \circ x = x$, since (u, v) is a pullback, and $u \circ x$ is a factorization of x through k. This factorization is unique since k is a monomorphism.   □

Proposition 5.6.4 *Consider a localization $r \dashv i \colon \mathscr{A} \rightleftarrows \mathscr{B}$ of a finitely complete category $\mathscr{B}$ in which every arrow has a strong-epi–mono factorization. Write Σ for the class of those morphisms $f \in \mathscr{B}$ inverted by the reflection r.*

(1) A morphism $f \in \mathscr{B}$ is in Σ iff its image is in Σ and the equalizer of its kernel pair is in Σ.

(2) An object $A \in \mathscr{B}$ is in $\mathscr{A}$ iff for every monomorphism $u \in \Sigma$, $u \perp A$.

Proof Let us use the notation of diagram 5.60. Since r preserves finite limits, $r(i)$ is a monomorphism (see 2.9.3), $\big(r(u), r(v)\big)$ is the kernel pair of $r(p)$ and $r(k)$ is the equalizer of $\big(r(u), r(v)\big)$. On the other hand since $r \dashv i$, the morphism $r(p)$ is a strong epimorphism (see 4.3.9).

By 4.3.6, $r(f)$ is a strong epimorphism iff $r(i)$ is a strong epimorphism, i.e. iff $r(i)$ is an isomorphism. On the other hand $r(f)$ is a monomorphism iff $r(u) = r(v)$ (see 2.5.6), i.e. iff $r(k)$ is an isomorphism (see 2.4.5). Thus $r(f)$ is an isomorphism if and only if both $r(i)$ and $r(k)$ are isomorphisms.

We know that every object $A \in \mathscr{A}$ obeys $f \perp A$, for every $f \in \Sigma$; see 5.4.4. Conversely choose $A \in \mathscr{B}$ such that $f \perp A$, for every monomorphism f of Σ. Now $\mathscr{A}$ is replete by definition (see 3.5.2) and $rA \in \mathscr{A}$; thus it suffices to prove that $\eta_A \colon A \longrightarrow ir(A)$ is an isomorphism. Let us consider diagram 5.61, where (j, p) is the strong-epi–mono factorization of η_A (see 4.4.3), (u, v) is the kernel pair of p and k is the equalizer of (u, v). Since $r(\eta_A)$ is an isomorphism with inverse $\varepsilon_{r(A)}$ (see 3.4.1 and 3.1.5), $j \in \Sigma$ and $k \in \Sigma$ by the first part of the proof. On the other hand $u \circ k = 1_A = v \circ k$ (see 5.6.3); since $k \perp A$, the uniqueness condition in the definition of orthogonality implies $u = v$. But then p is a monomorphism (see 2.5.6) and thus also an isomorphism (see 4.3.6). Then η_A

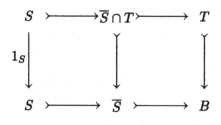

Diagram 5.62

is isomorphic to the monomorphism $j \in \Sigma$, and η_A is a monomorphism belonging to Σ. This implies $\eta_A \perp A$ and thus the existence of a unique $w: ir(A) \longrightarrow A$ such that $w \circ \eta_A = 1_A$. From $\eta_A \circ w \circ \eta_A = 1_{ir(A)} \circ \eta_A$, we deduce $\eta_A \circ w = 1_{ir(A)}$ by the universal property of η_A; see 3.1.1. Thus η_A is an isomorphism with inverse w (see 1.9.3). $\qquad\square$

5.7 Universal closure operations

Definition 5.7.1 *Consider a finitely complete category $\mathscr{B}$. A universal closure operation on $\mathscr{B}$ consists in giving, for every subobject $S \rightarrowtail B$ in $\mathscr{B}$, another subobject $\overline{S} \rightarrowtail B$ called "the closure of S in B"; these assignments have to satisfy the following properties, where S, T are subobjects of B and $f: A \longrightarrow B$ is a morphism of $\mathscr{B}$:*

(1) $S \subseteq \overline{S}$;
(2) $S \subseteq T \Rightarrow \overline{S} \subseteq \overline{T}$;
(3) $\overline{\overline{S}} = \overline{S}$;
(4) $f^{-1}(\overline{S}) = \overline{f^{-1}(S)}$.

Proposition 5.7.2 *Consider a finitely complete category $\mathscr{B}$ provided with a universal closure operation. Given subobjects S, T of $B \in \mathscr{B}$, one has*

(1) $\overline{B} = B$,
(2) $\overline{S \cap T} = \overline{S} \cap \overline{T}$.

Proof Since $B \subseteq \overline{B}$ and, of course, $\overline{B} \subseteq B$ as subobjects of B, one has $\overline{B} = B$.

Now let us first observe that given $S \subseteq T \subseteq B$ diagram 5.62 where both squares are pullbacks indicates that $\overline{S} \cap T$ is the closure of S in T; see 5.7.1.(4). Thus the closure of S in T is smaller than the closure $\overline{S}$ of S in B.

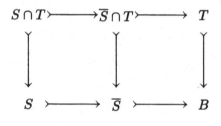

Diagram 5.63

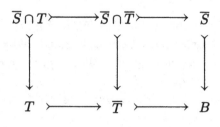

Diagram 5.64

Given now arbitrary subobjects $T, S \subseteq B$, the relations $T \cap S \subseteq T$, $T \cap S \subseteq S$ imply $\overline{T \cap S} \subseteq \overline{T}$ and $\overline{T \cap S} \subseteq \overline{S}$, thus $\overline{T \cap S} \subseteq \overline{T} \cap \overline{S}$. On the other hand diagram 5.63, where both squares are pullbacks, indicates that $\overline{S} \cap T$ is the closure of $S \cap T$ in T; see 5.7.1.(4). Since the closure in T is smaller than the closure in B (previous step of the proof) one gets $\overline{S} \cap T \subseteq \overline{S \cap T}$. Analogously diagram 5.64, where both squares are pullbacks, indicates that $\overline{S} \cap \overline{T}$ is the closure of $\overline{S} \cap T$ in $\overline{S}$. Since the closure in $\overline{S}$ is smaller than the closure in B, $\overline{S} \cap \overline{T} \subseteq \overline{\overline{S} \cap T} \subseteq \overline{\overline{S \cap T}} = \overline{S \cap T}$.   □

Definition 5.7.3 *Consider a finitely complete category $\mathscr{B}$ provided with a universal closure operation.*

(1) A subobject $S \rightarrowtail B$ is dense when $\overline{S} = B$;

(2) a subobject $S \rightarrowtail B$ is closed when $\overline{S} = S$.

Proposition 5.7.4 *Consider a finitely complete category $\mathscr{B}$ provided with a universal closure operation, a morphism $f: A \longrightarrow B$ and a subobject S of B; see diagram 5.65.*

(1) If S is dense in B, $f^{-1}(S)$ is dense in A;

(2) if S is closed in B, $f^{-1}(S)$ is closed in A.

Proof If $\overline{S} = B$, then $\overline{f^{-1}(S)} = f^{-1}(\overline{S}) = f^{-1}(B) = A$, which proves the first assertion.

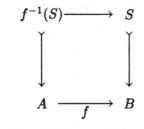

Diagram 5.65

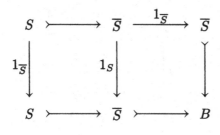

Diagram 5.66

If $\overline{S} = S$, then $\overline{f^{-1}(S)} = f^{-1}(\overline{S}) = f^{-1}(S)$, which proves the second assertion. □

Corollary 5.7.5 *Consider a finitely complete category $\mathscr{B}$ provided with a universal closure operation. If $S \subseteq B$ is any subobject, S is dense in $\overline{S}$.*

Proof Just consider diagram 5.66 where both squares are pullbacks, showing that $\overline{S}$ is the closure of S in $\overline{S}$. □

Corollary 5.7.6 *Consider a finitely complete category $\mathscr{B}$ provided with a universal closure operation. If $S \subseteq B$ is both closed and dense, then $S = B$.*

Proof One has $\overline{S} = S$ and $\overline{S} = B$. □

Proposition 5.7.7 *Consider a finitely complete category $\mathscr{B}$ provided with a universal closure operation. Given subobjects $S \subseteq T \subseteq B$, the following conditions are equivalent:*
(1) S is dense in T and T is dense in B;
(2) S is dense in B.

Proof Assume (1). We have observed in the proof of 5.7.2 that the closure of S in T (which is T) is smaller than the closure of S in B; thus $T \subseteq \overline{S}$. Therefore $B = \overline{T} \subseteq \overline{\overline{S}} = \overline{S}$, thus $B = \overline{S}$.

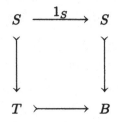

$$S \xrightarrow{\ 1_S\ } S$$

Diagram 5.67

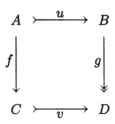

$$A \xrightarrow{\ u\ } B$$

Diagram 5.68

Assume (2). Since $S \subseteq T$, we get $B = \overline{S} \subseteq \overline{T}$, thus $B = \overline{T}$ and T is dense in B. On the other hand diagram 5.67 is a pullback and since S is dense in B, S is dense in T; see 5.7.4. $\qquad\square$

Corollary 5.7.8 *Consider a finitely complete category $\mathscr{B}$ provided with a universal closure operation. In a commutative square $g \circ u = v \circ f$ (see diagram 5.68), if u is a dense monomorphism, g is a strong epimorphism and v is a monomorphism, then v is a dense monomorphism.*

Proof We consider the closure $\overline{C}$ of C and diagram 5.69, where the squares are pullbacks and the monomorphism i is the unique factorization of f and u through the pullback of v and g. The subobject d is closed by definition, thus b is closed (see 5.7.4). By definition u is dense, thus b is dense (see 5.7.7). Therefore b is an isomorphism (see 5.7.6) and $d \circ h \circ b^{-1} = g$. Since g is a strong epimorphism, the monomorphism d is also a strong epimorphism and thus an isomorphism (see 4.3.6). So v is isomorphic to c, which is dense (see 5.7.5). $\qquad\square$

Proposition 5.7.9 *Consider a finitely complete category $\mathscr{B}$ provided with a universal closure operation. Given subobjects $S \subseteq T \subseteq B$, the following implications hold.*

(1) If S is closed in T and T is closed in B, then S is closed in B.

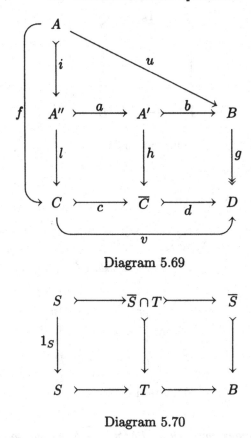

Diagram 5.69

Diagram 5.70

(2) If S is closed in B, then S is closed in T.

Proof Assume S is closed in T and T is closed in B. Consider diagram 5.70, where the squares are pullbacks and $\overline{S}$ is the closure of S in B. The subobject S is dense in $\overline{S}$ (see 5.7.5), thus $\overline{S} \cap T$ is dense in $\overline{S}$ (see 5.7.7); but since T is closed in B, $\overline{S} \cap T$ is also closed in $\overline{S}$ (see 5.7.4); therefore $\overline{S} \cap T = \overline{S}$. In the same way, S is dense in $\overline{S} \cap T$ because S is so in $\overline{S}$, and S is also closed in $\overline{S} \cap T$ because S is so in T; thus $S = \overline{S} \cap T$. Finally $S = \overline{S}$ and S is closed in B.

Now assume S is closed in B. Diagram 5.71 is a pullback and since S is closed in B, S is closed in T; see 5.7.4. $\qquad \square$

It is probably useful to dwell on the fact that, when $S \subseteq T \subseteq B$ and S is closed in B, T is in general not closed in B; see 5.9.6 for a counterexample.

Corollary 5.7.10 *Consider a finitely complete category $\mathscr{B}$ provided*

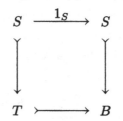

Diagram 5.71

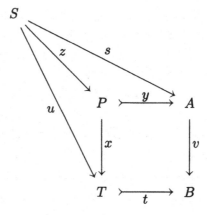

Diagram 5.72

with a universal closure operation. If $s\colon S{\rightarrowtail}A$ is a dense subobject of A and $t\colon T{\rightarrowtail}B$ is a closed subobject of B, then $s \perp t$.

Proof Consider diagram 5.72 where $t{\circ}u = v{\circ}s$. Computing the pullback (x,y) of (t,v), y is a closed monomorphism since t is (see 5.7.4). On the other hand y is dense since s is (see 5.7.7). Thus y is an isomorphism (see 5.7.6) and $x \circ y^{-1}$ is a factorization satisfying

$$x \circ y^{-1} \circ s = x \circ z = u, \quad t \circ x \circ y^{-1} = v \circ y \circ y^{-1} = v.$$

Since t is a monomorphism $x \circ y^{-1}$ is unique with these properties . $\quad\square$

Let us now indicate an interesting relation between localizations and universal closure operations.

Proposition 5.7.11 *Consider a localization $r \dashv i\colon \mathscr{A}{\rightleftarrows}\mathscr{B}$ of a finitely complete category $\mathscr{B}$. The localization induces a universal closure operation on $\mathscr{B}$, associating with a subobject $s\colon S{\rightarrowtail}B$ the subobject $\overline{S}{\rightarrowtail}B$ defined as $\eta_B^{-1}(ir(S))$, where η_B is the canonical morphism of the adjunction (see diagram 5.73).*

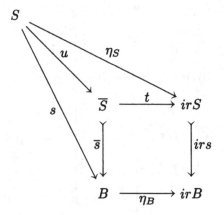

Diagram 5.73

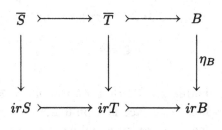

Diagram 5.74

Proof Considering diagram 5.73, one observes that irs is a monomorphism since r preserves finite limits (by assumption) and i preserves all limits (it has a left adjoint r; see 3.2.2). Thus $\bar{s}$ is a monomorphism (see 2.5.3). By naturality of η (see 3.1.5), the outer diagram is commutative from which u is such that $\bar{s} \circ u = s$, $t \circ u = \eta_S$; u is a monomorphism since s is. This defines the closure $\bar{S}$ of S and proves that $S \subseteq \bar{S}$.

If $S \subseteq T \subseteq B$, the pullbacks in diagram 5.74 indicate that $\bar{S} \subseteq \bar{T}$.

In 5.6.2 we have observed that $s = \bar{s} \circ u$ is the $(\mathcal{E}, \mathcal{M})$ factorization of s, thus $r(u)$ is an isomorphism. This proves that $ir(S)$ is isomorphic to $ir(\bar{S})$ and thus $\bar{S}$ is isomorphic to $\bar{\bar{S}}$.

Finally given a morphism $f: A \longrightarrow B$, consider diagram 5.75 where the right face is the construction of $\bar{S}$ and the left face is the construction of $\overline{f^{-1}(S)}$. The bottom face is commutative by naturality of η; see 3.1.5. The whole back face is the pullback of f and $s = \bar{s} \circ u$, producing $f^{-1}(s) = \bar{s'} \circ v$ and y. Since v is a dense monomorphism (see 5.7.5) and $\bar{s}$ is a closed monomorphism, the relation $\bar{s} \circ (uy) = (f\bar{s'}) \circ v$ implies

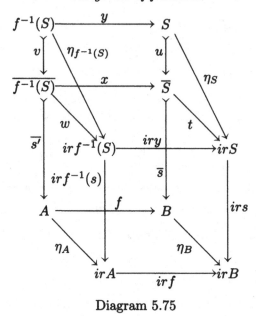

Diagram 5.75

the existence of a unique x such that $x \circ v = u \circ y$, $\bar{s} \circ x = f \circ \bar{s'}$; see 5.7.10. Since the whole back face is a pullback by definition, the front face, which is its image by ir, is a pullback as well. The left lateral "square" is the pullback defining $\overline{f^{-1}(S)}$ and in the same way the right lateral "square" is the pullback defining $\overline{S}$. Since the whole diagram is commutative, the associativity properties of pullbacks (see 2.5.9) imply that $(\bar{s'}, x)$ is indeed the pullback of $(f, \bar{s})$, proving $f^{-1}(\overline{S}) = \overline{f^{-1}(S)}$. $\square$

Examples 5.7.12

5.7.12.a In the category Top of topological spaces and continuous mappings, a monomorphism is just a continuous injection $s\colon S \rightarrowtail A$; see 1.7.7.b. Define the closure of s as the subspace $s(S) \hookrightarrow A$, where $s(S)$ is provided with the induced topology. It is straightforward to observe that this is a universal closure operation.

5.7.12.b In the category Ab of abelian groups, define the closure of a subgroup $S \subseteq A$ by

$$\overline{S} = \{a \in A \mid \exists k \in \mathbb{N} \quad n^k a \in S\}$$

where $n \neq 0$ is some fixed natural number. This defines a universal closure operation.

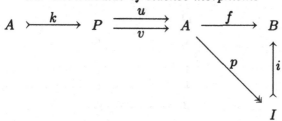

Diagram 5.76

5.7.12.c In the category **Cat** of small categories and functors, define the closure of a subcategory $\mathscr{S} \subseteq \mathscr{C}$ as the full subcategory $\overline{\mathscr{S}}$ generated by $\mathscr{S}$. This is a universal closure operation.

5.8 The calculus of bidense morphisms

To emphasize the relations between universal closure operations and calculus of fractions, let us consider a finitely complete category in which every arrow f factors as $f = i \circ p$, with i a monomorphism and p a strong epimorphism (see 4.4). As in section 5.6 we are interested in the image i of f and in the equalizer k of the kernel pair (u, v) of f (see 5.6.3), as in diagram 5.76. Since i is a monormophism, (u, v) is also the kernel pair of p.

Definition 5.8.1 *Consider a finitely complete category $\mathscr{B}$ with strong-epi–mono factorizations. Given a universal closure operation on $\mathscr{B}$, a morphism $f\colon A \longrightarrow B$ is bidense when its image is dense and the equalizer of its kernel pair is dense.*

Proposition 5.8.2 *Consider a finitely complete category $\mathscr{B}$ admitting strong-epi–mono factorizations. Given a universal closure operation on $\mathscr{B}$, a monomorphism is dense if and only if it is bidense.*

Proof With the previous notation, if f is a monomorphism, p is both a monomorphism and a strong epimorphism, thus it is an isomorphism (see 4.3.6). On the other hand, $u = v = 1_A$ (see 2.5.6) and thus $k = 1_A$, which is in any case dense. On the other hand since p is an isomorphism, f is dense if and only if i is dense. $\qquad\square$

Proposition 5.8.3 *Consider a finitely complete category $\mathscr{B}$ admitting strong-epi–mono factorizations. Given a localization $r \dashv i\colon \mathscr{A} \underset{\longrightarrow}{\overset{\longleftarrow}{}} \mathscr{B}$, consider the corresponding universal closure operation as in 5.7.11. A morphism $f \in \mathscr{B}$ is inverted by the reflection r if and only if it is bidense.*

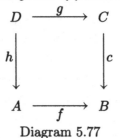

Diagram 5.77

Proof By 5.6.4, it suffices to prove that a monomorphism $s\colon S \rightarrowtail B$ is inverted by r if and only if it is dense. Considering diagram 5.73, if $ir(s)$ is an isomorphism, $\bar{s}$ is an isomorphism and $\overline{S} = B$, thus S is dense in B. Conversely, suppose $\overline{S} = B$. In 5.6.2 we have proved that $s = \bar{s} \circ u$ is the $(\mathcal{E}, \mathcal{M})$-factorization of s associated with the localization. Thus $u \in \mathcal{E}$ and $r(u)$ is an isomorphism. Since $\bar{s}$ is an isomorphism by assumption, $ir(s) = ir(\bar{s}) \circ ir(u)$ is an isomorphism. $\qquad\square$

In general, a universal closure operation has no reason to be induced by a localization as in proposition 5.7.11. We shall nevertheless investigate this question a little bit more in some particular cases of interest. In the rest of this section, we shall freely use the notions of *regular* and *locally presentable* category which will be studied systematically in chapters 2 and 5 of volume 2.

First of all, when strong-epi–mono factorizations exist and are stable under pullbacks (this is the essence of the definition of a regular category; see 2.2.2, volume 2), the class of bidense morphisms has good properties.

Proposition 5.8.4 *Consider a finitely complete regular category $\mathscr{B}$ provided with a universal closure operation. The class Σ of bidense morphisms has a right calculus of fractions. Moreover, this class Σ satisfies the following additional properties:*

(1) every isomorphism is bidense;

(2) if two sides of a commutative triangle are bidense, the third side is bidense as well;

(3) bidense morphisms are stable under pullbacks, i.e. given a pullback in $\mathscr{B}$ as in diagram 5.77, if f is bidense, g is bidense as well.

Proof By 10.2.2, every morphism has a strong-epi–mono factorization and the pullback of a strong epimorphism is still a strong epimorphism.

With the notation of diagram 5.76, if f is an isomorphism we can choose $i = 1_A$, $u = v = 1_A$, $k = 1_A$. Thus i and k are dense (see 5.7.3)

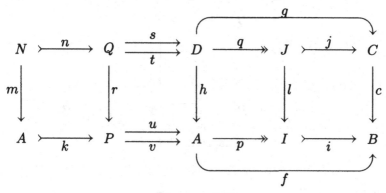

Diagram 5.78

and the isomorphism f is bidense. In particular this implies condition 5.2.3.(1).

Next let us consider f, i, p, u, v, k as above and in addition diagram diagram 5.78, where the last two squares on the right are pullbacks, (s, t) is the kernel pair of q and $n = \mathsf{Ker}(s, t)$. Observe that $p \circ h \circ s = l \circ q \circ s = l \circ q \circ t = p \circ h \circ t$, which yields a unique r such that $h \circ s = u \circ r$ and $h \circ t = v \circ r$. Moreover, $u \circ r \circ n = h \circ s \circ n = h \circ t \circ n = v \circ r \circ n$, from which there is a unique m such that $k \circ m = r \circ n$. Let us prove that (m, n) is the pullback of (k, r). Given x, y such that $r \circ x = k \circ y$, the relations

$$q \circ s \circ x = q \circ t \circ x, \quad h \circ s \circ x = u \circ r \circ x = u \circ k \circ y = v \circ k \circ y = v \circ r \circ x = h \circ t \circ x$$

imply $s \circ x = t \circ x$, because (q, h) is a pullback. Since $n = \mathsf{Ker}(s, t)$, there exists a unique z such that $n \circ z = x$; moreover $k \circ m \circ z = r \circ n \circ z = r \circ x = k \circ y$ so that $m \circ z = y$ because k is a monomorphism. Finally z is the expected factorization and it is necessarily unique because n is a monomorphism.

Suppose f is bidense. Then j and n are dense monomorphisms since i and k are (see 5.7.4). Moreover q is a strong epimorphism since p is (see 10.2.2). All these observations prove that pulling back the bidense morphism f along an arbitrary morphism c, one gets another bidense morphism g. In particular, this implies condition 5.2.3.(3).

Next let us check condition 5.2.3.(4). Let us consider $g, h \colon C \rightrightarrows A$ in $\mathscr{B}$ and a bidense morphism $f \colon A \longrightarrow B$ such that $f \circ g = f \circ h$. In diagram 5.79, (u, v) is the kernel pair of f and $k = \mathsf{Ker}(u, v)$. From $f \circ g = f \circ h$ we find a unique r such that $u \circ r = g$, $v \circ r = h$. Computing the pullback (t, s) of (k, r) we obtain a monomorphism s, which is dense

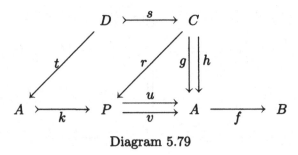

Diagram 5.79

since k is such that

$$g \circ s = u \circ r \circ s = u \circ k \circ t = t = v \circ k \circ t = v \circ r \circ s = h \circ s.$$

By 5.8.2 the dense monomorphism s is bidense, which concludes the proof of condition 5.2.3.(4).

Now consider a composite $g \circ f$. We refer to diagram 5.80 where $f = i \circ p$, $g = j \circ q$, $q \circ i = n \circ r$ are image factorizations. The kernel pair of f is (u, v), (s, t) is that of g and (a, b) that of $g \circ f$. Moreover, $l = \mathsf{Ker}\,(s, t)$, $k = \mathsf{Ker}\,(u, v)$ and $m = \mathsf{Ker}\,(a, b)$.

Observe that $g \circ f \circ a = g \circ f \circ b$ implies the existence of a unique w such that $s \circ w = f \circ a$, $t \circ w = f \circ b$. We construct the pullback (d, z) of (w, l). On the other hand $g \circ f \circ u = g \circ f \circ v$ implies the existence of a unique h such that $s \circ h = f \circ u$, $t \circ h = f \circ v$. Observe also that

$$s \circ l \circ f = f = f \circ u \circ k = s \circ h \circ k, \quad t \circ l \circ f = f = f \circ v \circ k = t \circ h \circ k,$$

from which $l \circ f = h \circ k$ since (s, t) is a pullback. Observe next that

$$f \circ a \circ d = s \circ w \circ d = s \circ l \circ z = z = t \circ l \circ z = t \circ w \circ d = f \circ b \circ d,$$

from which there is a unique c such that $u \circ c = a \circ d$, $v \circ c = b \circ d$. Finally

$$s \circ w \circ m = f \circ a \circ m = f = s \circ l \circ f, \quad t \circ w \circ m = f \circ b \circ m = f = t \circ l \circ f,$$

from which $w \circ m = l \circ f$ since (s, t) is a pullback. This implies the existence of a unique θ such that $d \circ \theta = m$ and $z \circ \theta = f$.

Let us now suppose that f and g are bidense. Since i is dense and q is a strong epimorphism, n is dense (see 5.7.8). Since j and n are dense, $j \circ n$ is dense (see 5.7.7), and this is the image of $g \circ f$.

Observe now that

$$u \circ c \circ \theta = a \circ d \circ \theta = a \circ m = 1_A = u \circ k, \quad v \circ c \circ \theta = b \circ d \circ \theta = b \circ m = 1_A = v \circ k,$$

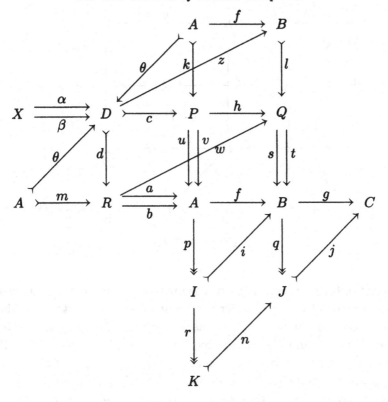

Diagram 5.80

from which $c \circ \theta = k$ because (u, v) is a pullback. Moreover c is a mono-morphism because $c \circ \alpha = c \circ \beta$ implies

$$a \circ d \circ \alpha = u \circ c \circ \alpha = u \circ c \circ \beta = a \circ d \circ \beta, \quad b \circ d \circ \alpha = v \circ c \alpha = v \circ c \circ \beta = b \circ d \circ \beta,$$

from which $d \circ \alpha = d \circ \beta$, since (a, b) is a pullback. But d is a monomor-phism since l is, thus $\alpha = \beta$. The monomorphism d is dense since l is (see 5.7.4). But θ is dense since k is and $c \circ \theta = k$; see 5.7.7. Thus $m = d \circ \theta$ is dense (see 5.7.7) and $g \circ f$ is bidense. This implies in particular condition 5.2.3.(2).

Now suppose that $g \circ f$ and f are bidense. Since $g \circ f$ is bidense, $j \circ n$ is dense and thus j is dense (see 5.7.7). On the other hand m is dense, thus θ and d are dense (see 5.7.7). If we can prove that w is bidense, then $w \circ d$ will be bidense and thus $l \circ z = w \circ d$ will be as well. But since $l \circ z$ factors through l, the subobject l contains the dense image of $l \circ z$ and thus l is dense (see 5.7.7). Thus g will be bidense as long as w is.

To prove that w is bidense, consider diagram 5.81 where (e, ε) is the

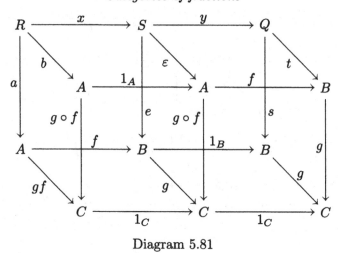

Diagram 5.81

pullback of $(g, g \circ f)$ and x, y are the obvious factorizations making the diagram commutative. Consider the square $f \circ \varepsilon = t \circ y$. It is a pullback because given $f \circ \sigma = t \circ \tau$, one has $g \circ f \sigma = g \circ t \circ \tau = g \circ s \circ \tau$ from which there is a unique ρ such that $\varepsilon \circ \rho = \sigma$ and $e \circ \rho = s \circ \tau$. This implies $s \circ y \circ \rho = e \circ \rho = s \circ \tau$ and $t \circ y \circ \rho = f \circ \varepsilon \circ \rho = f \circ \sigma = t \circ \tau$, thus $y \circ \rho = \tau$ since (s, t) is a pullback. This factorization ρ is unique because, given ρ' such that $\varepsilon \circ \rho' = \sigma$ and $y \circ \rho' = \tau$, one deduces $e \circ \rho' = s \circ y \circ \rho' = s \circ \tau$, from which $\rho = \rho'$ by uniqueness of ρ. Since f is bidense, y is bidense (previous part of the proof). A perfectly analogous proof on the left-hand cube yields the bidenseness of x. Since $s \circ y \circ x = f \circ a$ and $t \circ y \circ x = f \circ b$, one has $w = y \circ x$. Thus w is bidense as composite of two bidense morphisms (previous part of the proof).

Finally let us suppose that $g \circ f$ and g are bidense. We again refer to diagram 5.80. From $g \circ f \circ u = g \circ f \circ v$ we conclude that u, v are coequalized by the bidense morphism $g \circ f$; as we have seen previously in the proof, this implies the existence of a dense monomorphism $\xi \colon Z \longrightarrow A$ such that $u \circ \xi = v \circ \xi$. But then ξ factors through the equalizer $k = \mathsf{Ker}\,(u, v)$ and since ξ is dense, k is dense as well (see 5.7.7).

To prove that the image i of f is dense, consider diagram 5.82 obtained from the two previous diagrams and where all three squares are pullbacks. Observe that $f \circ \varepsilon \circ \lambda = s \circ y \circ \lambda = s \circ l \circ \mu = t \circ l \circ \mu = t \circ y \circ \lambda$. Since l is dense, λ is (see 5.7.4). Applying previous parts of the proof several times, $t \circ y$ is bidense because $g \circ f$ is, and ε is bidense because g is. Therefore $\varepsilon \circ \lambda$ and $t \circ y \circ \lambda$ are bidense. Since $f \circ (\varepsilon \circ \lambda) = t \circ y \circ \lambda$ and $\varepsilon \circ \lambda$ are bidense, f is also. $\qquad\square$

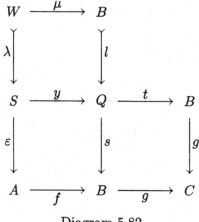

Diagram 5.82

Given a universal closure operation on a category $\mathscr{B}$, the corresponding bidense morphisms constitute in general a proper class Σ, so that the corresponding category of fractions $\mathscr{B}[\Sigma^{-1}]$ has no reason to exist (see section 5.2). But when $\mathscr{B}$ has a "sufficiently good family of generators", the size problem disappears. The precise notion we need is that of a *locally presentable category* as studied in sections 5.1, 5.2 of volume 2. Such a category is in particular complete and cocomplete (see 5.2.8, volume 2) and admits a dense family of presentable generators (see 5.2.5, volume 2).

Proposition 5.8.5 *Consider a locally α-presentable category $\mathscr{B}$ provided with a universal closure operation. A monomorphism $s\colon S \rightarrowtail B$ is dense for the closure operation iff, for every α-presentable object P and every morphism $f\colon P \longrightarrow B$,*

$$f^{-1}(s)\colon f^{-1}(S) \rightarrowtail P$$

is a dense monomorphism.

Proof By 5.7.4, if s is dense, so is $f^{-1}(s)$. Conversely B can be written as an α-filtered colimit $B = \operatorname{colim}_i P_i$, where each P_i is α-presentable (see 5.2.5 of volume 2). Consider the pullbacks given by diagram 5.83, where the σ_i's are the canonical morphisms of the colimit. Given $p\colon P_j \longrightarrow P_i$ such that $\sigma_i \circ p = \sigma_j$, we have $s \circ \tau_j = \sigma_j \circ s_j = \sigma_i \circ p \circ s_j$, from which there is a unique $q\colon S_j \longrightarrow S_i$ such that $\tau_i \circ q = \tau_j$, $s_i \circ q = p \circ s_j$. This defines a new α-filtered diagram with vertices S_i and, since α-filtered colimits are universal in $\mathscr{B}$ (see 5.2.8 of volume 2), $S = \operatorname{colim}_i S_i$.

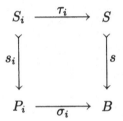

Diagram 5.83

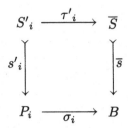

Diagram 5.84

An analogous argument can be developed replacing S by $\overline{S}$, considering now the pullbacks of diagram 5.84. By 5.7.1.(4), $S'_i = \sigma_i^{-1}(\overline{S}) = \overline{\sigma_i^{-1}(S)} = \overline{S_i} = P_i$, since by assumption s_i is dense. But then each s'_i is an isomorphism and the morphisms

$$(s'_i)^{-1}\colon P_i \longrightarrow S'_i$$

induce a factorization $\tilde{s}\colon \operatorname{colim}_i P_i \longrightarrow \overline{S}$ such that $\tilde{s} \circ \sigma_i = \tau'_i \circ (s'_i)^{-1}$. The relations

$$\overline{s} \circ \tilde{s} \circ \sigma_i = \overline{s} \circ \tau'_i \circ (s'_i)^{-1} = \sigma_i \circ s'_i \circ (s'_i)^{-1} = \sigma_i,$$
$$\tilde{s} \circ \overline{s} \circ \tau'_i = \tilde{s} \circ \sigma_i \circ s'_i = \tau'_i \circ (s'_i)^{-1} \circ s'_i = \tau'_i,$$

imply $\overline{s} \circ \tilde{s} = 1_B$ and $\tilde{s} \circ \overline{s} = 1_{\overline{S}}$. Thus $\overline{S} = B$ and S is dense in B. $\qquad\square$

Proposition 5.8.6 *Consider a locally α-presentable regular category $\mathscr{B}$ provided with a universal closure operation. For an object $A \in \mathscr{B}$, the following conditions are equivalent:*

(1) for every bidense morphism f, $f \perp A$;

(2) for every dense monomorphism s, $s \perp A$;

(3) for every α-presentable object P and every dense subobject $S \rightarrowtail P$ of P, $s \perp A$.

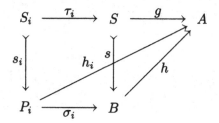

Diagram 5.85

Proof (1) ⇒ (2) and (2) ⇒ (3) are obvious. Let us prove (3) ⇒ (2). We consider an object A satisfying (3) and a dense monomorphism $s: S \rightarrowtail B$. We construct σ_i, s_i, τ_i as in the proof of 5.8.5, yielding diagram 5.85. Consider now an arbitrary morphism $g: S \longrightarrow A$. Each s_i is dense and thus we find a unique $h_i: P_i \longrightarrow A$ such that $g \circ \tau_i = h_i \circ s_i$. Given $p: P_j \longrightarrow P_i$ such that $\sigma_i \circ p = \sigma_j$, there exists $q: S_j \longrightarrow S_i$ such that $\tau_i \circ q = \tau_j$, $s_i \circ q = p \circ s_j$ (see proof of 5.8.5). Therefore

$$h_i \circ p \circ s_j = h_i \circ s_i \circ q = g \circ \tau_i \circ q = g \circ \tau_j = h_j \circ s_j$$

from which $h_i \circ p = h_j$ by uniqueness of h_j. So the morphisms h_i constitute a cocone, from which there is a unique factorization h through $B = \operatorname{colim}_i P_i$, with the property $h \circ \sigma_i = h_i$.

Since $S = \operatorname{colim}_i S_i$, the relations

$$h \circ s \circ \tau_i = h \circ \sigma_i \circ s_i = h_i \circ s_i = g \circ \tau_i$$

imply $h \circ s = g$. If h' is another morphism such that $h' \circ s = g$, one gets

$$h' \circ \sigma_i \circ s_i = h' \circ s \circ \tau_i = g \circ \tau_i = h_i \circ s_i,$$

from which $h' \circ \sigma_i = h_i$, by the uniqueness condition in the definition of $s_i \perp A$. But then $h' \circ \sigma_i = h_i = h \circ \sigma_i$, from which $h = h'$ since $B = \operatorname{colim}_i P_i$. This concludes the proof of (3) ⇒ (2).

Now let us prove (2) ⇒ (1). We consider a bidense morphism f, its strong-epi–mono factorization $f = i \circ p$, its kernel pair (u, v) and $k = \mathsf{Ker}\,(u, v)$. We choose $A \in \mathscr{B}$ such that $s \perp A$, for every dense monomorphism A. Finally we consider $g: B \longrightarrow A$ and diagram 5.86. Since $u \circ k = v \circ k$, one has $g \circ u \circ k = g \circ v \circ k$. Since $k \perp A$, this implies $g \circ u = g \circ v$. But since i is a monomorphism, (u, v) is also the kernel pair of the strong, thus regular epimorphism p (see 10.1.5). Since $p = \mathsf{Coker}\,(u, v)$ (see 2.5.7) and $g \circ u = g \circ v$, we get a unique q such that $g = q \circ p$. Since $i \perp A$, this provides a unique h such that $h \circ i = q$. This

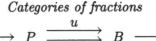
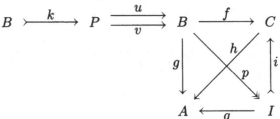

Diagram 5.86

implies $h \circ f = h \circ i \circ p = q \circ p = g$. If h' is another morphism such that $h' \circ f = g$, then $h' \circ i \circ p = h' \circ f = g = q \circ p$, from which $h' \circ i = q$ since p is an epimorphism. By uniqueness of h, $h = h'$. $\qquad\square$

Corollary 5.8.7 *Consider a locally presentable regular category $\mathscr{B}$ provided with a universal closure operation. Write Σ for the corresponding class of bidense morphisms and $\mathscr{B}_\Sigma$ for the full subcategory of those objects $A \in \mathscr{B}$ such that $f \perp A$, for every bidense morphism f. Under these conditions, $\mathscr{B}_\Sigma$ is reflective in $\mathscr{B}$ and each bidense morphism is inverted by the reflection.*

Proof By 5.8.6, it suffices to require orthogonality with respect to the dense monomorphisms $s: S \rightarrowtail P$, with P α-presentable. Since there is just a set of such dense monomorphisms (see 4.5.15, this volume and 5.2.1, volume 2), the existence of a reflection $r \dashv i: \mathscr{B}_\Sigma \xrightarrow{\longleftarrow} \mathscr{B}$ follows from 5.4.7.

By 5.4.4, each bidense morphism is inverted by the reflection r. $\qquad\square$

It should be clear that the previous corollary does not at all state that Σ is precisely the class of all morphisms inverted by the reflection. As a consequence, $\mathscr{B}_\Sigma$ has a priori no reason to be a localization of $\mathscr{B}$. An additional condition on Σ is required for that, as attested by the next proposition.

Proposition 5.8.8 *Consider a locally presentable regular category $\mathscr{B}$ provided with a universal closure operation. The following conditions are equivalent:*

(1) *the universal closure operation on $\mathscr{B}$ is that induced by a localization $r \dashv i: \mathscr{A} \xrightarrow{\longleftarrow} \mathscr{B}$ of $\mathscr{B}$, as in 5.7.11;*

(2) *the class Σ of bidense morphisms is closed under colimits (see 5.4.9).*

Under these conditions, the bidense morphisms are exactly those inverted by the reflection r.

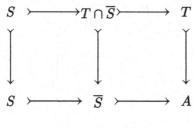

Diagram 5.87

Proof If (1) holds, Σ is the class of those morphisms inverted by the reflection (see 5.8.3) and thus is closed under colimits (see 5.4.10).

Assume now condition (2) of the statement. Corollary 5.8.7 implies the existence of a reflection $r \dashv i: \mathscr{B}_\Sigma \overset{\longleftarrow}{\underset{\longrightarrow}{}} \mathscr{B}$, where $A \in \mathscr{B}_\Sigma$ precisely when $f \perp A$ for every bidense morphism f. Moreover, the class Σ of bidense morphisms is contained in the class $\mathcal{E}$ of all those morphisms inverted by the reflection r. But condition (2) of our statement, together with 5.8.4, allows us to apply 5.4.10 and conclude that $\mathcal{E} \subseteq \Sigma$. Thus finally $\mathcal{E} = \Sigma$. Since Σ has a right calculus of fractions (see 5.8.4), the reflection is a localization (see 5.6.1). The universal closure operation associated with that localization has the same bidense morphisms as the original closure operation (the morphisms inverted by r), thus both closure operations have the same dense monomorphisms (see 5.8.2). But given a subobject $S \rightarrowtail A$, $\overline{S} \rightarrowtail A$ is the biggest subobject of A in which S is dense. Indeed, $S \rightarrowtail \overline{S}$ is dense (see 5.7.5) and if $S \leq T \leq A$ with S dense in T, diagram 5.87, where both squares are pullbacks, indicates that $T \cap \overline{S} \rightarrowtail T$ is closed, since $\overline{S} \rightarrowtail A$ is (see 5.7.4). On the other hand, $T \cap \overline{S} \rightarrowtail T$ is dense, since $S \rightarrowtail T$ is (see 5.7.7). Therefore $T \cap \overline{S} = T$ (see 5.7.6) and $T \subseteq \overline{S}$. Thus $\overline{S}$ is indeed the biggest subobject of A in which S is dense, so that two universal closure operations with the same dense monomorphisms must coincide. $\square$

A useful lemma, when trying to apply 5.8.8, is given by the following result.

Lemma 5.8.9 *Consider a locally α-presentable regular category $\mathscr{B}$ provided with a universal closure operation. The class of bidense morphisms is closed under α-filtered colimits.*

Proof Consider an α-filtered category $\mathscr{D}$, two functors $F, G: \mathscr{D} \rightrightarrows \mathscr{B}$ and a natural transformation $\alpha: F \Rightarrow G$ such that, for every $D \in \mathscr{D}$, α_D is bidense. In the category of functors $\mathsf{Fun}(\mathscr{D}, \mathscr{B})$ let us consider

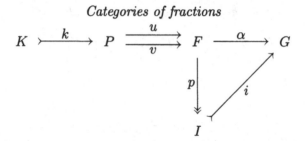

Diagram 5.88

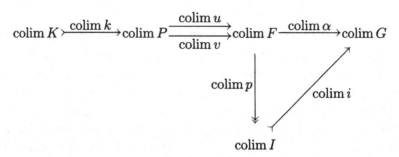

Diagram 5.89

the situation of diagram 5.88, where (u, v) is the kernel pair of α, $k = \mathsf{Ker}\,(u, v)$ and $p = \mathsf{Coker}\,(u, v)$; all these constructions are performed pointwise in $\mathscr{B}$; see 2.15.1. From $\alpha \circ u = \alpha \circ v$ we get a unique factorization i such that $i \circ p = \alpha$ and we know i is a monomorphism, since $\mathsf{Fun}(\mathscr{D}, \mathscr{B})$ is regular; see 2.1.4 of volume 2. Now α-filtered colimits commute in $\mathscr{B}$ with α-limits (see 3.2.8 of volume 2), thus in particular with finite limits; but they commute also with coequalizers (see 2.12.11). Therefore we get an analogous diagram 5.89, with $(\mathrm{colim}\,u, \mathrm{colim}\,v)$ the kernel pair of $\mathrm{colim}\,\alpha$, $\mathrm{colim}\,k$ its equalizer and $\mathrm{colim}\,p$ its coequalizer, $\mathrm{colim}\,i$ a monomorphism. To conclude that $\mathrm{colim}\,\alpha$ is bidense, it suffices to prove that $\mathrm{colim}\,i$ and $\mathrm{colim}\,k$ are dense monomorphisms, i.e. that an α-filtered colimit of dense monomorphisms is again a dense monomorphism. We write out the proof in the case of the monomorphisms $i_D \colon ID \longrightarrow GD$.

Consider diagram 5.90 where σ_D, τ_D are the canonical morphisms of the colimits, $\overline{\mathrm{colim}\,I}$ is the closure of $\mathrm{colim}\,I$ in $\mathrm{colim}\,G$ and the square is a pullback. We get a factorization l_D through the pullback and since i_D is dense, j_D is dense as well (see 5.7.7). But since v is closed by construction, j_D is closed (see 5.7.4). Therefore j_D is an isomorphism (see 5.7.6). Now since α-filtered colimits are universal in $\mathscr{B}$ (see 5.2.8, volume 2), $\overline{\mathrm{colim}\,I} = \mathrm{colim}\,P_D$ and finally $v = \mathrm{colim}\,j_D$. Since each j_D

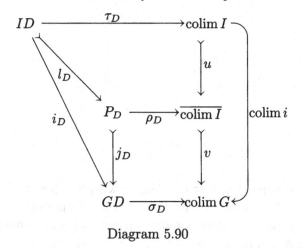

Diagram 5.90

is an isomorphism, v is an isomorphism and finally $\overline{\operatorname{colim} I} = \operatorname{colim} G$, proving that $\operatorname{colim} i$ is dense. $\qquad\square$

Corollary 5.8.10 *Consider a locally α-presentable regular category $\mathscr{B}$ provided with a universal closure operation. The following conditions are equivalent:*

(1) the universal closure operation on $\mathscr{B}$ is that induced by a localization $r \dashv i \colon \mathscr{A} \xrightarrow{} \mathscr{B}$ of $\mathscr{B}$, as in 5.7.11;

(2) the class Σ of bidense morphisms is closed under α-colimits (see 5.4.9).

Under these conditions, the bidense morphisms are exactly those inverted by the reflection r.

Proof By 5.8.8 it remains to show that condition (2) implies in fact that bidense morphisms are stable under all (small) colimits. Every colimit is an α-filtered colimit of α-generated colimits (this is the α-version of proposition 2.13.7). Since an α-generated colimit can be constructed from α-coproducts and coequalizers (this is the α-version of proposition 2.8.5), our assumption implies that bidense morphisms are stable under α-generated colimits. Combining this with 5.8.9, we conclude that bidense morphisms are stable under all small colimits. $\qquad\square$

When condition 5.8.10.(2) is satisfied by every universal closure operation on $\mathscr{B}$, we get at once a bijection between the localizations of $\mathscr{B}$ and the universal closure operations on $\mathscr{B}$. Rather strong conditions are needed in general on a category $\mathscr{B}$ as in 5.8.10 to get such a bijection; we shall prove it is the case for locally finitely presentable abelian categories

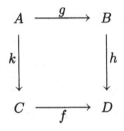

Diagram 5.91

(see 1.13.5, volume 2) and for Grothendieck toposes (see 3.5.7, volume 3). More generally, it is the case for the exact locally presentable categories in which a coproduct of monomorphisms is still a monomorphism (see 5.9.5). In 9.3.9, volume 3, we shall also prove the indicated bijection in the case of an (elementary) topos $\mathscr{B}$, but following completely different arguments: a topos is in general not locally presentable.

5.9 Exercises

5.9.1 Consider a finitely complete category $\mathscr{C}$ provided with a system $(\mathcal{E}, \mathcal{M})$ of factorization. Prove that

(1) if $f \circ g \in \mathcal{M}$ and f is a monomorphism, then $g \in \mathcal{M}$,
(2) the class $\mathcal{M}$ is stable under pullbacks, i.e. in the pullback of diagram 5.91, if $f \in \mathcal{M}$, then $g \in \mathcal{M}$.

5.9.2 Consider a small category $\mathscr{C}$, a set Σ of morphisms of $\mathscr{C}$ and the corresponding category of fractions $\varphi \colon \mathscr{C} \longrightarrow \mathscr{C}[\Sigma^{-1}]$. Prove the following:

(1) if φ is faithful, every morphism of Σ must be both a monomorphism and an epimorphism;
(2) if Σ admits a left calculus of fractions and all the morphisms of Σ are monomorphisms, then φ is faithful.

5.9.3 Consider a category $\mathscr{C}$ and a class Σ of morphisms of $\mathscr{C}$, which admits a left calculus of fractions and is such that the corresponding category $\varphi \colon \mathscr{C} \longrightarrow \mathscr{C}[\Sigma^{-1}]$ of fractions exists. Write Γ for the class of all morphisms inverted by the functor φ. Prove the following:

(1) $f \in \Gamma$ iff there are morphisms $v, w \in \mathscr{C}$ such that $v \circ f \in \Sigma$ and $w \circ v \in \Sigma$;
(2) Γ admits a left calculus of fractions.

5.9.4 Let $\mathscr{C}$ be a finitely complete category provided with a universal closure operation. Prove that the closure of an equivalence relation is still an equivalence relation (see 2.5.2, volume 2, for the definition of an equivalence relation).

5.9.5 Consider a locally presentable category $\mathscr{C}$ which is exact in the sense of 2.6.1, volume 2. Suppose that in $\mathscr{C}$, a coproduct of monomorphisms is again a monomorphism. Given a universal closure operation on $\mathscr{C}$, prove that the class of bidense morphisms is stable under colimits [Hint: consider separately the cases of (bi)dense monomorphisms and of bidense strong epimorphisms, of coproducts and of coequalizers; apply 5.9.4]. Deduce the existence of a bijection between the universal closure operations on $\mathscr{C}$ and the localizations of $\mathscr{C}$; see **Borceux and Veit**.

5.9.6 For the universal closure operation defined in 5.7.12.b, prove that (0) is closed in $\mathbb{Z}$, but not every subgroup of $\mathbb{Z}$ is closed.

6

Flat functors and Cauchy completeness

6.1 Exact functors

Extending a classical terminology for abelian categories (see 1.11.2, volume 2), we define

Definition 6.1.1 *Consider two finitely complete categories $\mathscr{A}, \mathscr{B}$. A functor $F\colon \mathscr{A} \longrightarrow \mathscr{B}$ is left exact when it preserves finite limits.*

Proposition 6.1.2 *Let $F\colon \mathscr{A} \longrightarrow \mathsf{Set}$ be a functor defined on a finitely complete category $\mathscr{A}$. The following conditions are equivalent:*

(1) F is left exact,

(2) the category $\mathsf{Elts}(F)$ of elements of F is cofiltered (see 1.6.4);

(3) F is a filtered colimit of representable functors.

Proof Let us recall that $\mathsf{Elts}(F)$ is defined in the following way.

- Objects: pairs (A, a) where $A \in |\mathscr{A}|$ and $a \in FA$.
- Arrows: $f\colon (A, a) \longrightarrow (A', a')$ is an arrow $f\colon A \longrightarrow A'$ in $\mathscr{A}$ such that $Ff(a) = a'$.

First we prove (1) $\Rightarrow$ (2). If $\mathbf{1} \in \mathscr{A}$ is the terminal object, $F(\mathbf{1})$ is the singleton $\{*\}$ so that $(\mathbf{1}, *)$ is an object of $\mathsf{Elts}(F)$.

Given (A, a) and (A', a') in $\mathsf{Elts}(F)$, one has $(a, a') \in FA \times FA' \cong F(A \times A')$, yielding an object $(A \times A', (a, a'))$ of $\mathsf{Elts}(F)$. The two projections of the product give the required arrows

$$(A, a) \xleftarrow{\ p_A\ } (A \times A', (a, a')) \xrightarrow{\ p_{A'}\ } (A', a').$$

If $f, g\colon (A, a) \rightrightarrows (A', a')$ are two morphisms of $\mathsf{Elts}(F)$, consider the equalizer $k\colon K \rightarrowtail A$ of the pair f, g in $\mathscr{A}$. Since $Fk = \mathsf{Ker}\,(Ff, Fg)$ in Set, the relation $Ff(a) = a' = Fg(a)$ implies $a \in FK$, yielding an object

250

$(K, a) \in \mathsf{Elts}(F)$ and a morphism $k: (K, a) \longrightarrow (A, a)$ which equalizes f and g.

To prove $(2) \Rightarrow (3)$, let us recall that F is the colimit object of the following composite:

$$\mathsf{Elts}(F) \xrightarrow{\ \phi\ } \mathscr{A} \xrightarrow{\ Y\ } \mathsf{Fun}(\mathscr{A}, \mathsf{Set}),$$

where $\phi(A, a) = A$ is the obvious forgetful functor and $Y(A) = \mathscr{A}(A, -)$ is the contravariant Yoneda embedding. Since the functor Y is contravariant, $(2) \Rightarrow (3)$ follows at once.

Finally we prove $(3) \Rightarrow (1)$. Suppose we are given a cofiltered category $\mathscr{C}$ and a functor θ such that F is the colimit object of the composite

$$\mathscr{C} \xrightarrow{\ \theta\ } \mathscr{A} \xrightarrow{\ Y\ } \mathsf{Fun}(\mathscr{A}, \mathsf{Set})$$

which we write $F = \mathrm{colim}\,_C \mathscr{A}(\theta C, -)$ for short. Consider a finite category $\mathscr{D}$ and a functor $\psi: \mathscr{D} \longrightarrow \mathscr{A}$. This yields a bifunctor

$$\Gamma: \mathscr{C} \times \mathscr{D} \longrightarrow \mathsf{Set}, \quad \Gamma(C, D) = \mathscr{A}(\theta C, \psi D).$$

Since representable functors commute with limits (see 2.9.4) and finite limits commute with filtered colimits in the category of sets (see 2.13.4), one deduces

$$
\begin{aligned}
F(\lim_D \psi D) &= \mathrm{colim}\,_C \mathscr{A}(\theta C, \lim_D \psi D) \\
&= \mathrm{colim}\,_C \lim_D \mathscr{A}(\theta C, \psi D) \\
&= \lim_D \mathrm{colim}\,_C \mathscr{A}(\theta C, \psi D) \\
&= \lim_D F(\psi D). \qquad \square
\end{aligned}
$$

Assuming now that $\mathscr{A}$ is small, we can consider the covariant Yoneda embedding

$$Y: \mathscr{A} \longrightarrow \mathsf{Fun}(\mathscr{A}^*, \mathsf{Set}), \quad Y(A) = \mathscr{A}(-, A).$$

Anticipating the results of section 6.2, proposition 6.1.2 can be completed as follows.

Proposition 6.1.3 *Let* $F: \mathscr{A} \longrightarrow \mathsf{Set}$ *be a functor defined on a small finitely complete category* $\mathscr{A}$. *The following conditions are equivalent:*

(1) F is left exact;

(2) the left Kan extension $\mathrm{Lan}_Y F$ *of* F *along the covariant Yoneda embedding is left exact.*

Proof   By 2.15.5, Y preserves limits. By 3.7.3, $(\operatorname{Lan}_Y F) \circ Y = F$. Therefore (2) certainly implies (1).

Conversely suppose $F \colon \mathscr{A} \longrightarrow \mathsf{Set}$ left exact and choose a contravariant functor $G \in \mathsf{Fun}(\mathscr{A}^*, \mathsf{Set})$. By 3.8.1 we have

$$(\operatorname{Lan}_Y F)(G) \cong (\operatorname{Lan}_{Y^*} G)(F).$$

To avoid heavy notation, let us make the convention that (A, a) runs through the category of elements of F. Using the axiom system of universes, let us assume that "set" means "belonging to some universe $\mathscr{U}$" and "class" implies "belonging to some universe $\mathscr{V} \ni \mathscr{U}$". The categories Set, $\mathsf{Fun}(\mathscr{A}^*, \mathsf{Set})$, and $\mathsf{Fun}\big(\mathsf{Fun}(\mathscr{A}^*, \mathsf{Set}), \mathsf{Set}\big)$ are then $\mathscr{V}$-small categories. Writing

$$\mathrm{ev}_A \colon \mathsf{Fun}(\mathscr{A}^*, \mathsf{Set}) \longrightarrow \mathsf{Set}$$

for the evaluation functor at $A \in |\mathscr{A}|$, the pointwise nature of colimits in $\mathsf{Fun}\big(\mathsf{Fun}(\mathscr{A}^*, \mathsf{Set}), \mathsf{Set}\big)$ implies (see 2.15.12)

$$\begin{aligned}
(\operatorname{Lan}_Y F)(G) &\cong (\operatorname{Lan}_{Y^*} G)(F) \\
&\cong \operatorname{colim}_{(A,a)} G(A) \\
&\cong \operatorname{colim}_{(A,a)} \big(\mathrm{ev}_A(G)\big) \\
&\cong \big(\operatorname{colim}_{(A,a)} \mathrm{ev}_A\big)(G).
\end{aligned}$$

But the evaluation functor ev_A is left exact since, by the Yoneda lemma (see 1.3.3), it is represented by $\mathscr{A}(A, -)$; see 2.9.4. On the other hand the category of elements of F is cofiltered, since F is left exact (see 6.1.2). Thus $\operatorname{Lan}_Y F \cong \operatorname{colim}_{(A,a)} \mathrm{ev}_A$ is a filtered colimit of left exact functors and therefore is left exact (see 6.2.2). $\square$

Proposition 6.1.4 *Consider a functor $F \colon \mathscr{A} \longrightarrow \mathscr{B}$ with $\mathscr{A}, \mathscr{B}$ finitely complete. The following conditions are equivalent:*

(1) F *is left exact;*

(2) $\forall B \in \mathscr{B}$ *the functor $\mathscr{B}(B, F-) \colon \mathscr{A} \longrightarrow \mathsf{Set}$ is left exact.*

Proof   (1) $\Rightarrow$ (2) since $\mathscr{B}(B, F-) = \mathscr{B}(B, -) \circ F$ and both functors F, $\mathscr{B}(B, -)$ are left exact (see 2.9.4).

Conversely suppose that each $\mathscr{B}(B, F-)$ is left exact and consider a functor $G \colon \mathscr{D} \longrightarrow \mathscr{A}$, with $\mathscr{D}$ finite. Write $\big(L, (p_D)_{D \in \mathscr{D}}\big)$ for the limit of G. We must prove that $\big(FL, (Fp_D)_{D \in \mathscr{D}}\big)$ is the limit of $F \circ G$. Given a cone $(q_D \colon B \longrightarrow FGD)_{D \in \mathscr{D}}$ on FG, we have in fact a compatible family

$\left(q_D \in \mathscr{B}(B, FGD)\right)_{D \in \mathscr{D}}$. By construction of limits in Set (see section 2.8), this compatible family corresponds to a unique element

$$q \in \lim_D \mathscr{B}(B, FG-) \cong \mathscr{B}(B, FL),$$

where the isomorphism holds since $\mathscr{B}(B, F-)$ preserves finite limits by assumption. In this way we have obtained a unique $q\colon B \longrightarrow FL$ such that $Fp_D \circ q = q_D$ for each $D \in \mathscr{D}$. $\qquad\square$

6.2 Left exact reflection of a functor

Now let us study more intensively the category $\mathsf{Lex}(\mathscr{A}, \mathsf{Set})$ of left exact functors on a small finitely complete category $\mathscr{A}$. The morphisms of $\mathsf{Lex}(\mathscr{A}, \mathsf{Set})$ are just the natural transformations, so that $\mathsf{Lex}(\mathscr{A}, \mathsf{Set})$ is a full subcategory of $\mathsf{Fun}(\mathscr{A}, \mathsf{Set})$. We intend to prove it is a reflective subcategory.

Proposition 6.2.1 *Let $\mathscr{A}$ be a small finitely complete category $\mathscr{A}$. The category $\mathsf{Lex}(\mathscr{A}, \mathsf{Set})$ is complete and limits are computed pointwise.*

Proof Given a functor $H\colon \mathscr{X} \longrightarrow \mathsf{Lex}(\mathscr{A}, \mathsf{Set})$, with $\mathscr{X}$ a small category, compute its pointwise limit $L\colon \mathscr{A} \longrightarrow \mathsf{Set}$. We must prove that L is left exact. Consider a functor $K\colon \mathscr{Y} \longrightarrow \mathscr{A}$ with $\mathscr{Y}$ finite; applying 2.12.1 and 2.15.2 we get

$$\begin{aligned}
L(\lim_Y KY) &\cong (\lim_X HX)(\lim_Y KY) \\
&\cong \lim_X(HX)(\lim_Y KY) \\
&\cong \lim_X \lim_Y (HX)(KY) \\
&\cong \lim_Y \lim_X (HX)(KY) \\
&\cong \lim_Y L(KY),
\end{aligned}$$

since each HX is by assumption left exact. (We have written $\lim_X$ and $\lim_Y$ for short to denote the limits of functors defined respectively on $\mathscr{X}$ or $\mathscr{Y}$.) $\qquad\square$

Proposition 6.2.2 *Let $\mathscr{A}$ be a small finitely complete category. The category $\mathsf{Lex}(\mathscr{A}, \mathsf{Set})$ has filtered colimits: they are computed pointwise, are universal and commute with finite limits.*

Proof Given a functor $H\colon \mathscr{X} \longrightarrow \mathsf{Lex}(\mathscr{A}, \mathsf{Set})$ with $\mathscr{X}$ a filtered category, compute its pointwise colimit $L\colon \mathscr{A} \longrightarrow \mathsf{Set}$. We must prove that L is left exact. Consider a functor $K\colon \mathscr{Y} \longrightarrow \mathscr{A}$ with $\mathscr{Y}$ finite; applying

2.13.4 and 2.15.2 and using abbreviations analogous to that of 6.2.1, we get

$$L(\lim_Y KY) \cong (\operatorname{colim}_X HX)(\lim_Y KY)$$
$$\cong \operatorname{colim}_X (HX)(\lim_Y KY)$$
$$\cong \operatorname{colim}_X \lim_Y (HX)(KY)$$
$$\cong \lim_Y \operatorname{colim}_X (HX)(KY)$$
$$\cong \lim_Y L(KY).$$

Finite limits and filtered colimits are computed pointwise in the category $\mathsf{Lex}(\mathscr{A}, \mathsf{Set})$; their commutativity and the universality of filtered colimits are thus a consequence of the corresponding results in Set (see 2.13.4 and 2.14.2). $\qquad\square$

Proposition 6.2.3 *Let $\mathscr{A}$ be a small finitely complete category. The contravariant Yoneda embedding*

$$Y\colon \mathscr{A} \longrightarrow \mathsf{Lex}(\mathscr{A}, \mathsf{Set}); \quad A \mapsto \mathscr{A}(A, -)$$

transforms finite limits into finite colimits. When $\mathscr{A}$ is finitely cocomplete, Y transforms also finite colimits into finite limits.

Proof Let $H\colon \mathscr{X} \longrightarrow \mathscr{A}$ be a functor defined on a finite category $\mathscr{X}$; write $\big(L, (p_X)_{X\in\mathscr{X}}\big)$ for its limit in $\mathscr{A}$. We know $\big(YL, (Yp_X)_{X\in\mathscr{X}}\big)$ is a cocone on $Y \circ H$, because Y is contravariant. Consider another cocone

$$\big(\alpha_X\colon \mathscr{A}(HX, -) \Rightarrow F\big)_{X\in\mathscr{X}}$$

where F is left exact. Applying the Yoneda lemma (see 1.3.3), this corresponds to a compatible family of elements $\big(a_X \in F(HX)\big)_{X\in\mathscr{X}}$. Since F is left exact, $\big(FL, (Fp_X)_{X\in\mathscr{X}}\big)$ is a finite limit in Set. By construction of a limit in Set (see section 2.8), this yields a unique element $a \in F(L)$ such that $(Fp_X)(a) = a_X$ for each $X \in \mathscr{X}$. By the Yoneda lemma again, a corresponds to a natural transformation $\alpha\colon \mathscr{A}(L, -) \Rightarrow F$ which by the naturality of the Yoneda isomorphisms satisfies $\alpha \circ \mathscr{A}(p_X, -) = \alpha_X$. The uniqueness of such a factorization α is an immediate consequence of the uniqueness of the corresponding element a. The last statement is a direct consequence of 2.9.5, 2.15.2 and 6.2.1. $\qquad\square$

Proposition 6.2.4 *Let $\mathscr{A}$ be a small finitely complete category. The category $\mathsf{Lex}(\mathscr{A}, \mathsf{Set})$ is cocomplete.*

Proof The first proof one can think of is showing that $\mathsf{Lex}(\mathscr{A}, \mathsf{Set})$ is a reflective subcategory of the category $\mathsf{Fun}(\mathscr{A}, \mathsf{Set})$ of all functors from

$\mathscr{A}$ to Set, from which Lex($\mathscr{A}$, Set) is cocomplete (see 3.5.4 and 2.15.2). Considering 6.2.1 and the adjoint functor theorem (see 3.3.3), it remains to prove the solution set condition. Given two functors $F, G\colon \mathscr{A} \rightrightarrows$ Set with F left exact and a natural transformation $\alpha\colon G \Rightarrow F$, one considers first the set S_0 of all elements of the form $\alpha_A(x)$ for all $A \in \mathscr{A}$ and $x \in GA$. It is quite straightforward to construct the smallest left exact subfunctor $H \subseteq F$ containing all the elements of S_0:

- given a compatible family $(x_i)_{i \in I}$ of elements in S_0 along a finite diagram $(D_i)_{i \in I}$ in $\mathscr{A}$, one adds to S_0 the corresponding element $x \in \lim_i (FD_i) = F(\lim_i D_i)$; this yields a bigger set S_0' of elements in F;
- given an element $y \in FA$ in S_0' and an arrow $f\colon A \longrightarrow B$ in $\mathscr{A}$, one adds to S_0' the element $(Ff)(y)$; this yields a bigger set S_1 of elements in F.

One repeats the same operations on S_1 to get a bigger set S_2 and so on by induction on the integers. Finally one considers $S = \bigcup_{n \in \mathbb{N}} S_n$ and since this union is filtered, it follows easily that S is exactly the set of all elements $y \in H(A)$, $A \in \mathscr{A}$, for some left exact subfunctor $H \subseteq F$ containing all the original elements $\alpha_A(x)$; in particular α factors through H. Some routine arithmetic on infinite cardinals shows that the cardinality of $\coprod_{A \in \mathscr{A}} H(A)$ is smaller than some cardinal β depending only on the cardinality of the set of arrows of $\mathscr{A}$ and that of the set $\coprod_{A \in \mathscr{A}} G(A)$; the crucial point is that the cardinal β does not depend at all on F. Straightforward cardinality arguments show that (up to isomorphisms) there is just a set of left exact functors H such that $\coprod_{A \in \mathscr{A}} H(A)$ has cardinality less than β. The previous development lets us then conclude that those H's constitute a solution set for G.

Instead of developing the details of the proof we have just sketched, we prefer an alternative "constructive proof" avoiding cardinal arithmetic. The advantages of such a "constructive proof" are its possible generalizations, for example in the context of toposes (see chapter 6, volume 3).

An arbitrary colimit can be written as a filtered colimit of finitely generated colimits (see 2.13.7). Since filtered colimits exist (see 6.2.2), it suffices to prove the existence of finitely generated colimits. But finitely generated colimits can be computed from finite coproducts and coequalizers (see 2.8.5).

If **1** is the terminal object of $\mathscr{A}$, the representable functor $\mathscr{A}(\mathbf{1}, -)$ is the initial object of Lex($\mathscr{A}$, Set); see 6.2.3.

If $F, G\colon \mathscr{A} \rightrightarrows \mathsf{Set}$ are left exact, consider the categories of elements of F and G and the corresponding forgetful functors $\phi\colon \mathsf{Elts}(F) \longrightarrow \mathscr{A}$, $\psi\colon \mathsf{Elts}(G) \longrightarrow \mathscr{A}$. Since the categories $\mathsf{Elts}(F)$ and $\mathsf{Elts}(G)$ are both cofiltered (see 6.1.2), the product category $\mathsf{Elts}(F) \times \mathsf{Elts}(G)$ is obviously cofiltered. Let us consider the composite

$$\mathsf{Elts}(F) \times \mathsf{Elts}(G) \xrightarrow{\phi \times \psi} \mathscr{A} \xrightarrow{Y} \mathsf{Fun}(\mathscr{A}, \mathsf{Set})$$

where $(\phi \times \psi)\big((A, a), (B, b)\big) = \phi(A, a) \times \psi(B, b)$ and $Y(A) = \mathscr{A}(A, -)$. Since the domain category is cofiltered,

$$\operatorname{colim}\big(Y \circ (\phi \times \psi)\big) = \big(H, (s_{(A,a)(B,b)})\big)$$

where H is left exact (see 6.1.2). But the morphisms

$$\mathscr{A}(A, -) \xrightarrow{\mathscr{A}(p_A, -)} \mathscr{A}(A \times B, -) \xrightarrow{s_{(A,a)(B,b)}} H$$

constitute a cocone on $Y \circ \phi$, just because the morphisms $s_{(A,a)(B,b)}$ constitute a cocone on $Y \circ (\phi \times \psi)$; this induces a factorization $s_1\colon F \longrightarrow H$ through $F = \operatorname{colim}(Y \circ \phi)$. In the same way we get $s_2\colon G \longrightarrow H$. Let us prove that (H, s_1, s_2) is the coproduct of F, G in $\mathsf{Lex}(\mathscr{A}, \mathsf{Set})$.

Choose a left exact functor K and natural transformations $\alpha\colon F \Rightarrow K$, $\beta\colon G \Rightarrow K$. The composites

$$\mathscr{A}(A, -) \xrightarrow{s_{(A,a)}} F \xrightarrow{\alpha} K, \quad \mathscr{A}(B, -) \xrightarrow{s_{(B,b)}} G \xrightarrow{\beta} K$$

yield a factorization

$$\mathscr{A}(A \times B, -) \xrightarrow{\cong} \mathscr{A}(A, -) \amalg \mathscr{A}(B, -) \xrightarrow{\gamma_{(A,a)(B,b)}} K;$$

see 6.2.3. Since $s_{(A,a)}$ and $s_{(B,b)}$ constitute cocones, $\gamma_{(A,a)(B,b)}$ is a cocone on $Y \circ (\phi \times \psi)$, from which we may obtain a unique factorization $\gamma\colon H \longrightarrow K$ such that

$$\gamma \circ s_{(A,a)(B,b)} \cong \gamma_{(A,a)(B,b)}.$$

The relations

$$\gamma \circ s_1 \circ s_{(A,a)} = \gamma \circ s_{(A,a)(B,b)} \circ \mathscr{A}(p_A, -)$$
$$\cong \gamma_{(A,a)(B,b)} \circ \mathscr{A}(p_A, -)$$
$$= \alpha \circ s_{(A,a)}$$

yield $\gamma \circ s_1 = \alpha$. In the same way, $\gamma \circ s_2 = \beta$. The uniqueness of γ results immediately from that of $\gamma_{(A,a)(B,b)}$.

To prove the existence of coequalizers in $\mathsf{Lex}(\mathscr{A}, \mathsf{Set})$, let us first consider two natural transformations $\alpha, \beta \colon \mathscr{A}(C, -) \rightrightarrows F$ with F left exact. Let us write F as the filtered colimit

$$F = \operatorname{colim}_{(A,a)\in\mathrm{ELTS}(F)} \mathscr{A}(A, -).$$

By the Yoneda lemma (see 1.3.3), α, β correspond to elements

$$a, b \in FC = \operatorname{colim}_{(A,a)} \mathscr{A}(A, C).$$

a is thus the equivalence class of some element $f \colon A_1 \longrightarrow C$ for an index (A_1, a_1), while b is the equivalence class of some element $g \colon A_2 \longrightarrow C$ for an index (A_2, a_2). Since the colimit is filtered, there is no restriction in choosing $(A_1, a_1) = (A_2, a_2)$. In this way we have obtained

$$\mathscr{A}(C, -) \underset{\mathscr{A}(g,-)}{\overset{\mathscr{A}(f,-)}{\rightrightarrows}} \mathscr{A}(A_1, -) \xrightarrow{s_{(A_1,a_1)}} F,$$

with $s_{(A_1,a_1)} \circ \mathscr{A}(f, -) = \alpha$ and $s_{(A_1,a_1)} \circ \mathscr{A}(g, -) = \beta$. By 6.2.3, the following diagram is a coequalizer in $\mathsf{Lex}(\mathscr{A}, \mathsf{Set})$,

$$\mathscr{A}(C, -) \underset{\mathscr{A}(g,-)}{\overset{\mathscr{A}(f,-)}{\rightrightarrows}} \mathscr{A}(A_1, -) \xrightarrow{\mathscr{A}(k,-)} \mathscr{A}(K, -),$$

where $k = \mathsf{Ker}\,(f, g)$.

Now given any morphism $a \colon (A_3, a_3) \longrightarrow (A_1, a_1)$, we can consider in the same way the coequalizer

$$\mathscr{A}(C, -) \underset{\mathscr{A}(g \circ a,-)}{\overset{\mathscr{A}(f \circ a,-)}{\rightrightarrows}} \mathscr{A}(A_3, -) \xrightarrow{\mathscr{A}(k_a,-)} \mathscr{A}(K_a, -),$$

where $k_a = \mathsf{Ker}\,(f \circ a, g \circ a)$. Given an additional morphism b,

$$(A_4, a_4) \xrightarrow{\ b\ } (A_3, a_3) \xrightarrow{\ a\ } (A_1, a_1),$$

the relation $f \circ a \circ b \circ k_{aob} = g \circ a \circ b \circ k_{aob}$ implies the existence of a unique factorization $b' \colon K_{aob} \longrightarrow b \circ K_a$ such that $k_a \circ b' = b \circ k_{aob}$. In this way we have defined three functors from $\mathsf{Elts}(F)/(A_1, a_1)$ to $\mathsf{Lex}(\mathscr{A}, \mathsf{Set})$:

- the constant functor on $\mathscr{A}(C, -)$;
- the functor mapping $((A_3, a_3), a)$ to $\mathscr{A}(A_3, -)$;
- the functor mapping $((A_3, a_3), a)$ to $\mathscr{A}(K_a, -)$.

The obvious functor

$$\pi \colon \mathsf{Elts}(F)/(A_1, a_1) \longrightarrow \mathsf{Elts}(F), \quad ((A_3, a_3), a) \mapsto (A_3, a_3)$$

is cofinal, just because $\mathsf{Elts}(F)$ is cofiltered (see 2.11.2). The colimit of the composite $Y \circ \phi \circ \pi$ is thus again F (see 2.11.1). Observe moreover that $\mathsf{Elts}(F)/(A_1, a_1)$ is still cofiltered, which implies that the three functors we have just described have a colimit (see 6.2.2). Computing those colimits thus yields a diagram in $\mathsf{Lex}(\mathscr{A}, \mathsf{Set})$,

$$\mathscr{A}(C, -) \underset{\beta}{\overset{\alpha}{\rightrightarrows}} F \xrightarrow{\gamma} Q.$$

Finally $\gamma = \mathsf{Coker}\,(\alpha, \beta)$, by the interchange property of colimits (see 2.12.1), thus (α, β) indeed have a coequalizer.

Now consider two arbitrary morphisms $\alpha, \beta \colon G \rightrightarrows F$ in $\mathsf{Lex}(\mathscr{A}, \mathsf{Set})$ and write G as a filtered colimit

$$G = \mathrm{colim}_{\,(B,b) \in \mathsf{Elts}(G)}\,\mathscr{A}(B, -).$$

For each index (B, b) we have a diagram

$$\mathscr{A}(B, -) \xrightarrow{s_{(B,b)}} G \underset{\beta}{\overset{\alpha}{\rightrightarrows}} F \xrightarrow{q_{(B,b)}} Q_{(B,b)},$$

where $q_{(B,b)} = \mathsf{Coker}\,(\alpha s_{(B,b)}, \beta s_{(B,b)})$. Observe that given a morphism $f \colon (B, b) \longrightarrow (B', b')$ in $\mathsf{Elts}(G)$, the relations

$$\begin{aligned}
q_{(B,b)} \circ \alpha \circ s_{(B',b')} &= q_{(B,b)} \circ \alpha \circ s_{(B,b)} \circ \mathscr{A}(f, -) \\
&= q_{(B,b)} \circ \beta \circ s_{(B,b)} \circ \mathscr{A}(f, -) \\
&= q_{(B,b)} \circ \beta \circ s_{(B',b')}
\end{aligned}$$

imply the existence of a unique morphism $Q_f \colon Q_{(B',b')} \longrightarrow Q_{(B,b)}$ such that $Q_f \circ q_{(B',b')} = q_{(B,b)}$. This defines a contravariant functor

$$Q \colon \mathsf{Elts}(G) \longrightarrow \mathsf{Lex}(\mathscr{A}, \mathsf{Set}),$$

which has a colimit $\left(H, (\sigma_{(B,b)})\right)$ in $\mathsf{Lex}(\mathscr{A}, \mathsf{Set})$ since $\mathsf{Elts}(G)$ is cofiltered (see 6.2.2). With the previous notations, observe that

$$\sigma_{(B,b)} \circ q_{(B,b)} = \sigma_{(B,b)} \circ Q_f \circ q_{(B',b')} = \sigma_{(B',b')} \circ q_{(B',b')}.$$

Thus if two objects (B, b), (B', b') of $\mathsf{Elts}(G)$ are connected by a morphism, the composites $\sigma_{(B,b)} \circ q_{(B,b)}$ and $\sigma_{(B',b')} \circ q_{(B',b')}$ are equal. Since $\mathsf{Elts}(G)$ is cofiltered, it is connected (see example 2.6.7.e) and finally the composites $\sigma_{(B,b)} \circ q_{(B,b)}$ and $\sigma_{(B',b')} \circ q_{(B',b')}$ are equal for every two objects of $\mathsf{Elts}(G)$. This defines a morphism

$$h \colon F \longrightarrow H, \quad h = \sigma_{(B,b)} \circ q_{(B,b)},$$

and we shall prove that $h = \mathsf{Coker}\,(\alpha, \beta)$ in $\mathsf{Lex}(\mathscr{A}, \mathsf{Set})$.

First of all one has

$$h \circ \alpha \circ s_{(B,b)} = \sigma_{(B,b)} \circ q_{(B,b)} \circ \alpha \circ s_{(B,b)}$$
$$= \sigma_{(B,b)} \circ q_{(B,b)} \circ \beta \circ s_{(B,b)}$$
$$= h \circ \beta \circ s_{(B,b)},$$

from which $h \circ \alpha = h \circ \beta$, since the $s_{(B,b)}$ constitute a colimit cocone. Next given $K \in \mathsf{Lex}(\mathscr{A}, \mathsf{Set})$ and $k \colon F \longrightarrow K$ such that $k \circ \alpha = k \circ \beta$, the relation $k \circ \alpha \circ s_{(B,b)} = k \circ \beta \circ s_{(B,b)}$ implies the existence of a unique $l_{(B,b)} \colon Q_{(B,b)} \longrightarrow K$ such that $l_{(B,b)} \circ q_{(B,b)} = k$. Observe that given $f \colon (B,b) \longrightarrow (B',b')$, the relations

$$l_{(B,b)} \circ Q_f \circ q_{(B',b')} = l_{(B,b)} \circ q_{(B,b)} = k = l_{(B',b')} \circ q_{(B',b')}$$

imply $l_{(B,b)} \circ Q_f = l_{(B',b')}$, since $q_{(B',b')}$ is an epimorphism. Thus the morphisms $l_{(B,b)}$ constitute a cocone on Q, from which there is a unique $l \colon H \longrightarrow K$ such that $l \circ \sigma_{(B,b)} = l_{(B,b)}$ for each (B,b). In particular

$$l \circ h = l \circ \sigma_{(B,b)} \circ q_{(B,b)} = l_{(B,b)} \circ q_{(B,b)} = k.$$

The uniqueness of l follows immediately from that of $l_{(B,b)}$. $\qquad\square$

Theorem 6.2.5 *Let $\mathscr{A}$ be a small category. The category $\mathsf{Lex}(\mathscr{A}, \mathsf{Set})$ of left exact functors is reflective in the category $\mathsf{Fun}(\mathscr{A}, \mathsf{Set})$ of all functors.*

Proof A first try, reducing the problem to the adjoint functor theorem (see 3.3.3), has been sketched at the beginning of the previous proof. Let us instead give a constructive proof.

Consider an arbitrary functor $F \colon \mathscr{A} \longrightarrow \mathsf{Set}$. Define $\left(\widetilde{F}, \left(s_{(A,a)} \right) \right)$ to be the colimit of the composite

$$\mathsf{Elts}(F) \overset{\phi}{\longrightarrow} \mathscr{A} \overset{Y}{\longrightarrow} \mathsf{Lex}(\mathscr{A}, \mathsf{Set})$$

where $\mathsf{Elts}(F)$ is the category of elements of F and ϕ is the corresponding forgetful functor (see 6.2.4). We know by 2.15.6 that F itself can be obtained via the colimit $\left(F, (\sigma_{(A,a)}) \right)$ of the composite

$$\mathsf{Elts}(F) \overset{\phi}{\longrightarrow} \mathscr{A} \overset{Y}{\longrightarrow} \mathsf{Lex}(\mathscr{A}, \mathsf{Set}) \overset{i}{\longrightarrow} \mathsf{Fun}(\mathscr{A}, \mathsf{Set})$$

where $Y(A) = \mathscr{A}(A, -)$ and i is the canonical inclusion. The morphisms $s_{(A,a)}$ constitute a cocone on $iY\phi$ in $\mathsf{Fun}(\mathscr{A}, \mathsf{Set})$, from which a unique factorization $\varphi \colon F \longrightarrow \widetilde{F}$ such that $\varphi \circ \sigma_{(A,a)} = s_{(A,a)}$ for every object $(A,a) \in \mathsf{Elts}(F)$. We shall prove that $(\widetilde{F}, \varphi)$ is the left exact reflection of F.

If G is left exact and $\psi\colon F \Rightarrow G$ is a natural transformation, the composites $\psi \circ \sigma_{(A,a)}$ constitute a cocone on $Y \circ \phi$ in $\mathsf{Lex}(\mathscr{A}, \mathsf{Set})$, from which there is a unique $\theta\colon \widetilde{F} \Rightarrow G$ such that $\theta \circ s_{(A,a)} = \psi \circ \sigma_{(A,a)}$ for every index (A, a). In particular

$$\theta \circ \varphi \circ \sigma_{(A,a)} = \theta \circ s_{(A,a)} = \psi \circ \sigma_{(A,a)},$$

from which $\theta \circ \varphi = \psi$ since the morphisms $(\sigma_{(A,a)})_{(A,a)}$ constitute a colimit cocone in the category $\mathsf{Fun}(\mathscr{A}, \mathsf{Set})$. Now if $\theta'\colon \widetilde{F} \Rightarrow G$ is another natural transformation such that $\theta' \circ \varphi = \psi$, the relations

$$\theta' \circ s_{(A,a)} = \theta' \circ \varphi \circ \sigma_{(A,a)} = \psi \circ \sigma_{(A,a)} = \theta \circ s_{(A,a)}$$

imply $\theta = \theta'$, since the morphisms $s_{(A,a)}$ constitute a colimit cocone in $\mathsf{Lex}(\mathscr{A}, \mathsf{Set})$.   □

6.3 Flat functors

Some functors are bound to preserve limits, like representable functors (see 2.9.4), functors having a left adjoint (see 3.2.2) or covariant Yoneda embeddings (see 2.15.5). Most people think of the notion of *flat functor* as that of a functor *which would preserve finite limits if they existed*; this intuition can in some way be justified by 6.3.7. But even if no one considers the previous sentence as a definition (of course!), it is nevertheless misleading and can give a truncated intuition of what a *flat functor* actually is. For example the covariant Yoneda embeddings we have just mentioned are not flat in general (see 6.7.10). It is true that a flat functor preserves all finite limits which turn out to exist (see 6.7.5) and it is also true that *being flat* reduces to *preserving finite limits* when the categories considered do have finite limits. But when not all finite limits exist, being flat is more subtle than the rough idea of *preserving finite limits if they existed*. With a look on 6.1.2 and 6.1.4 we define

Definition 6.3.1 *For an arbitrary category $\mathscr{A}$, a functor $F\colon \mathscr{A} \longrightarrow \mathsf{Set}$ is flat when the category $\mathsf{Elts}(F)$ of elements of F is cofiltered. Given an arbitrary functor $F\colon \mathscr{A} \longrightarrow \mathscr{B}$, F is flat when for each object $B \in \mathscr{B}$, the functor $\mathscr{B}(B, F-)\colon \mathscr{A} \longrightarrow \mathsf{Set}$ is flat.*

It follows immediately from 6.1.2 and 6.1.4 that

Proposition 6.3.2 *Let $F\colon \mathscr{A} \longrightarrow \mathscr{B}$ be a functor, with $\mathscr{A}$, $\mathscr{B}$ finitely complete. The following conditions are equivalent:*

(1) F is left exact;

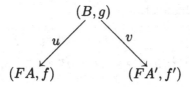

Diagram 6.1

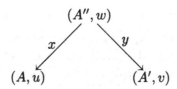

Diagram 6.2

(2) F is flat. □

Proposition 6.3.3 *Let $F: \mathscr{A} \longrightarrow \mathscr{B}$ and $G: \mathscr{B} \longrightarrow \mathscr{C}$ be flat functors. The composite $G \circ F$ is a flat functor as well.*

Proof Fix an object $C \in \mathscr{C}$ and consider the functor $\mathscr{C}(C, GF-)$. The category of elements of $\mathscr{C}(C, G-)$ is non-empty, which means the existence of $B \in \mathscr{B}$ and $f: C \longrightarrow GB$. The category of elements of $\mathscr{B}(B, F-)$ is non-empty, which means the existence of $A \in \mathscr{A}$ and $g: B \longrightarrow FA$. This yields the composite $(Gg) \circ f: C \longrightarrow GFA$, thus a pair $(A, (Gg) \circ f)$ in the category of elements of $\mathscr{C}(C, GF-)$.

Next consider two pairs (A, f), (A', f') in the category of elements of $\mathscr{C}(C, GF-)$; thus

$$f: C \longrightarrow GFA, \quad f': C \longrightarrow GFA'$$

are arrows of $\mathscr{C}$. This yields the pairs (FA, f) and (FA', f') in the category of elements of $\mathscr{C}(C, G-)$. By cofilteredness, we find (B, g) in this same category and morphisms u, v as in diagram 6.1, i.e. in the category $\mathscr{B}$, morphisms $u: B \longrightarrow FA$, $v: B \longrightarrow FA'$ such that $(Gu) \circ g = f$ and $(Gv) \circ g = f'$. The two pairs (A, u), (A', v) are now objects in the category of elements of $\mathscr{B}(B, F-)$. By cofilteredness we find (A'', w) in this same category and morphisms x, y as in diagram 6.2, thus in the category $\mathscr{A}$, morphisms $x: A'' \longrightarrow A$ and $y: A'' \longrightarrow A'$ such that $(Fx) \circ w = u$, $(Fy) \circ w = v$. Finally the pair $(A'', (Gw) \circ g)$ is in the category of elements of $\mathscr{C}(C, GF-)$ and the relations

$$(GFx) \circ Gw \circ g = Gu \circ g = f,$$
$$(GFy) \circ Gw \circ g = Gv \circ g = f',$$

show that we have produced morphisms

$$(A, f) \xleftarrow{\quad x \quad} (A'', (Gw) \circ g) \xrightarrow{\quad y \quad} (A', f')$$

as required.

Finally consider $x, y: (A, f) \rightrightarrows (A', f')$ in the category of elements of $\mathscr{C}(C, GF-)$; thus $x, y: A \rightrightarrows A'$ are such that $(GFx) \circ f = f' = (GFy) \circ f$. This yields the morphisms Fx, Fy in the category of elements of $\mathscr{C}(C, G-)$:

$$(B, g) \xrightarrow{\quad u \quad} (FA, f) \underset{Fy}{\overset{Fx}{\rightrightarrows}} (FA', f').$$

By cofilteredness, we get an arrow u in this same category, such that $(Fx) \circ u = (Fy) \circ u$; observe that $(Gu) \circ g = f$. Writing $(Fx) \circ u = v = (Fy) \circ u$, we now have two arrows $x, y: (A, u) \rightrightarrows (A', v)$ in the category of elements of $\mathscr{B}(B, F-)$. By cofilteredness we get an arrow $z: (A'', w) \longrightarrow (A, u)$ in this same category, such that $x \circ z = y \circ z$; notice that $(Fz) \circ w = u$. The relation $GFz \circ Gw \circ g = Gu \circ g = f$ indicates that we have produced a morphism $z: (A'', (Gw) \circ g) \longrightarrow (A, f)$ in the category of elements of $\mathscr{C}(C, GF-)$, with $x \circ z = y \circ z$. □

Proposition 6.3.4 *Given a category $\mathscr{A}$, every representable functor*

$$\mathscr{A}(A, -): \mathscr{A} \longrightarrow \mathsf{Set}$$

is flat.

Proof   The object $(A, 1_A)$ is initial in the category of elements of $\mathscr{A}(A, -)$. Indeed given any other object (A', f) of this category, a morphism $g: (A, 1_A) \longrightarrow (A', f)$ is an arrow $g: A \longrightarrow A'$ such that $g \circ 1_A = f$; f is of course the unique such arrow. A category with an initial object is obviously cofiltered. □

Proposition 6.3.5 *Let $F: \mathscr{A} \longrightarrow \mathscr{B}$ be a functor with a left adjoint. Then F is flat.*

Proof   We must prove that $\mathscr{B}(B, F-)$ is flat for each $B \in \mathscr{B}$; see 6.3.1. Since $\mathscr{B}(B, F-) \cong \mathscr{A}(GB, -)$ by 3.1.5, the result follows at once from 6.3.4. □

Proposition 6.3.6 *Given a small category $\mathscr{A}$, the category* $\mathsf{Flat}(\mathscr{A}, \mathsf{Set})$
of flat functors from $\mathscr{A}$ to Set has pointwise filtered colimits.

Proof Consider a filtered small category $\mathscr{D}$ and a functor

$$H \colon \mathscr{D} \longrightarrow \mathsf{Flat}(\mathscr{A}, \mathsf{Set}).$$

Write $\left(F, (s_D)_{D \in \mathscr{D}}\right)$ for the (pointwise) colimit of the composite

$$\mathscr{D} \xrightarrow{\ H\ } \mathsf{Flat}(\mathscr{A}, \mathsf{Set}) \xrightarrow{\ i\ } \mathsf{Fun}(\mathscr{A}, \mathsf{Set});$$

thus $s_D \colon HD \longrightarrow F$. We must prove that F is flat.

As $\mathscr{D}$ is non-empty, choose $D \in \mathscr{D}$. Since HD is flat, $\mathsf{Elts}(HD)$ is non-empty and we find $(A, a) \in \mathsf{Elts}(HD)$, i.e. $a \in HD(A)$. So $s_D(a) \in FA$ and $\left(A, s_D(a)\right) \in \mathsf{Elts}(F)$.

Next consider (A, a) and (B, b) in $\mathsf{Elts}(F)$. One has

$$a \in FA = \operatorname{colim}(HD)(A),$$

thus $a = s_{DA}(a')$ for some $a' \in (HD)(A)$; in the same way $b = s_{D'B}(b')$ for some $b' \in (HD')(B)$. Since $\mathscr{D}$ is filtered, there is no restriction in choosing $D = D'$. So (A, a') and (B, b') are objects of $\mathsf{Elts}(HD)$, which is cofiltered since HD is flat. We can thus choose (C, c) in $\mathsf{Elts}(HD)$, together with

$$u \colon (C, c) \longrightarrow (A, a'), \quad v \colon (C, c) \longrightarrow (B, b'),$$

yielding $(HD)(u)(c) = a'$ and $(HD)(v)(c) = b'$. One has $c \in (HD)(C)$, thus $s_{DC}(c) \in FC$ and by the naturality of s_D,

$$F(u)\left(s_{DC}(c)\right) = \left(s_{DA} \circ HD(u)\right)(c) = s_{DA}(a') = a.$$

So u is a morphism in $\mathsf{Elts}(F)$:

$$u \colon \left(C, s_{DC}(c)\right) \longrightarrow (A, a), \quad v \colon \left(C, s_{DC}(c)\right) \longrightarrow (B, b),$$

and similarly for v.

Finally choose $u, v \colon (A, a) \rightrightarrows (B, b)$ in $\mathsf{Elts}(F)$, i.e. $a \in FA$, $b \in FB$ and $(Fu)(a) = b = (Fv)(a)$. As before write $a = s_{DA}(a')$, $b = s_{DB}(b')$ for some $D \in \mathscr{D}$, $a' \in (HD)(A)$, $b' \in (HD)(B)$. One has

$$\left(s_{DB} \circ HD(u)\right)(a') = (Fu \circ s_{DA})(a') = (Fu)(a) = b$$

and analogously for v. Thus

$$\left(s_{DB} \circ HD(u)\right)(a') = b = \left(s_{DB} \circ HD(v)\right)(a')$$

and the filteredness of $\mathscr{D}$ implies the existence of $d\colon D \longrightarrow D'$ such that

$$((Hd)_B \circ HD(u))(a') = ((Hd)_B \circ HD(v))(a');$$

see 2.13.3. Write b' for that element of $HD'(B)$. Since the morphisms s_D constitute a cocone,

$$s_{D'B}(b') = (s_{D'B} \circ (Hd)_B \circ (HD)(u))(a') = (s_{DB} \circ (HD)(u))(a') = b.$$

By the naturality of Hd we also have

$$b' = ((Hd)_B \circ HD(u))(a') = (HD'(u) \circ (Hd)_A)(a')$$

and similarly for v. This implies

$$HD'(u)((Hd)_A(a')) = b' = HD'(v)((Hd)_A(a')),$$

so that $u, v\colon \big(A, (Hd)_A(a')\big) \rightrightarrows (B, b')$ are morphisms in $\mathsf{Elts}(HD')$. But $\mathsf{Elts}(HD')$ is cofiltered since HD' is flat, thus we can choose

$$w\colon (C, c) \longrightarrow \big(A, (Hd)_A(a')\big)$$

in $\mathsf{Elts}(HD')$ such that $u \circ w = v \circ w$. Since $c \in (HD')(C)$, one has $s_{D'C}(c) \in F(C)$ and $w\colon (C, c) \longrightarrow (A, a)$ is a morphism in $\mathsf{Elts}(F)$ because

$$\begin{aligned} F(w)(c) &= \big(F(w) \circ s_{D'C}\big)(c) = \big(s_{D'A} \circ HD'(w)\big)(c) \\ &= \big(s_{D'A} \circ (Hd)_A\big)(a') = s_{DA}(a') = a. \end{aligned}$$

On the other hand we know already that $u \circ w = v \circ w$. $\qquad\square$

As a consequence, we can generalize 6.1.2 to the case of flat functors.

Proposition 6.3.7 *Let $F\colon \mathscr{A} \longrightarrow \mathsf{Set}$ be a functor defined on an arbitrary category $\mathscr{A}$. The following conditions are equivalent:*
(1) F is flat;
(2) F is a filtered colimit of representable functors.

Proof $(1) \Rightarrow (2)$ is immediate by 6.2.1 and 2.15.6. Conversely, suppose we are given a cofiltered category $\mathscr{D}$ and a functor ψ such that F is the colimit object of the composite

$$\mathscr{D} \xrightarrow{\ \psi\ } \mathscr{A} \xrightarrow{\ Y\ } \mathsf{Fun}(\mathscr{A}, \mathsf{Set}).$$

We shall prove that $\mathsf{Elts}(F)$ is cofiltered.

First of all $\mathscr{D}$ is not empty; thus choose $D_0 \in \mathscr{D}$. Since $F(\psi D_0) = \mathrm{colim}_D \mathscr{A}(\psi D, \psi D_0)$, it suffices to choose $D = D_0$ in order to get an element $[1_{\psi D_0}] \in F(\psi D_0)$, thus an object $(\psi D_0, [1_{\psi D_0}])$ of $\mathsf{Elts}(F)$.

Next choose (A, a) and (A', a') in $\mathsf{Elts}(F)$. Since $a \in \operatorname{colim}_D(\psi D, A)$ and $a' \in \operatorname{colim}_D(\psi D, A')$, a is represented by $g: \psi D_1 \longrightarrow A$ and a' by $g': \psi D_2 \longrightarrow A'$. Since $\mathscr{D}$ is cofiltered, choose $D_3 \in \mathscr{D}$ with $d: D_3 \longrightarrow D_1$ and $d': D_3 \longrightarrow D_2$. Observe that $F(\psi D_3) = \operatorname{colim}_D(\psi D, \psi D_3)$, so that choosing $D = D_3$ we find an element $a_3 = [1_{\psi D_3}] \in F(\psi D_3)$; this gives an object $(\psi D_3, a_3)$ of $\mathsf{Elts}(F)$. Moreover

$$F(g \circ \psi d)(a_3) = [g \circ \psi d \circ 1_{\psi D_3}] = [g] = a$$

and in the same way $F(g' \circ \psi d')(a_3) = a'$. This yields two morphisms of $\mathsf{Elts}(F)$

$$(A, a) \xleftarrow{\;g \circ \psi d\;} (\psi D_3, a_3) \xrightarrow{\;g' \circ \psi d'\;} (A', a')$$

as required.

Now consider $f, g: (A, a) \rightrightarrows (A', a')$ in $\mathsf{Elts}(F)$. Again a is represented by some $h: \psi D_1 \longrightarrow A$ and a' by some $h': \psi D_2 \longrightarrow A'$. The equality $(Ff)(a) = a'$ means the existence of $d_1: D_2 \longrightarrow D_1$ in $\mathscr{D}$ such that $f \circ h \circ \psi d_1 = h' \circ \psi d_1$. In the same way $(Fg)(a) = a'$ implies the existence of $d_2: D_3 \longrightarrow D_1$ in $\mathscr{D}$ such that $g \circ h \circ \psi d_2 = h' \circ \psi d_2$. Since $\mathscr{D}$ is cofiltered, we can find $D_4 \in \mathscr{D}$ and $d_3: D_4 \longrightarrow D_2$, $d_4: D_4 \longrightarrow D_3$ such that $d_1 \circ d_3 = d_2 \circ d_4$. Putting $d = d_1 \circ d_3$ we obtain

$$g \circ h \circ \psi(d) = g \circ h \circ \psi(d_2 \circ d_4) = h' \circ \psi(d_2 \circ d_4)$$
$$= h' \circ \psi(d_1 \circ d_3) = f \circ h \circ \psi(d_1 \circ d_3) = f \circ h \circ \psi(d).$$

From the relation $F(\psi D_4) = \operatorname{colim}_D(\psi D, \psi D_4)$, we deduce the existence of an element $a_4 = [1_{\psi D_4}] \in F(\psi D_4)$. The relation

$$F(h \circ \psi d)(a_4) = [h \circ \psi d \circ 1_{\psi_4}] = [h] = a$$

shows that we have defined a morphism

$$h \circ \psi d: (\psi A_4, a_4) \longrightarrow (A, a)$$

which equalizes f and g. This concludes the proof of $(2) \Rightarrow (1)$. $\qquad\square$

When the category $\mathscr{A}$ is small, we can complete the statement of the previous proposition in a way which explains the precise relation between flatness and left exactness.

Proposition 6.3.8 *Let* $F: \mathscr{A} \longrightarrow \mathsf{Set}$ *be a functor defined on a small category* $\mathscr{A}$. *The following conditions are equivalent:*

(1) F *is flat;*

(2) *the left Kan extension* $\operatorname{Lan}_Y F$ *of* F *along the covariant Yoneda embedding* $Y: \mathscr{A} \longrightarrow \mathsf{Fun}(\mathscr{A}^*, \mathsf{Set})$ *is left exact.*

Proof The proof of $(1) \Rightarrow (2)$ given in 6.1.3 applies here, just replacing "left exact" by "flat" (see 6.3.4 and 6.3.8 instead of 2.9.4 and 6.2.2).

Conversely, assume condition 2. Since the Yoneda embedding is full and faithful, $(\mathrm{Lan}_Y F) \circ Y = F$; see 3.7.3. We must prove the cofilteredness of the category of elements of F.

The category of elements of the constant functor $\Delta \colon \mathscr{A}^* \longrightarrow \mathsf{Set}$ on the singleton is just $\mathscr{A}$. Since Δ is the terminal object of $\mathsf{Fun}(\mathscr{A}^*, \mathsf{Set})$ and $\mathrm{Lan}_Y F$ is exact,

$$\{*\} = (\mathrm{Lan}_Y F)(\Delta) = \mathrm{colim}\,(F \circ 1_{\mathscr{A}}) = \mathrm{colim}\,F.$$

Since $\mathrm{colim}\,F$ is not empty, there exists at least an object $A \in \mathscr{A}$ with FA non-empty (see 2.8.1); choosing $a \in FA$ we get an object (A, a) in the category $\mathsf{Elts}(F)$ of elements of F.

Choose now two objects (A, a), (B, b) in $\mathsf{Elts}(F)$. Since $\mathrm{Lan}_Y F$ preserves finite products, applying 3.8.1 we get

$$\begin{aligned} FA \times FB &\cong (\mathrm{Lan}_Y F)(YA) \times (\mathrm{Lan}_Y F)(YB) \\ &\cong (\mathrm{Lan}_Y F)(YA \times YB) \\ &\cong \big(\mathrm{Lan}_{Y^*}(YA \times YB)\big)(F) \\ &\cong \mathrm{colim}_{(X,x)}\big(\mathscr{A}(X, A) \times \mathscr{A}(X, B)\big), \end{aligned}$$

where (X, x) runs through $\mathsf{Elts}(F)$. The pair (a, b) is in $FA \times FB$, thus is represented by a pair

$$(f, g) \in \mathscr{A}(X, A) \times \mathscr{A}(X, B)$$

for some index $(X, x) \in \mathsf{Elts}(F)$. To prove that

$$(A, a) \xleftarrow{\quad f \quad} (X, x) \xrightarrow{\quad g \quad} (B, b)$$

are morphisms in $\mathsf{Elts}(F)$, it suffices to observe that since (f, g) is representative of (a, b), then $(Ff)(x) = a$, $(Fg)(x) = b$. Indeed the previous colimit construction is obviously natural in A, B and, writing $s^A_{(X,x)}$ for the canonical morphisms of the colimit, one has the commutative diagram 6.3. By construction of the Yoneda isomorphisms (see 1.3.3), $x = s^X_{(X,x)}(1_X)$. Therefore

$$(Ff)(x) = \Big(Ff \circ s^X_{(X,x)}\Big)(1_X) = \Big(s^A_{(X,x)} \circ \mathscr{A}(1, f)\Big)(1_X) = s^A_{(X,x)}(f) = a.$$

Now given two arrows $f, g\colon (A, a) \rightrightarrows (B, b)$ in $\mathsf{Elts}(F)$, let us consider the equalizer diagram

$$K \rightarrowtail \xrightarrow{\quad k \quad} YA \underset{Yg}{\overset{Yf}{\rightrightarrows}} YB$$

$$\mathscr{A}(X,X)\xrightarrow{s^X_{(X,x)}} FX$$

$$\mathscr{A}(1,f)\Big\downarrow \qquad\qquad \Big\downarrow Ff$$

$$\mathscr{A}(X,A)\xrightarrow[s^A_{(X,x)}]{} FA$$

Diagram 6.3

in $\mathsf{Fun}(\mathscr{A}^*,\mathsf{Set})$. Again applying 3.8.1 and the fact that equalizers are computed pointwise in $\mathsf{Fun}(\mathscr{A}^*,\mathsf{Set})$ (see 2.15.2), we get

$$\begin{aligned}
\mathsf{Ker}\,(Ff,Fg) &\cong \mathsf{Ker}\,\big((\mathsf{Lan}_Y F)(Yf),(\mathsf{Lan}_Y F)(Yg)\big)\\
&\cong (\mathsf{Lan}_Y F)\big(\mathsf{Ker}\,(Yf,Yg)\big)\\
&\cong (\mathsf{Lan}_Y F)(K)\\
&\cong (\mathsf{Lan}_{Y^*} K)(F)\\
&\cong \operatorname{colim}_{(X,x)} K(X)\\
&\cong \operatorname{colim}_{(X,x)} \mathsf{Ker}\,\big(\mathscr{A}(X,f),\mathscr{A}(X,g)\big),
\end{aligned}$$

where (X,x) runs through the category of elements of F. Since f,g are arrows of $\mathsf{Elts}(F)$, one has $(Ff)(a)=b=(Fg)(a)$; thus $a\in\mathsf{Ker}\,(Ff,Fg)$. Via the previous isomorphism, a can be represented by some element $\alpha\in KX$. But via the injection $k_X\colon KX \rightarrowtail \mathscr{A}(X,A)$, this element α can be seen as an arrow $h=k_X(\alpha)\colon X\longrightarrow A$. As previously the fact that h represents a means exactly that $(Fh)(x)=a$. This yields a morphism $h\colon (X,x)\longrightarrow (A,a)$ in $\mathsf{Elts}(F)$ with the property $f\circ h=g\circ h$, since h is in the kernel KX of $\mathscr{A}(X,f),\,\mathscr{A}(X,g)$. $\qquad\square$

6.4 The relevance of regular cardinals

At this stage it is time to introduce an easy generalization of various previous results. Many of our results deal with "finiteness". For example "finite limits in Set commute with filtered colimits". Why does such a property hold just for finite limits and not for arbitrary limits? Analysing the proof of 2.13.4, one realizes immediately that it is due to the finiteness requirements in the definition of filtered category: sending two (thus finitely many) objects to the same one; coequalizing two (thus finitely many) parallel arrows.

Being finite means being strictly less than $\aleph_0 = \#\mathbb{N}$, where $\#$ means "cardinality of" and $\aleph$ is just the usual "aleph" notation of set theory (see **Bell and Machover**). One could imagine replacing $\aleph_0$ by an arbitrary cardinal α, then replacing "finite" by "strictly less than α". A problem which will arise immediately is that no generalization can be found of the very useful fact that

a finite union of finite sets is again finite.

In fact, the correct attitude is to replace $\aleph_0$ by a cardinal α which ensures the generalization of the previous property.

Definition 6.4.1 *An infinite cardinal α is regular when it satisfies*

$$\left(\#I < \alpha \ \text{ and } \ \forall i \in I \ \#X_i < \alpha\right) \Rightarrow \# \left(\bigcup_{i \in I} X_i\right) < \alpha$$

where I, X_i are arbitrary sets.

It is useful to recall the following result of set theory, attesting that there are "enough" regular cardinals:

given a set $(\alpha_i)_{i \in I}$ of cardinals, there exists a regular cardinal α such that for every $i \in I$, $\alpha_i < \alpha$.

Let us first generalize the considerations of section 2.13.

Definition 6.4.2 *Let α be a regular cardinal. A category $\mathscr{C}$ is α-filtered when*

(1) there exists at least one object in $\mathscr{C}$,

(2) given a set I with $\#I < \alpha$ and a family $(C_i \in \mathscr{C})_{i \in I}$ of objects of $\mathscr{C}$, there exist an object $C \in \mathscr{C}$ and morphisms $f_i \colon C_i \longrightarrow C$ in $\mathscr{C}$,

(3) given a set I with $\#I < \alpha$ and a family $(f_i \colon C \longrightarrow C')_{i \in I}$ in $\mathscr{C}$, there exist an object $C'' \in \mathscr{C}$ and a morphism $f \colon C' \longrightarrow C''$ such that $f \circ f_i = f \circ f_j$, for all indices i, j.

Definition 6.4.3 *Let α be a regular cardinal.*

(1) By an α-filtered colimit in a category $\mathscr{C}$, we mean the colimit of a functor $F \colon \mathscr{D} \longrightarrow \mathscr{C}$ where the category $\mathscr{D}$ is α-filtered.

(2) By an α-limit in a category $\mathscr{C}$, we mean the limit of a functor $F \colon \mathscr{D} \longrightarrow \mathscr{C}$ where $\mathscr{D}$ is a small category and $\#\mathrm{Ar}(\mathscr{D}) < \alpha$, where $\mathrm{Ar}(\mathscr{D})$ indicates the set of arrows of $\mathscr{D}$.

In general, we shall just write $\#\mathscr{D} < \alpha$ to indicate that the small category $\mathscr{D}$ has a set $\mathrm{Ar}(\mathscr{D})$ of arrows of cardinality strictly less than α.

As a consequence, the cardinal of the set $|\mathscr{D}|$ of objects of $\mathscr{D}$ is a fortiori strictly less than α.

Observe now that choosing $\alpha = \aleph_0 = \#\mathbb{N}$, to be strictly less than α means just being finite. Thus definitions 6.4.2, 6.4.3 describe in this case filtered categories, filtered colimits and finite limits in the usual sense (see 2.13.2).

Just replacing "finite" everywhere in the proof by "strictly less than α", our lemma 2.13.2 yields immediately

Lemma 6.4.4 *Let α be a regular cardinal and $\mathscr{C}$ an α-filtered category. For every category $\mathscr{D}$ such that $\#\mathscr{D} < \alpha$ and every functor $F \colon \mathscr{D} \longrightarrow \mathscr{C}$, there exists a cocone on F.* $\qquad\square$

Since an α-filtered category is a fortiori filtered, our proposition 2.13.3 applies in particular to α-filtered colimits of sets. Finally, again replacing "finite" by "strictly less than α" in the proof of theorem 2.13.4, we obtain

Theorem 6.4.5 *Let α be a regular cardinal. In the category of sets and mappings, α-limits commute with α-filtered colimits.* $\qquad\square$

Next we generalize the results of sections 6.1–6.3. Again the proofs are obtained by replacing "finite" by "strictly less than α".

Definition 6.4.6 *Let α be a regular cardinal. Consider two α-complete categories $\mathscr{A}$, $\mathscr{B}$. A functor $F \colon \mathscr{A} \longrightarrow \mathscr{B}$ is α-left-exact when it preserves α-limits.*

Proposition 6.4.7 *Let α be a regular cardinal and $F \colon \mathscr{A} \longrightarrow \mathsf{Set}$ be a functor defined on an α-complete category $\mathscr{A}$. The following conditions are equivalent:*

(1) F is α-left-exact;

(2) the category $\mathsf{Elts}(F)$ of elements of F is α-cofiltered;

(3) F is an α-filtered colimit of representable functors.

Moreover when $\mathscr{A}$ is small, those conditions are also equivalent to

(4) the left Kan extension $\mathrm{Lan}_Y F$ of F along the covariant Yoneda embedding $Y \colon \mathscr{A} \longrightarrow \mathsf{Fun}(\mathscr{A}^*, \mathsf{Set})$ is α-left-exact. $\qquad\square$

Proposition 6.4.8 *Consider a functor $F \colon \mathscr{A} \longrightarrow \mathscr{B}$ between α-complete categories $\mathscr{A}$, $\mathscr{B}$. The following conditions are equivalent:*

(1) F is α-left-exact;

(2) $\forall B \in \mathscr{B}$ the functor $\mathscr{B}(B, F-) \colon \mathscr{B} \longrightarrow \mathsf{Set}$ is α-left-exact. $\qquad\square$

Proposition 6.4.9 *Let $\mathscr{A}$ be a small α-complete category. The notation $\mathsf{Lex}_\alpha(\mathscr{A}, \mathsf{Set})$ indicates the category of α-left-exact functors from $\mathscr{A}$ to Set.*

(1) $\mathsf{Lex}_\alpha(\mathscr{A}, \mathsf{Set})$ *is complete and limits are computed pointwise;*

(2) $\mathsf{Lex}_\alpha(\mathscr{A}, \mathsf{Set})$ *has pointwise and universal α-filtered colimits and they commute with α-limits;*

(3) *the contravariant Yoneda embedding*

$$\mathscr{A} \longrightarrow \mathsf{Lex}_\alpha(\mathscr{A}, \mathsf{Set}), \quad A \mapsto \mathscr{A}(A, -)$$

transforms α-limits into α-colimits;

(4) $\mathsf{Lex}_\alpha(\mathscr{A}, \mathsf{Set})$ *is cocomplete;*

(5) $\mathsf{Lex}_\alpha(\mathscr{A}, \mathsf{Set})$ *is reflective in* $\mathsf{Fun}(\mathscr{A}, \mathsf{Set})$. $\qquad\square$

Now comes the case of flat functors.

Definition 6.4.10 *Let α be a regular cardinal.*

(1) *A functor $F\colon \mathscr{A} \longrightarrow \mathsf{Set}$ is α-flat when its category of elements is α-cofiltered.*

(2) *A functor $F\colon \mathscr{A} \longrightarrow \mathscr{B}$ is α-flat when, for every $B \in \mathscr{B}$, the functor $\mathscr{B}(B, F-)\colon \mathscr{A} \longrightarrow \mathsf{Set}$ is α-flat.*

Proposition 6.4.11 *Given a functor $F\colon \mathscr{A} \longrightarrow \mathscr{B}$ between α-complete categories, for some regular cardinal α, the following two conditions are equivalent:*

(1) *F is α-left-exact;*

(2) *F is α-flat.* $\qquad\square$

Proposition 6.4.12 *Let α be a regular cardinal:*

(1) *the composite of two α-flat functors is α-flat;*

(2) *representable functors are α-flat;*

(3) *functors with a left adjoint are α-flat.* $\qquad\square$

Proposition 6.4.13 *Let α be a regular cardinal and $F\colon \mathscr{A} \longrightarrow \mathsf{Set}$ a functor. The following conditions are equivalent:*

(1) *F is α-flat;*

(2) *F is an α-filtered colimit of representable functors.*

Moreover when $\mathscr{A}$ is small, those conditions are also equivalent to

(3) *the left Kan extension $\mathsf{Lan}_Y F$ of F along the covariant Yoneda embedding $Y\colon \mathscr{A} \longrightarrow \mathsf{Fun}(\mathscr{A}^*, \mathsf{Set})$ is α-left-exact.* $\qquad\square$

Corollary 6.4.14 *Let α be a regular cardinal and $\mathscr{A}$ a small category. The category* $\mathsf{Flat}_\alpha(\mathscr{A}, \mathsf{Set})$ *of α-flat functors from $\mathscr{A}$ to* Set *has pointwise α-filtered colimits.* $\qquad\square$

Finally let us generalize the definition of a functor which preserves "all small limits".

Definition 6.4.15 *A functor* $F: \mathscr{A} \longrightarrow \mathscr{B}$ *is absolutely flat when it is α-flat for every regular cardinal α.*

By 6.4.12 we conclude that representable functors are absolutely flat and so are those admitting a left adjoint.

6.5 The splitting of idempotents

Definition 6.5.1 *In a category $\mathscr{C}$, a morphism $e: C \longrightarrow C$ is idempotent when $e \circ e = e$.*

Proposition 6.5.2 *In a category $\mathscr{C}$, consider a retract $r, i: R \overset{\longleftarrow}{\longrightarrow} C$, i.e. $r \circ i = 1_R$. Under these conditions, $e = i \circ r$ is idempotent.*

Proof We have $e \circ e = i \circ r \circ i \circ r = i \circ 1_R \circ r = i \circ r = e$. $\qquad\square$

Definition 6.5.3 *In a category $\mathscr{C}$, an idempotent $e: C \longrightarrow C$ splits when there exists a retract $r, i: R \overset{\longleftarrow}{\longrightarrow} C$ of C such that $i \circ r = e$.*

Often we shall use the expression "an idempotent" or "a split idempotent" just to abbreviate the language.

Proposition 6.5.4 *The following conditions are equivalent for an idempotent $e: C \longrightarrow C$ of a category $\mathscr{C}$:*

(1) e splits as $e = i \circ r$, with $r, i: R \overset{\longleftarrow}{\longrightarrow} C$, $r \circ i = 1_R$;
(2) the equalizer $\mathsf{Ker}\,(e, 1_C)$ *exists;*
(3) the coequalizer $\mathsf{Coker}\,(e, 1_C)$ *exists.*

Moreover, under these conditions, $i = \mathsf{Ker}\,(e, 1_C)$, $r = \mathsf{Coker}\,(e, 1_C)$ and this equalizer and this coequalizer are absolute.

Proof Assuming (1) and considering diagram 6.4, we conclude by 2.10.2 that r is the absolute coequalizer of the pair $(e, 1_C)$. Conversely if $r = \mathsf{Coker}\,(e, 1_C)$, the relation $e \circ e = e = e \circ 1_C$ implies the existence of a unique i such that $i \circ r = e$. Since $r \circ ir = r \circ e = r$ and r is an epimorphism, $r \circ i = 1_R$.

The equivalence (1) $\Leftrightarrow$ (2) follows by duality. $\qquad\square$

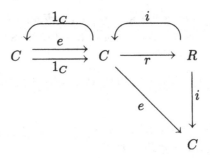

Diagram 6.4

Proposition 6.5.4 shows that splitting of idempotents is some weak form of completeness or cocompleteness. In fact, this result can be made much more precise. We recall from 2.10.1 that a colimit is absolute when it is preserved by all functors. We would like to make precise when a small category has "all small absolute colimits". For this consider the covariant Yoneda embedding

$$Y\colon \mathscr{C} \longrightarrow \mathsf{Fun}(\mathscr{C}^*, \mathsf{Set}), \quad C \mapsto \mathscr{C}(-, C).$$

Like every functor, Y preserves all absolute colimits. Therefore it is sensible to define:

Definition 6.5.5 *A small category $\mathscr{C}$ has all small absolute colimits when, given a small category $\mathscr{D}$ and a functor $F\colon \mathscr{D} \longrightarrow \mathscr{C}$, if the colimit of the composite*

$$\mathscr{D} \xrightarrow{\;\;F\;\;} \mathscr{C} \xrightarrow{\;\;Y\;\;} \mathsf{Fun}(\mathscr{C}^*, \mathsf{Set})$$

is absolute, then the colimit of F exists and is absolute.

Lemma 6.5.6 *Let $\mathscr{C}$ be a small category in which every idempotent splits. Then each retract of a representable functor $\mathscr{C}(-, C)$ is itself representable.*

Proof Consider $\rho, \iota\colon F \xrightarrow{\;\longleftarrow\;} \mathscr{C}(-, C)$, with $\rho \circ \iota = 1_F$. The natural transformation $\iota \circ \rho\colon \mathscr{C}(-, C) \Rightarrow \mathscr{C}(-, C)$ is representable as $\iota \circ \rho = \mathscr{C}(-, e)$ for some morphism $e\colon C \longrightarrow C$, just because the Yoneda embedding is full. But from $(\iota \circ \rho) \circ (\iota \circ \rho) = \iota \circ \rho$ we deduce $\mathscr{C}(-, e \circ e) = \mathscr{C}(-, e)$, which implies $e \circ e = e$ since the Yoneda embedding is faithful (see 1.5.2). The idempotent e splits in $\mathscr{C}$ as $e = r \circ i$, with $r, i\colon R \xrightarrow{\;\longleftarrow\;} C$ and $i \circ r = 1_R$. This implies that $\mathscr{C}(-, r)$, $\mathscr{C}(-, i)$ constitute a splitting of the idempotent $\mathscr{C}(-, e)$. But such a splitting is unique up to isomorphism (see 6.5.4), thus F is isomorphic to $\mathscr{C}(-, R)$. $\qquad\square$

$$\mathscr{D} \xrightarrow{\;\;F\;\;} \mathscr{C} \xrightarrow{\;\;Y\;\;} \mathsf{Fun}(\mathscr{C}^*, \mathsf{Set})$$

$$G \Big\downarrow \qquad\qquad \Big\downarrow G_A$$

$$\mathscr{A} \xrightarrow[\;\;\mathscr{A}(-,A)\;\;]{} \mathsf{Set}^*$$

Diagram 6.5

Proposition 6.5.7 *Let $\mathscr{C}$ be a small category. The following conditions are equivalent:*

(1) In $\mathscr{C}$, all idempotents split;

(2) $\mathscr{C}$ has all small absolute colimits.

Proof   Suppose idempotents split. With the notation of 6.5.5, write $(L, (s_D)_{D \in \mathscr{D}})$ for the colimit of $Y \circ F$ and suppose this colimit is absolute. This colimit is in particular preserved by the representable functor $\mathsf{Nat}(L, -)$, thus

$$\mathsf{Nat}(L, L) \cong \mathrm{colim}\,_D \mathsf{Nat}\big((L, \mathscr{C}(-, FD)\big).$$

A colimit in Set can be constructed as a quotient of a coproduct (see 2.8.1); thus the natural transformation $1_L \colon L \Rightarrow L$ corresponds to the equivalence class of some natural transformation $\beta \colon L \Rightarrow \mathscr{C}(-, FD)$, for some $D \in \mathscr{D}$, with the property $s_D \circ \beta = 1_L$. By lemma 6.5.6, L is representable as $\mathscr{C}(-, R)$ and since the Yoneda embedding is full and faithful (see 1.5.2), the cone $(s_D)_{D \in \mathscr{D}}$ is induced by a cone $(\sigma_D \colon FD \longrightarrow R)_{D \in \mathscr{D}}$ in $\mathscr{C}$. The universal property of $(L, (s_D)_{D \in \mathscr{D}})$ restricted to cocones with representable vertex implies that $(R, (\sigma_D)_{D \in \mathscr{D}})$ is the colimit of F.

By construction, Y preserves the colimit of F. On the other hand given an arbitrary functor $G \colon \mathscr{C} \longrightarrow \mathscr{A}$, we must prove that G preserves the colimit of F. For every $A \in \mathscr{A}$ let us consider the Kan extension G_A of $\mathscr{A}(G-, A)$ along the Yoneda embedding (see 3.7.1), as in diagram 6.5. By 3.7.3, $G_A \circ Y = \mathscr{A}(G-, A)$ because Y is full and faithful. And since the colimit of $Y \circ F$ is absolute,

$$\mathscr{A}\big(G(\mathrm{colim}\,_D FD), A\big) \cong G_A \circ Y(\mathrm{colim}\,_D FD)$$
$$\cong G_A\big(\mathrm{colim}\,_D Y(FD)\big)$$
$$\cong \lim_D G_A Y(FD)$$
$$\cong \lim_D \mathscr{A}(GFD, A)$$

(we have written the equalities in Set, not in Set*; notice that $\mathscr{A}(-, A)$ and G_A are contravariant). Now choose a cocone $(u_D \colon GFD \longrightarrow A)_{D \in \mathscr{D}}$ in $\mathscr{A}$. This is just a compatible family in the limit in Set and so this corresponds to a unique morphism $u \colon G(R) \longrightarrow A$ such that $u \circ G(\sigma_D) = u_D$ for every $D \in \mathscr{D}$.

Conversely if $\mathscr{C}$ has all small absolute colimits and $e \colon C \longrightarrow C$ is idempotent, the corresponding natural transformation

$$\mathscr{C}(-, e) \colon \mathscr{C}(-, C) \Rightarrow \mathscr{C}(-, C)$$

is idempotent and thus splits, since $\mathsf{Fun}(\mathscr{C}^*, \mathsf{Set})$ is cocomplete (see 2.15.2 and 6.5.4). By 6.5.4 again, $\mathscr{C}(-, e)$ and $\mathscr{C}(-, 1_C)$ have an absolute coequalizer, so that by assuption, $e, 1_C$ have an absolute coequalizer in $\mathscr{C}$. Proposition 6.5.4 again implies that e splits. $\qquad\square$

Proposition 6.5.7 suggests considering the splitting of idempotents as some (weak) intrinsic notion of cocompleteness. The following terminology will be justified in 6.8.9, volume 2.

Definition 6.5.8 *A category $\mathscr{C}$ is Cauchy complete when all idempotents of $\mathscr{C}$ split.*

We shall now prove the existence of a "Cauchy completion" for every small category.

Proposition 6.5.9 *Every small category $\mathscr{C}$ can be embedded as a full subcategory in a Cauchy complete small category $\overline{\mathscr{C}}$. Moreover,*

(1) *given a functor $F \colon \mathscr{C} \longrightarrow \mathscr{D}$ where $\mathscr{D}$ is Cauchy complete, F extends uniquely (up to isomorphism) as a functor $\overline{F} \colon \overline{\mathscr{C}} \longrightarrow \mathscr{D}$,*

(2) *given another functor $G \colon \mathscr{C} \longrightarrow \mathscr{D}$, its extension $\overline{G} \colon \overline{\mathscr{C}} \longrightarrow \mathscr{D}$ and a natural transformation $\alpha \colon F \Rightarrow G$, α extends uniquely to a natural transformation $\overline{\alpha} \colon \overline{F} \Rightarrow \overline{G}$,*

(3) *the inclusion $\mathscr{C} \hookrightarrow \overline{\mathscr{C}}$ is an equivalence of categories iff, on $\mathscr{C}$, every retract of a representable functor is itself representable.*

Proof   Given a small category $\mathscr{C}$, the category $\mathsf{Fun}(\mathscr{C}^*, \mathsf{Set})$ of contravariant functors $\mathscr{C} \longrightarrow \mathsf{Set}$ is complete (see 2.15.2), thus Cauchy complete (see 6.5.4). For each object $C \in \mathscr{C}$ and each idempotent $e \colon C \longrightarrow C$ in $\mathscr{C}$, choose a splitting $r_e, i_e \colon R_e \underset{\longleftarrow}{\overset{\longrightarrow}{}} \mathscr{C}(-, C)$, with the condition that $r_{1_C} = 1_{\mathscr{C}(-, C)} = i_{1_C}$. Observe that all splittings of this kind for the same e are isomorphic (see 6.5.4), but making a choice allows us to speak of the small category $\overline{\mathscr{C}}$ as generated by all the chosen objects R_e (there is just a set of them since there is just a set of idempotents in the small category $\mathscr{C}$). The category $\mathscr{C}$ is thus contained as a full subcategory in $\overline{\mathscr{C}}$, via the Yoneda embedding.

Let us prove that $\overline{\mathscr{C}}$ is Cauchy complete. Indeed for every idempotent $e\colon C \longrightarrow C$ in $\mathscr{C}$ and every idempotent $\varepsilon\colon R_e \longrightarrow R_e$ in $\mathsf{Fun}(\mathscr{C}^*, \mathsf{Set})$, ε factors as $\iota \circ \rho$ for some retract $\rho, \iota\colon R \xleftarrow{\quad} R_e$ of ε in $\mathsf{Fun}(\mathscr{C}^*, \mathsf{Set})$. Therefore $\rho \circ r_e, i_e \circ \iota\colon R \xleftarrow{\quad} \mathscr{C}(-, C)$ is a retract of $\mathscr{C}(-, C)$, thus is isomorphic to some $r_{e'}, i_{e'}\colon R_{e'} \xleftarrow{\quad} \mathscr{C}(-, C)$ in $\overline{\mathscr{C}}$. So up to isomorphism, ρ, ι are in $\overline{\mathscr{C}}$ and finally ε splits in $\overline{\mathscr{C}}$.

Now consider a functor $F\colon \mathscr{C} \longrightarrow \mathscr{D}$ with $\mathscr{D}$ Cauchy complete. For every idempotent $e\colon C \longrightarrow C$ in $\mathscr{C}$, $Fe\colon FC \longrightarrow FC$ is idempotent in $\mathscr{D}$ and thus splits as $Fe = j_e \circ s_e$ with $s_e, j_e\colon S_e \xleftarrow{\quad} FC$ a retract of FC. We put $\overline{F}(R_e) = S_e$. If $\alpha\colon R_e \longrightarrow R_{e'}$ is a morphism in $\overline{\mathscr{C}}$ with $e\colon C \longrightarrow C$, $e'\colon C' \longrightarrow C'$ idempotents of $\mathscr{C}$,

$$\alpha = 1_{R_{e'}} \circ \alpha \circ 1_{R_e} = r_{e'} \circ i_{e'} \circ \alpha \circ r_e \circ i_e,$$

and the composite

$$\mathscr{C}(-, C) \xrightarrow{\ r_e\ } R_e \xrightarrow{\ \alpha\ } R_{e'} \xrightarrow{\ i_{e'}\ } \mathscr{C}(-, C')$$

has the form $\mathscr{C}(-, f)$ for a unique $f\colon C \longrightarrow C'$, just because the Yoneda embedding is full and faithful (see 1.5.2). We define

$$\overline{F}(\alpha) = s_{e'} \circ F(f) \circ j_e.$$

It is a straightforward matter to check the functoriality of $\overline{F}$ defined in this way. Moreover, $\overline{F}$ extends F. Indeed $\overline{F}(\mathscr{C}(-, C)) = F(C)$ just because $R_{1_C} = C$ by convention. Moreover given $f\colon C \longrightarrow C'$ and putting $\alpha = \mathscr{C}(-, f)$ in the previous argument, one gets $e = r_e = i_e = 1_C$ and $e' = r_{e'} = i_{e'} = 1_{C'}$ by convention, so that $\overline{F}(\mathscr{C}(-, f)) = F(f)$.

Let us now consider another functor $G\colon \mathscr{C} \longrightarrow \mathscr{D}$ and "its" extension $\overline{G}\colon \overline{\mathscr{C}} \longrightarrow \mathscr{D}$. Given a natural transformation $\beta\colon F \Rightarrow G$, every extension $\overline{\beta}\colon \overline{F} \Rightarrow \overline{G}$ of β satisfies, just by naturality,

$$\begin{aligned}
\overline{\beta}_{R_e} &= \overline{\beta}_{R_e} \circ \overline{F}(r_e) \circ \overline{F}(i_e) \\
&= \overline{G}(r_e) \circ \overline{\beta}_{\mathscr{C}(-, C)} \circ \overline{F}(i_e) \\
&= \overline{G}(r_e) \circ \beta_C \circ \overline{F}(i_e).
\end{aligned}$$

This proves the uniqueness of a natural transformation $\overline{\beta}\colon \overline{F} \Rightarrow \overline{G}$ extending β. It is straightforward to check that the previous formula indeed defines a natural transformation $\overline{\beta}$ extending β.

The last assertion is obvious by construction of $\overline{\mathscr{C}}$. $\qquad\qquad\square$

With a view to establishing the major property of the Cauchy completion (theorem 6.5.11), we need the following lemma.

Lemma 6.5.10 *Consider a small category $\mathscr{C}$ and the corresponding category* $\mathsf{Fun}(\mathscr{C}^*, \mathsf{Set})$ *of contravariant functors to* Set. *For a contravariant functor* $F\colon \mathscr{C}^* \longrightarrow \mathsf{Set}$, *the following conditions are equivalent:*
(1) F is a retract of a representable functor;
(2) the functor $\mathsf{Nat}(F, -)\colon \mathsf{Fun}(\mathscr{C}^*, \mathsf{Set}) \longrightarrow \mathsf{Set}$ *preserves colimits.*

Proof Using the axiom system of universes (see section 1.1), we assume that all categories considered are small with respect to some sufficiently big universe $\mathscr{V}$, including Set. This allows us to consider categories like

$$\mathsf{Fun}\big(\mathsf{Fun}(\mathscr{C}^*, \mathsf{Set}), \mathsf{Set}\big).$$

Suppose $\rho, \iota\colon F \underset{\longleftarrow}{\overset{\longrightarrow}{}} \mathscr{C}(-, C)$, $\rho \circ \iota = 1_F$ is a retract. By the Yoneda lemma (see 1.3.3), the functor

$$\mathsf{Nat}(\mathscr{C}(-, C), -)\colon \mathsf{Fun}(\mathscr{C}^*, \mathsf{Set}) \longrightarrow \mathsf{Set}$$

is just the functor "evaluation at C", which preserves colimits since colimits in $\mathsf{Fun}(\mathscr{C}^*, \mathsf{Set})$ are computed pointwise (see 2.15.2). Since F is a retract of $\mathscr{C}(-, C)$, we get another retract in $\mathsf{Fun}\big((\mathscr{C}^*, \mathsf{Set}), \mathsf{Set}\big)$

$$\mathsf{Nat}(F, -) \xrightarrow[\mathsf{Nat}(\rho, -)]{\overset{\mathsf{Nat}(\iota, -)}{\longleftarrow}} \mathsf{Nat}(\mathscr{C}(-, C), -)$$

Therefore $\mathsf{Nat}(\iota, -)$ is the (absolute) coequalizer of $\mathsf{Nat}(\mathscr{C}(-, e), -)$ and $\mathsf{Nat}(\mathscr{C}(-, 1_C), -)$. By the interchange property for colimits (see 2.12.1), the coequalizer $\mathsf{Nat}(F, -)$ preserves colimits since $\mathsf{Nat}(\mathscr{C}(-, C), -)$ does.

Conversely suppose $\mathsf{Nat}(F, -)$ preserves colimits. We can write F itself as the colimit of the composite

$$\mathsf{Elts}(F) \xrightarrow{\phi} \mathscr{C} \xrightarrow{Y} \mathsf{Fun}(\mathscr{C}^*, \mathsf{Set}),$$

where $\mathsf{Elts}(F)$ is the category of elements of F (see 2.15.6). By assumption we thus have:

$$\mathsf{Nat}(F, F) \cong \mathsf{Nat}\big(F, \mathrm{colim}_{(C,c)} \mathscr{C}(-, C)\big)$$
$$\cong \mathrm{colim}_{(C,c)} \mathsf{Nat}\big(F, \mathscr{C}(-, C)\big).$$

Since a colimit of sets is a quotient of a coproduct (see 2.8.2), the identity on F corresponds to the equivalence class of some $\sigma\colon F \Rightarrow \mathscr{C}(-, C)$, for some index (C, c), with the property $s_{(C,c)} \circ \sigma = 1_F$, where $s_{(C,c)}$ is the canonical morphism of the colimit. This proves that F is a retract of $\mathscr{C}(-, C)$. $\qquad\qquad\square$

The reader should now observe that since $\overline{F}$ is just defined up to isomorphism in 6.5.9, "the" Cauchy completion $\overline{\mathscr{C}}$ of a small category $\mathscr{C}$ is just defined up to equivalence, not up to isomorphism. This is good enough for most purposes and we shall speak freely of "the" Cauchy completion. Here is its main property.

Theorem 6.5.11 *Given two small categories $\mathscr{A}, \mathscr{B}$, the following conditions are equivalent:*

(1) the categories $\mathsf{Fun}(\mathscr{A}^*, \mathsf{Set})$ and $\mathsf{Fun}(\mathscr{B}^*, \mathsf{Set})$ are equivalent;

(2) $\mathscr{A}$ and $\mathscr{B}$ have the same Cauchy completion.

In particular, given a small category $\mathscr{C}$ and its Cauchy completion $\overline{\mathscr{C}}$, the categories $\mathsf{Fun}(\mathscr{C}^*, \mathsf{Set})$ and $\mathsf{Fun}(\overline{\mathscr{C}}^*, \mathsf{Set})$ are equivalent.

Proof If $\mathsf{Fun}(\mathscr{A}^*, \mathsf{Set})$ is equivalent to $\mathsf{Fun}(\mathscr{B}^*, \mathsf{Set})$, the corresponding full subcategories of functors F such that $\mathsf{Nat}(F, -)$ preserves colimits are themselves equivalent. But those two categories are respectively equivalent to the Cauchy completions of $\mathscr{A}$ and $\mathscr{B}$ (see 6.5.10 and the construction of the Cauchy completion in 6.5.9).

For the converse, it suffices to prove that $\mathscr{A}$ and its Cauchy completion $\overline{\mathscr{A}}$ give rise to equivalent categories $\mathsf{Fun}(\mathscr{A}^*, \mathsf{Set})$ and $\mathsf{Fun}(\overline{\mathscr{A}}^*, \mathsf{Set})$. Indeed we have an obvious functor

$$\varphi \colon \mathsf{Fun}(\overline{\mathscr{A}}^*, \mathsf{Set}) \longrightarrow \mathsf{Fun}(\mathscr{A}^*, \mathsf{Set})$$

which is the composition with the inclusion $\mathscr{A}^* \subseteq \overline{\mathscr{A}}^*$. Since each functor $\mathscr{A} \longrightarrow \mathsf{Set}^*$ extends to a functor $\overline{\mathscr{A}} \longrightarrow \mathsf{Set}^*$ (see 6.5.9), each functor $\mathscr{A}^* \longrightarrow \mathsf{Set}$ is the restriction of a functor $\overline{\mathscr{A}}^* \longrightarrow \mathsf{Set}$; thus φ is surjective on the objects. It remains to prove that φ is full and faithful (see 3.4.3), but this is precisely the content of the last assertion in 6.5.9. $\square$

6.6 The more general adjoint functor theorem

This section presents an amazing generalization of the adjoint functor theorem (see 3.3.3).

Theorem 6.6.1 *Consider a functor $F \colon \mathscr{A} \longrightarrow \mathscr{B}$, with $\mathscr{A}$ a Cauchy complete category. The following conditions are equivalent:*

(1) F has a left adjoint;

(2) F is absolutely flat and satisfies the solution set condition for every object $B \in \mathscr{C}$ (see 3.3.2).

Proof The considerations following 6.4.15 and 3.3.2 prove (1) $\Rightarrow$ (2). Conversely fix an object $B \in \mathscr{B}$ and a corresponding solution set S_B. Write $\mathscr{E}_B$ for the category of elements of $\mathscr{B}(B, F-)$ and $\mathscr{S}_B$ for its small full subcategory generated by the objects (A, b), with $A \in S_B$ (and $b: B \longrightarrow FA$).

Choose a regular cardinal $\alpha > \#\mathscr{S}_B$. By α-flatness of F, we get an object $(Z, z) \in \mathscr{E}_B$ provided with morphisms $\alpha_{(X,x)}: (Z, z) \longrightarrow (X, x)$ for every $(X, x) \in \mathscr{S}_B$.

Choose a regular cardinal $\beta > \#\mathscr{E}_B((Z, z), (Z, z))$. By β-flatness of F, we get an object $(Y, y) \in \mathscr{E}_B$ and a morphism $u: (Y, y) \longrightarrow (Z, z)$ such that $f \circ u = g \circ u$ for every two endomorphisms $f, g: (Z, z) \rightrightarrows (Z, z)$. Applying the solution set condition, we choose now $(X, x) \in \mathscr{S}_B$ and a morphism $v: (X, x) \longrightarrow (Y, y)$.

Observe that we have obtained an endomorphism

$$u \circ v \circ \alpha_{(X,x)}: (Z, z) \longrightarrow (Z, z).$$

By definition of u, one has $u \circ v \circ \alpha_{(X,x)} \circ u = 1_{(Z,z)} \circ u$, thus also

$$u \circ v \circ \alpha_{(X,x)} \circ u \circ v \circ \alpha_{(X,x)} = u \circ v \circ \alpha_{(X,x)},$$

proving that $u \circ v \circ \alpha_{(X,x)}$ is idempotent. By Cauchy completeness of $\mathscr{A}$, we find a retract $r, i: (W, w) \rightleftarrows (Z, z)$, with $i \circ r = u \circ v \circ \alpha_{(X,x)}$, $r \circ i = 1_{(W,w)}$ and $w = F(r) \circ z$, $z = Fi \circ w$. We shall prove that (W, w) is the reflection of B along F.

One already has $w: B \longrightarrow FW$ by definition of $\mathscr{E}_B$. Given $V \in \mathscr{A}$ and $v': B \longrightarrow FV$ in $\mathscr{B}$, the pair (V, v') is an object of $\mathscr{E}_B$. By the solution set condition, choose an object $(U, u') \in \mathscr{S}_B$ and a morphism $l: (U, u') \longrightarrow (V, v')$. This yields a composite

$$(W, w) \xrightarrow{\quad i \quad} (Z, z) \xrightarrow{\alpha_{(U,u')}} (U, u') \xrightarrow{\quad l \quad} (V, v'),$$

i.e. a morphism $l \circ \alpha_{(U,u')} \circ i: W \longrightarrow V$ such that $F(l \circ \alpha_{(U,u')} \circ i) \circ w = v'$. We still have to show the uniqueness of such a factorization.

To do this let us prove first that every endomorphism

$$g: (W, w) \longrightarrow (W, w)$$

is necessarily the identity. Indeed since $i \circ g \circ r$ and $i \circ r$ are endomorphisms of (Z, z), one gets, by definition of u,

$$i \circ g \circ r = i \circ g \circ r \circ i \circ r = i \circ g \circ r \circ u \circ v \circ \alpha_{(X,x)}$$

$$= i \circ r \circ u \circ v \circ \alpha_{(X,x)} = i \circ r \circ i \circ r = i \circ 1_{(W,w)} \circ r.$$

Since i is a monomorphism and r is an epimorphism, $g = 1_{(W,w)}$.

Proving the uniqueness of the factorization $l \circ \alpha_{(U,u')} \circ i$ means proving the uniqueness of a morphism $(W,w) \longrightarrow (V,v')$. Choose a regular cardinal $\gamma > \#\mathscr{E}_B((W,w),(V,v'))$. By γ-flatness of F, there are an object $(T,t) \in \mathscr{E}_B$ and a morphism $m\colon (T,t) \longrightarrow (W,w)$ such that $f \circ m = g \circ m$ for every two morphisms $f, g\colon (W,w) \rightrightarrows (V,v')$. By the solution set condition choose $(S,s) \in \mathscr{S}_B$ and a morphism $n\colon (S,s) \longrightarrow (T,t)$. We get a composite

$$(W,w) \xrightarrow{\ i\ } (Z,z) \xrightarrow{\ \alpha_{(S,s)}\ } (S,s) \xrightarrow{\ n\ } (T,t) \xrightarrow{\ m\ } (W,w),$$

which is necessarily the identity, as we have proved. Given two morphisms $f, g\colon (W,w) \rightrightarrows (V,v')$, we thus have

$$f = f \circ m \circ n \circ \alpha_{(S,s)} \circ i$$
$$= g \circ m \circ n \circ \alpha_{(S,s)} \circ i$$
$$= g. \qquad \square$$

Observe that theorem 6.6.1 contains as a special case the general adjoint functor theorem proved in 3.3.3. Indeed when $\mathscr{A}$ is complete, it is Cauchy complete (see 6.5.4) and the absolute flatness of F means the preservation of α-limits for every cardinal α (see 6.4.11), thus the preservation of all small limits.

6.7 Exercises

6.7.1 Develop the details of the first proof suggested for 6.2.4 and indicate a lower bound for the cardinal β.

6.7.2 Given a small category $\mathscr{A}$, prove that the category $\mathsf{Flat}(\mathscr{A}, \mathsf{Set})$ of flat functors is (up to equivalence) the smallest full subcategory of $\mathsf{Fun}(\mathscr{A}, \mathsf{Set})$ containing all the representable functors and stable under filtered colimits.

6.7.3 Given a small category $\mathscr{A}$, prove that $\mathsf{Flat}(\mathscr{A}^*, \mathsf{Set})$ is the free cocompletion of $\mathscr{A}$ for filtered colimits. This means that given a category $\mathscr{B}$ with filtered colimits and a functor $F\colon \mathscr{A} \longrightarrow \mathscr{B}$, there exists a functor $\widetilde{F}\colon \mathsf{Flat}(\mathscr{A}^*, \mathsf{Set}) \longrightarrow \mathscr{B}$, unique up to isomorphism, which preserves filtered colimits and restricts to F via the covariant Yoneda embedding.

6.7.4 Given a small category $\mathscr{A}$, prove that $\mathsf{Fun}(\mathscr{A}^*, \mathsf{Set})$ is the free cocompletion of $\mathscr{A}$ for small colimits. This means that given a complete category $\mathscr{B}$ and a functor $F\colon \mathscr{A} \longrightarrow \mathscr{B}$ there exists a functor

$$\widetilde{F}\colon \mathsf{Fun}(\mathscr{A}^*, \mathsf{Set}) \longrightarrow \mathscr{B},$$

unique up to isomorphism, which preserves small colimits and restricts to F via the covariant Yoneda embedding.

6.7.5 Given a flat functor $F\colon \mathscr{A} \longrightarrow \mathscr{B}$ between arbitrary categories, prove that F preserves all finite limits which turn out to exist in $\mathscr{A}$.

6.7.6 Let $F\colon \mathscr{A} \longrightarrow \mathsf{Set}$ be a flat functor. Prove that the representable functor

$$\mathsf{Nat}(F, -)\colon \mathsf{Flat}(\mathscr{A}, \mathsf{Set}) \longrightarrow \mathsf{Set}$$

preserves filtered colimits if and only if F is a retract of some representable functor.

6.7.7 Prove that every retract of a representable functor is absolutely flat.

6.7.8 Prove that the covariant Yoneda embedding

$$Y_{\mathscr{A}}\colon \mathscr{A} \longrightarrow \mathsf{Flat}(\mathscr{A}^*, \mathsf{Set}), \quad A \mapsto \mathscr{A}(-, A)$$

corestricted to the category of flat functors is itself a flat functor.

6.7.9 If $\mathscr{A}$ is a small category, prove that $\mathsf{Flat}(\mathscr{A}, \mathsf{Set})$ has a terminal object iff $\mathscr{A}$ is cofiltered. In particular, $\mathsf{Flat}(\mathscr{A}, \mathsf{Set})$ is in general not finitely complete. [Hint: apply the Yoneda lemma.]

6.7.10 If $\mathscr{A}$ is a small discrete category with at least two objects, prove that the covariant Yoneda embedding

$$Y\colon \mathscr{A} \longrightarrow \mathsf{Fun}(\mathscr{A}^*, \mathsf{Set}), \quad A \mapsto \mathscr{A}(-, A)$$

is not flat.

6.7.11 Prove that the Kan extension theorem (see 3.7.2) is a corollary of the more general adjoint functor theorem (see 6.7.1), but not of the adjoint functor theorem (see 3.3.3).

7

Bicategories and distributors

Even if containing some interesting results (like 7.9.3, 7.9.4), this chapter is not mainly concerned with proving theorems. Its aim is essentially to discuss some basic structures which turn out to appear quite naturally in categorical constructions.

7.1 2-categories

A category consists of a class of objects connected with morphisms. But in some cases the morphisms themselves can be connected with some additional devices: in the category of small categories and functors, we can define natural transformations between functors; in the category of topological spaces and continuous mappings, we can define homotopies between continuous mappings; and so on. This observation is at the origin of the notion of a 2-category.

A category has been presented in 1.2.1 as a class $|\mathscr{A}|$ of objects together, for each pair A, B of objects, with a set $\mathscr{A}(A, B)$ of morphisms. The composition law was just a mapping

$$c_{ABC}\colon \mathscr{A}(A, B) \times \mathscr{A}(B, C) \longrightarrow \mathscr{A}(A, C)$$

for each triple A, B, C of objects, while the identity on an object A could be seen as a mapping

$$u_A\colon \mathbf{1} \longrightarrow \mathscr{A}(A, A)$$

where $\mathbf{1}$ is the singleton, i.e. the terminal object of the category of sets. The associativity axiom just expresses the equality

$$c_{ACD} \circ (c_{ABC} \times 1) = c_{ABD} \circ (1 \times c_{BCD})$$

(see diagram 7.1) for all objects A, B, C, D. The identity axiom expresses

$$\begin{array}{ccc}
\mathscr{A}(A,B)\times\mathscr{A}(B,C)\times\mathscr{A}(C,D) & \xrightarrow{1\times c_{BCD}} & \mathscr{A}(A,B)\times\mathscr{A}(B,D)\\
\downarrow{\scriptstyle c_{ABC}\times 1} & & \downarrow{\scriptstyle c_{ABD}}\\
\mathscr{A}(A,C)\times\mathscr{A}(C,D) & \xrightarrow[c_{ACD}]{} & \mathscr{A}(A,D)
\end{array}$$

Diagram 7.1

$$\begin{array}{ccccc}
\mathbf{1}\times\mathscr{A}(A,B) & \xleftarrow{\cong} & \mathscr{A}(A,B) & \xrightarrow{\cong} & \mathscr{A}(A,B)\times 1\\
\downarrow{\scriptstyle u_A\times 1} & & \| & & \downarrow{\scriptstyle 1\times u_B}\\
\mathscr{A}(A,A)\times\mathscr{A}(A,B) & \xrightarrow[c_{AAB}]{} & \mathscr{A}(A,B) & \xleftarrow[c_{ABB}]{} & \mathscr{A}(A,B)\times\mathscr{A}(B,B)
\end{array}$$

Diagram 7.2

the equalities

$$c_{AAB}\circ(u_A\times 1)\cong 1\cong c_{ABB}\circ(1\times u_B)$$

(see diagram 7.2) for all objects A, B.

In the case of the category $\mathscr{A} = \mathsf{Cat}$ of small categories, the set $\mathscr{A}(A,B)$ of functors from the category A to the category B can be given the structure of a category $\mathscr{A}(A,B)$, with natural transformations as arrows. With that example in mind, we make the following definition.

Definition 7.1.1 *A 2-category $\mathscr{A}$ consists of*

(1) a class $|\mathscr{A}|$,

(2) for each pair A, B of elements of $|\mathscr{A}|$, a small category $\mathscr{A}(A,B)$,

(3) for each triple A, B, C of elements of $|\mathscr{A}|$, a bifunctor

$$c_{ABC}\colon \mathscr{A}(A,B)\times\mathscr{A}(B,C)\longrightarrow\mathscr{A}(A,C),$$

(4) for each element $A\in|\mathscr{A}|$, a functor

$$u_A\colon \mathbf{1}\longrightarrow\mathscr{A}(A,A),$$

where $\mathbf{1}$ is the terminal object of the category of small categories.

These data are required to satisfy the following axioms.

(1) Associativity axiom: given four elements $A, B, C, D\in\mathscr{A}$, the following equality holds:

$$c_{ACD}\circ(c_{ABC}\times 1) = c_{ABD}\circ(1\times c_{BCD})$$

(see diagram 7.1).

(2) *Unit axiom: given two elements $A, B \in |\mathscr{A}|$, the following equalities hold:*

$$c_{AAB} \circ (u_A \times 1) \cong 1 \cong c_{ABB} \circ (1 \times u_B)$$

(see diagram 7.2).

Let us first fix the terminology. Given a 2-category $\mathscr{A}$:

- the elements of the set $|\mathscr{A}|$ are called "0-cells" or "objects"; we use capital letters $A, B, C, \ldots$ to denote them;
- the objects of the category $\mathscr{A}(A, B)$ are called "1-cells" or "arrows"; we use small letters $a, b, c, \ldots$ to denote them and we designate them in the usual form $f\colon A \longrightarrow B$;
- the arrows of the category $\mathscr{A}(A, B)$ are called "2-cells"; we use greek letters $\alpha, \beta, \gamma \ldots$ to denote them; we designate them as $\alpha\colon f \Rightarrow g$ or just $\alpha\colon f \longrightarrow g$ when no confusion can occur;
- $\beta \circ \alpha$ denotes the composite in the category $\mathscr{A}(A, B)$,

$$f \overset{\alpha}{\Longrightarrow} g \overset{\beta}{\Longrightarrow} h;$$

when no confusion can occur, we write simply $\beta\alpha$;
- $l \circ f$ denotes the image of the pair (f, l) of arrows under the composition functor c_{ABC},

$$A \overset{f}{\longrightarrow} B \overset{l}{\longrightarrow} C;$$

when no confusion can occur, we write simply lf;
- $\varphi * \alpha$ denotes the image of the pair (α, φ) of 2-cells under the composition functor c_{ABC},

$$A \underset{g}{\overset{f}{\rightrightarrows}} {\Downarrow\alpha}\ B \underset{m}{\overset{l}{\rightrightarrows}} {\Downarrow\varphi}\ C;$$

- $1_A\colon A \longrightarrow A$ denotes the image of the unique object of $\mathbf{1}$ under the unit functor u_A; we write just i_A instead of i_{1_A} to denote the unit on the arrow 1_A in the category $\mathscr{A}(A, A)$.

It follows immediately from the axioms that the objects and the morphisms constitute a category, with the morphisms 1_A as identity arrows. Observe also that given the situation

$$A \overset{f}{\underset{h}{\overset{g}{\rightrightarrows}}} {\Downarrow\alpha \atop \Downarrow\beta}\ B \overset{l}{\underset{n}{\overset{m}{\rightrightarrows}}} {\Downarrow\varphi \atop \Downarrow\psi}\ C,$$

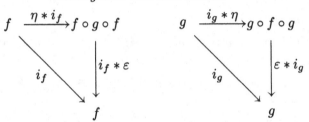

Diagram 7.3

by functoriality of c_{ABC},

$$
\begin{aligned}
(\psi * \beta) \circ (\varphi * \alpha) &= c_{ABC}(\beta, \psi) \circ c_{ABC}(\alpha, \varphi) \\
&= c_{ABC}((\beta, \psi) \circ (\alpha, \varphi)) \\
&= c_{ABC}(\beta \circ \alpha, \psi \circ \varphi) \\
&= (\beta \circ \alpha) * (\psi \circ \varphi).
\end{aligned}
$$

This formula is called the "interchange law" (see 1.3.5).

In a 2-category, it makes sense to speak of "adjoint arrows" or the "Kan extension of an arrow", when this exists. It suffices to transpose the classical definitions valid for ordinary categories (see 3.1.4, 3.4.4 and 3.7.1).

Definition 7.1.2 In a 2-category $\mathscr{A}$, consider arrows $f, g \colon A \underset{f}{\overset{g}{\leftrightarrows}} B$. They constitute an *adjoint pair of arrows* when there exist 2-cells
$$\eta \colon i_B \Rightarrow f \circ g, \quad \varepsilon \colon g \circ f \Rightarrow i_A$$
such that the following equalities between 2-cells hold:
$$(i_f * \varepsilon) \circ (\eta * i_f) = i_f, \quad (\varepsilon * i_g) \circ (i_g * \eta) = i_g;$$
see diagram 7.3. When η and ε are isomorphic 2-cells, the adjunction is called an *equivalence*.

Definition 7.1.3 In a 2-category $\mathscr{A}$, consider two arrows $f \colon A \longrightarrow C$, $g \colon A \longrightarrow B$. The *Kan extension* of f along g, when it exists, is a pair (h, α) where

(1) $h \colon B \longrightarrow C$ is an arrow and $\alpha \colon f \Rightarrow h \circ g$ is a 2-cell,
(2) given any pair (k, β) with $k \colon B \longrightarrow C$ an arrow and $\beta \colon f \Rightarrow k \circ g$ a 2-cell, there exists a unique 2-cell $\gamma \colon h \Rightarrow k$ such that
$$(\gamma * i_g) \circ \alpha = \beta.$$

While in an ordinary category most diagrams in which we are interested are commutative, very often in a 2-category one considers non-commutative diagrams of arrows "filled in" with 2-cells. Consider for

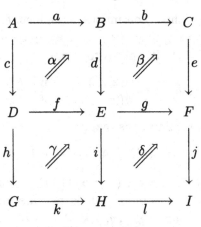

Diagram 7.4

example the situation of diagram 7.4 with no commutativity conditions at all. The squares are filled in with 2-cells

$$\alpha: f \circ c \Rightarrow d \circ a, \quad \beta: g \circ d \Rightarrow e \circ b, \quad \gamma: k \circ h \Rightarrow i \circ f, \quad \delta: l \circ i \Rightarrow j \circ g.$$

This allows us to "fill in" the outer square with a composite 2-cell

$$(i_j * \beta * i_a) \circ (\delta * \alpha) \circ (i_l * \gamma * i_c): l \circ k \circ h \circ c \Rightarrow j \circ e \circ b \circ a.$$

In fact one could combine the four 2-cells in many different ways, but the associativity rules and the interchange law imply immediately that all those composites are equal. There exists a general theorem (the pasting theorem) attesting this fact for a rather arbitrary diagram (see **Kelly, 1980**).

Examples 7.1.4

7.1.4.a A basic example of a 2-category is indeed obtained by choosing small categories as the objects, functors as the arrows and natural transformations as the 2-cells. The various compositions are those described in sections 1.2 and 1.3.

7.1.4.b Let us now choose as objects the topological spaces and as arrows the continuous mappings between them. Given two continuous mappings $f, g: A \rightrightarrows B$, a homotopy $\alpha: f \Rightarrow g$ is a continuous mapping

$$\alpha: I \times A \longrightarrow B$$

where $I = [0, 1]$ is the unit interval, $\alpha(0, a) = f(a)$ and $\alpha(1, a) = g(a)$.

If we consider now three continuous mappings $f, g, h\colon A \longrightarrow B$ and two homotopies $\alpha\colon f \Rightarrow g$, $\beta\colon g \Rightarrow h$, it is easy to construct a "composite homotopy" $\beta \circ \alpha\colon f \Rightarrow h$. It suffices to define

$$(\beta \circ \alpha)(t, a) = \begin{cases} \alpha(2t, a) & \text{if} \quad t \leq 1/2, \\ \beta(2t - 1, a) & \text{if} \quad t \geq 1/2, \end{cases}$$

and one indeed gets a continuous mapping

$$\beta \circ \alpha\colon I \times A \longrightarrow B$$

such that $(\beta \circ \alpha)(0, a) = f(a)$, $(\beta \circ \alpha)\left(\frac{1}{2}, a\right) = g(a)$, $(\beta \circ \alpha)(1, a) = h(a)$. Unfortunately this composition is not associative. Indeed consider four continuous mappings $f, g, h, k\colon A \longrightarrow B$ and three homotopies $\alpha\colon f \Rightarrow g$, $\beta\colon g \Rightarrow h$, $\gamma\colon h \Rightarrow k$; the composite

$$\gamma \circ (\beta \circ \alpha)\colon f \Rightarrow k$$

satisfies $\bigl(\gamma \circ (\beta \circ \alpha)\bigr)\left(\frac{1}{2}, a\right) = h(a)$, while the composite

$$(\gamma \circ \beta) \circ \alpha\colon f \Rightarrow k$$

is such that $\bigl((\gamma \circ \beta) \circ \alpha\bigr)\left(\frac{1}{2}, a\right) = g(a)$. Nevertheless the two homotopies $\gamma \circ (\beta \circ \alpha)$ and $(\gamma \circ \beta) \circ \alpha$ are easily checked to be themselves homotopic as continuous mappings:

$$\gamma \circ (\beta \circ \alpha), (\gamma \circ \beta) \circ \alpha\colon I \times A \rightrightarrows B.$$

Now consider the following situation:

$$A \xrightarrow[\substack{g}]{\substack{f}} \Downarrow\alpha\; B \xrightarrow[\substack{q}]{\substack{p}} \Downarrow\varphi\; C$$

with A, B, C topological spaces, f, g, p, q continuous mappings and α, φ homotopies. The composite

$$I \times A \xrightarrow{\;\alpha\;} B \xrightarrow{\;p\;} C$$

defines a homotopy $p \circ f \Rightarrow p \circ g$ while the composite

$$I \times A \xrightarrow{\;1 \times g\;} I \times B \xrightarrow{\;\varphi\;} C$$

defines a homotopy $p \circ g \Rightarrow q \circ g$. Using the composition of homotopies defined previously, we get a composite homotopy

$$\varphi * \alpha = \bigl(\varphi \circ (1 \times g)\bigr) \circ (p \circ \alpha)\colon p \circ f \Rightarrow q \circ g.$$

It is now a lengthy but straightforward calculation to check that both composition laws $\circ$ and $*$ of homotopies are compatible with the equivalence relation identifying two homotopic homotopies. Therefore we obtain corresponding composition laws on the homotopy classes of homotopies. Choosing the spaces as objects, the continuous mappings as arrows and the homotopy classes of homotopies as 2-cells, we now get a 2-category.

7.1.4.c The category Gr of groups has the groups as objects and the group homomorphisms as arrows. But given two groups G, H and two group homomorphisms $f, g \colon G \rightrightarrows H$, we can define a 2-cell $\alpha \colon f \Rightarrow g$ as an element $\alpha \in H$ such that for every element $x \in G$

$$f(x) \cdot \alpha = \alpha \cdot g(x)$$

where $\cdot$ denotes the composition law of the group H. In other words, a 2-cell $\alpha \colon f \Rightarrow g$ is an internal automorphism of H

$$H \longrightarrow H, \quad y \mapsto \alpha \cdot y \cdot \alpha^{-1},$$

transforming f into g.

It is obvious that given 2-cells $\alpha \colon f \Rightarrow g$ and $\beta \colon g \Rightarrow h$, the element $\beta \cdot \alpha$ is a 2-cell $\beta \circ \alpha \colon f \Rightarrow h$. This provides $\mathsf{Gr}(G, H)$ with the structure of a category or, more precisely, the structure of a groupoid (a category in which every arrow is an isomorphism). Observe that for every $f \colon G \longrightarrow H$, i_f is the unit element of H.

It remains to define the horizontal composition law on 2-cells. Thus we consider group homomorphisms $f, g \colon G \rightrightarrows H$ and $h, k \colon H \rightrightarrows K$, together with 2-cells $\alpha \colon f \Rightarrow g$, $\beta \colon h \Rightarrow k$. We get a composite 2-cell $\beta * \alpha \colon hf \Rightarrow kg$ by choosing the element $h(\alpha) \cdot \beta = \beta \cdot k(\alpha)$.

We leave to the reader the verification that we have indeed defined a 2-category.

7.1.4.d Every ordinary category $\mathscr{A}$ can be viewed as a 2-category with just the obvious 2-cells, i.e. each category $\mathscr{A}(A, B)$ is discrete.

7.2 2-functors and 2-natural transformations

When working with 2-categories, the functors and natural transformations which one considers had better be compatible with the given structures on 2-cells.

Definition 7.2.1 *Given two 2-categories $\mathscr{A}, \mathscr{B}$, a 2-functor $F \colon \mathscr{A} \longrightarrow \mathscr{B}$ consists in giving*

(1) for each object $A \in \mathscr{A}$, an object $FA \in \mathscr{B}$,

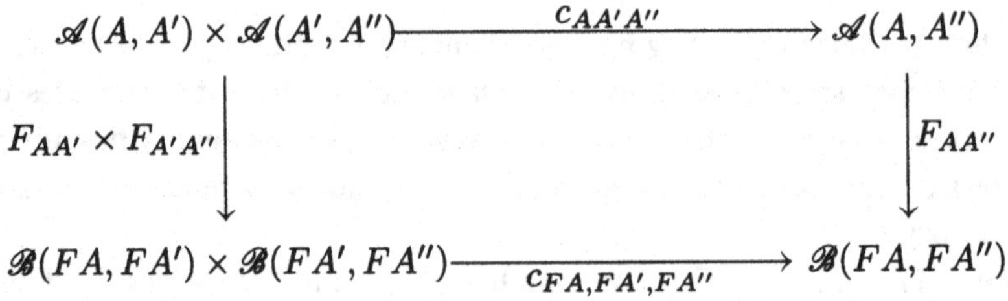

$$\mathscr{A}(A, A') \times \mathscr{A}(A', A'') \xrightarrow{\quad c_{AA'A''} \quad} \mathscr{A}(A, A'')$$

$$F_{AA'} \times F_{A'A''} \Big\downarrow \qquad\qquad\qquad\qquad \Big\downarrow F_{AA''}$$

$$\mathscr{B}(FA, FA') \times \mathscr{B}(FA', FA'') \xrightarrow{\quad c_{FA,FA',FA''} \quad} \mathscr{B}(FA, FA'')$$

Diagram 7.5

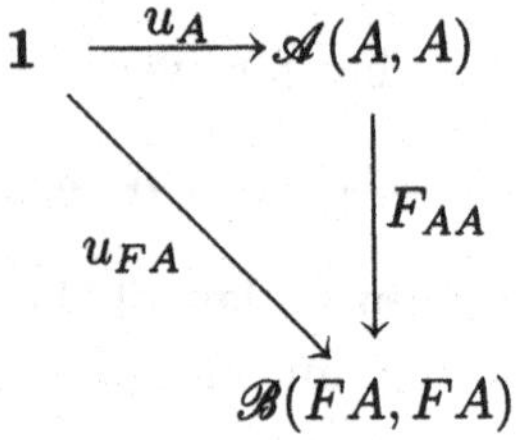

$$1 \xrightarrow{\quad u_A \quad} \mathscr{A}(A, A)$$

$$u_{FA} \searrow \qquad \Big\downarrow F_{AA}$$

$$\mathscr{B}(FA, FA)$$

Diagram 7.6

(2) for each pair of objects $A, A' \in \mathscr{A}$, a functor

$$F_{A,A'} \colon \mathscr{A}(A, A') \xrightarrow{\qquad} \mathscr{B}(FA, FA').$$

(For the sake of brevity, we often write F instead of $F_{A,A'}$). These data are required to satisfy the following axioms.

(1) Compatibility with composition: given three objects $A, A', A'' \in \mathscr{A}$, the following equality holds:

$$F_{AA''} \circ c_{AA'A''} = c_{FA,FA',FA''} \circ (F_{AA'} \times F_{A'A''})$$

(see diagram 7.5).

(2) Unit: for every object $A \in \mathscr{A}$, the following equality holds:

$$F_{AA} \circ u_A = u_{FA}$$

(see diagram 7.6).

Observe in particular that a 2-functor induces an ordinary functor between the underlying categories of objects and arrows.

In order to define the 2-natural transformations easily, observe that given a morphism $f \colon A' \longrightarrow A''$ in a 2-category $\mathscr{A}$, we get a functor

$$\mathscr{A}(A, f) \colon \mathscr{A}(A, A') \xrightarrow{\qquad} \mathscr{A}(A, A'')$$

for every object $A \in \mathscr{A}$, just by defining

$$\mathscr{A}(A, f)(g) = f \circ g, \quad \mathscr{A}(A, f)(\alpha) = i_f * \alpha.$$

In the same way one could define a functor

$$\mathscr{A}(f, A) \colon \mathscr{A}(A'', A) \xrightarrow{\qquad} \mathscr{A}(A', A).$$

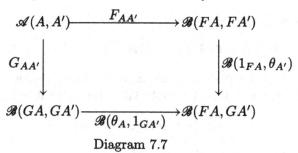

Diagram 7.7

Definition 7.2.2 *Consider two 2-categories $\mathscr{A}$, $\mathscr{B}$ and two 2-functors between them $F, G \colon \mathscr{A} \rightrightarrows \mathscr{B}$. A 2-natural transformation $\theta \colon F \Rightarrow G$ consists in giving, for each object $A \in \mathscr{A}$, an arrow $\theta_A \colon FA \longrightarrow GA$ such that the equality*

$$\mathscr{B}(1_{FA}, \theta_{A'}) \circ F_{AA'} = \mathscr{B}(\theta_A, 1_{GA'}) \circ G_{AA'}$$

holds for each pair of objects $A, A' \in \mathscr{A}$; see diagram 7.7.

In particular, every 2-natural transformation is also a natural transformation between the corresponding underlying functors.

We leave to the reader the straightforward proof that

Proposition 7.2.3 *Small 2-categories, 2-functors and 2-natural transformations themselves constitute a 2-category.* $\square$

In particular the notions of 2-adjunction, 2-equivalence or 2-Kan extension can immediately be obtained from definitions 7.1.2 and 7.1.3. The reader should observe that transposing those definitions will not cause any trouble when the 2-categories considered have proper classes of objects.

Examples 7.2.4

7.2.4.a Consider a small 2-category $\mathscr{A}$ and the 2-category Cat of small categories (see 7.1.4). Fixing an object $A \in \mathscr{A}$, we get at once a "representable" 2-functor

$$\mathscr{A}(A, -) \colon \mathscr{A} \longrightarrow \mathsf{Cat}$$

mapping $B \in \mathscr{A}$ to the category $\mathscr{A}(A, B)$, a morphism $f \colon B \longrightarrow C$ to the functor

$$\mathscr{A}(A, f) \colon \mathscr{A}(A, B) \longrightarrow \mathscr{A}(A, C),$$

$$\mathscr{A}(A, f)(g) = f \circ g, \quad \mathscr{A}(A, f)(\alpha) = i_f * \alpha,$$

and a 2-cell $\beta\colon f \Rightarrow f'$ to the natural transformation

$$\mathscr{A}(A,\beta)\colon \mathscr{A}(A,f) \Rightarrow \mathscr{A}(A,f'), \quad \mathscr{A}(A,\beta)_g = \beta * i_g.$$

7.2.4.b In the situation of example 7.2.4.a, consider another object $A' \in \mathscr{A}$ and an arrow $a\colon A \longrightarrow A'$. We obtain a 2-natural transformation

$$\mathscr{A}(a,-)\colon \mathscr{A}(A',-) \Rightarrow \mathscr{A}(A,-)$$

by defining, for every object $B \in \mathscr{A}$,

$$\mathscr{A}(a,-)_B\colon \mathscr{A}(A',B) \Rightarrow \mathscr{A}(A,B),$$

$$\mathscr{A}(a,-)_B(u) = u \circ a, \quad \mathscr{A}(a,-)_B(\alpha) = \alpha * i_a.$$

7.3 Modifications and n-categories

The basic example of a 2-category is that of ordinary categories, functors and natural transformations. Notice that the possibility of defining natural transformations between functors is directly related to the existence of arrows in the categories.

Consider now the 2-category of 2-categories, 2-functors and 2-natural transformations. The fact of having 2-cells in 2-categories now allows the definition of *modifications* between 2-natural transformations.

Definition 7.3.1 *Consider 2-categories $\mathscr{A}$, $\mathscr{B}$, 2-functors $F, G\colon \mathscr{A} \longrightarrow \mathscr{B}$ and 2-natural-transformations $\alpha, \beta\colon F \Rightarrow G$. A modification*

$$\Xi\colon \alpha \rightsquigarrow \beta$$

consists in giving, for every object $A \in \mathscr{A}$, a 2-cell

$$\Xi_A\colon \alpha_A \Rightarrow \beta_A,$$

in such a way that the following axiom is satisfied: for every pair of morphisms $f, g\colon A \rightrightarrows A'$ and every 2-cell $\alpha\colon f \Rightarrow g$ in $\mathscr{A}$, the equality

$$\Xi_{A'} * F\alpha = G\alpha * \Xi_A$$

holds in $\mathscr{B}$.

We consider 2-categories $\mathscr{A}, \mathscr{B}$, 2-functors $F, G\colon \mathscr{A} \rightrightarrows \mathscr{B}$, 2-natural transformations $\alpha, \beta, \gamma\colon F \Rightarrow G$ and two modifications $\Xi\colon \alpha \rightsquigarrow \beta$, $\Upsilon\colon \beta \rightsquigarrow \gamma$. One obviously gets a composite modification $\Upsilon \diamond \Xi\colon \alpha \rightsquigarrow \gamma$ by putting $(\Upsilon \diamond \Xi)_A = \Upsilon_A \circ \Xi_A$.

Now given 2-functors F, G, H,

$$F, G, H\colon \mathscr{A} \longrightarrow \mathscr{B},$$

2-natural transformations α, β, δ, ε,

$$\alpha, \beta \colon F \Rightarrow G, \quad \delta, \varepsilon \colon G \Rightarrow H,$$

and modifications Ξ, Ω,

$$\Xi \colon \alpha \rightsquigarrow \beta, \quad \Omega \colon \delta \rightsquigarrow \varepsilon,$$

one obviously gets a composite modification

$$\Omega \star \Xi \colon \delta \circ \alpha \rightsquigarrow \varepsilon \circ \beta$$

by putting $(\Omega \star \Xi)_A = \Omega_A \ast \Xi_A$.

Just because the composition laws on modifications are defined point-wise, it is straightforward to verify that the 2-functors from $\mathscr{A}$ to $\mathscr{B}$, the 2-natural transformations and the modifications have been given the structure of a 2-category 2-Fun($\mathscr{A}, \mathscr{B}$).

Given 2-categories $\mathscr{A}, \mathscr{B}, \mathscr{C}$ it is also easy to define a 2-functor of composition

$$c_{\mathscr{A}\mathscr{B}\mathscr{C}} \colon \text{2-Fun}(\mathscr{A}, \mathscr{B}) \times \text{2-Fun}(\mathscr{B}, \mathscr{C}) \longrightarrow \text{2-Fun}(\mathscr{A}, \mathscr{C}).$$

Given a pair (F, F') of 2-functors, we map it to the composite $F' \circ F$. Given another pair of this kind (G, G') and 2-natural transformations $\alpha \colon F \Rightarrow G$, $\alpha' \colon F' \Rightarrow G'$, we map the pair (α, α') to $\alpha' \ast \alpha$. Finally, given two other 2-natural transformations $\beta \colon F \Rightarrow G$, $\beta' \colon F' \Rightarrow G'$ and two modifications $\Xi \colon \alpha \rightsquigarrow \beta$, $\Xi' \colon \alpha' \rightsquigarrow \beta'$, the following relation holds, as Ξ' is a modification by definition:

$$\Xi'_{GA} \ast F'(\Xi_A) = G'(\Xi_A) \ast \Xi'_{FA}.$$

These composites define a 2-cell $(\alpha' \ast \alpha)_A \Rightarrow (\beta' \ast \beta)_A$ and finally a modification $\alpha' \ast \alpha \rightsquigarrow \beta' \ast \beta$, which we choose as the composite of the pair (Ξ, Ξ').

Now if $\mathscr{A}, \mathscr{B}, \mathscr{C}, \mathscr{D}$ are 2-categories, it is straightforward to verify the associativity axiom between the 2-functors of composition, i.e. the equality

$$c_{\mathscr{A}\mathscr{B}\mathscr{D}} \circ (1 \times c_{\mathscr{B}\mathscr{C}\mathscr{D}}) = c_{\mathscr{A}\mathscr{C}\mathscr{D}} \circ (c_{\mathscr{A}\mathscr{B}\mathscr{C}} \times 1);$$

see diagram 7.8. And it is just obvious that the identity modification on the identity 2-cell $i_A \colon 1_A \Rightarrow 1_A$ is an identity for the composition law on modifications.

Well, we have just sketched the construction of what is called a *3-category*.

$$1 \times c_{\mathscr{BCD}}$$

$$2\text{-Fun}[\mathscr{A},\mathscr{B}] \times 2\text{-Fun}[\mathscr{B},\mathscr{C}] \times 2\text{-Fun}[\mathscr{C},\mathscr{D}] \longrightarrow 2\text{-Fun}[\mathscr{A},\mathscr{B}] \times 2\text{-Fun}[\mathscr{B},\mathscr{D}]$$

$$c_{\mathscr{ABC}} \times 1 \Big\downarrow \qquad\qquad\qquad\qquad\qquad\qquad \Big\downarrow c_{\mathscr{ABD}}$$

$$2\text{-Fun}[\mathscr{A},\mathscr{C}] \times 2\text{-Fun}[\mathscr{C},\mathscr{D}] \xrightarrow{\;\;\;\; c_{\mathscr{ACD}} \;\;\;\;} 2\text{-Fun}[\mathscr{A},\mathscr{D}]$$

Diagram 7.8

Definition 7.3.2 *A 3-category consists of the following data:*

(1) *a class* $|\mathscr{A}|$;

(2) *for each pair* A, B *of elements of* $|\mathscr{A}|$, *a small 2-category* $\mathscr{A}(A,B)$;

(3) *for each triple* A, B, C *of elements of* $|\mathscr{A}|$, *a 2-functor*

$$c_{ABC} \colon \mathscr{A}(A,B) \times \mathscr{A}(B,C) \longrightarrow \mathscr{A}(A,C);$$

(4) *for each element* A *of* $|\mathscr{A}|$, *a 2-functor*

$$u_A \colon \mathbf{1} \longrightarrow \mathscr{A}(A,A)$$

where $\mathbf{1}$ *is the terminal 2-category (one object, one arrow, one 2-cell).*

These data must satisfy the following axioms.

(1) *Associativity axiom: given four elements* A, B, C, D *of* $\mathscr{A}$, *the following equality holds:*

$$c_{ABD} \circ (1 \times c_{BCD}) = c_{ACD} \circ (c_{ACD} \times 1)$$

(see diagram 7.1).

(2) *Unit axiom: given two elements* A, B *of* $\mathscr{A}$, *the following equalities hold:*

$$c_{AAB} \circ (u_A \times 1) \cong 1 \cong c_{ABB} \circ (1 \times u_B)$$

(see diagram 7.2).

Extending the terminology of section 7.1, we call the 2-cells of $\mathscr{A}(A,B)$ 3-cells of $\mathscr{A}$. The considerations preceding definition 7.3.2 have precisely proved

Proposition 7.3.3 *There exists a 3-category structure on the following data.*

- *Objects: the 2-categories.*
- *Arrows: the 2-functors.*
- *2-cells: the 2-natural transformations.*
- *3-cells: the modifications.* □

Clearly one could now define 3-functors, 3-natural transformations and 3-modifications. And the fact of having 3-cells in the 3-categories will allow the definition of "morphisms of 3-modifications". The 3-categories, 3-functors, 3-natural transformations, 3-modifications and morphisms of 3-modifications will now organize themselves in what is called a 4-category, whose definition can be easily adapted from the considerations of 7.3.1 and 7.3.2. The process can be iterated, yielding the notion of an n-category, for $n \in \mathbb{N}$, $n > 0$; those n-categories organize themselves in an $(n + 1)$-category. One could even define a 0-category as being a set and a 0-functor as being a mapping; applying the previous process yields the notions of 1-category and 1-functor, which are just the usual notions of category and functor.

7.4 2-limits and bilimits

Consider 2-categories $\mathscr{A}, \mathscr{B}$ with $\mathscr{A}$ small. For every object $B \in \mathscr{B}$ the constant functor

$$\Delta_B \colon \mathscr{A} \longrightarrow \mathscr{B}, \quad \Delta_B(A) = B, \quad \Delta_B(f) = 1_B, \quad \Delta_B(\alpha) = i_B,$$

is obviously a 2-functor.

Given a 2-functor $F \colon \mathscr{A} \longrightarrow \mathscr{B}$ and an object $B \in \mathscr{B}$, we shall write 2-Cone(B, F) to denote the category whose objects are the 2-natural transformations $\Delta_B \Rightarrow F$ (the "2-cones on F with vertex B") and whose morphisms are the modifications between them (see 7.3.1).

Definition 7.4.1 *We keep the notation we have just described. The 2-limit of F, if it exists, is a pair (L, π) where $L \in \mathscr{B}$ is an object of $\mathscr{B}$ and $\pi \colon \Delta_L \Rightarrow F$ is a 2-natural transformation such that the functor*

$$\mathscr{B}(B, L) \longrightarrow 2\text{-Cone}(B, F)$$

of composition with π is an isomorphism of categories, for each object $B \in \mathscr{B}$.

In more explicit terms, we have an arrow $\pi_A \colon L \longrightarrow FA$ for every $A \in \mathscr{A}$. Those arrows satisfy $Ff \circ \pi_A = \pi_{A'}$ for every $f \colon A \longrightarrow A'$ in $\mathscr{A}$, but also $F\alpha * i_{\pi_A} = i_{\pi_{A'}}$ for every $\alpha \colon f \Rightarrow g$ in $\mathscr{A}$. Now if the family $\sigma_A \colon B \longrightarrow FA$ has analogous properties, there exists a unique morphism $b \colon B \longrightarrow L$ such that $\pi_A \circ b = \sigma_A$ for each $A \in \mathscr{A}$. Given another family $\sigma'_A \colon B \longrightarrow FA$ with the same properties and the corresponding factorization $b' \colon B \longrightarrow L$, and given a family $\Xi_A \colon \sigma_A \Rightarrow \sigma'_A$ of 2-cells such that $F\alpha \circ \Xi_A = \Xi_{A'}$ for every α as mentioned above, there exists a unique 2-cell $\beta \colon b \Rightarrow b'$ such that $\Xi_A = i_{\pi_A} * \beta$.

As usual, the uniqueness of the factorization b in the previous discussion implies immediately:

Proposition 7.4.2 *If* (L, π) *and* (L', π') *are 2-limits of the same 2-functor* $F: \mathscr{A} \longrightarrow \mathscr{B}$*, there exists an isomorphism* $b: L \longrightarrow L'$ *such that* $\pi_A \circ b = \pi'_A$ *for each* $A \in \mathscr{A}$. $\qquad\qquad\square$

Examples 7.4.3

7.4.3.a   In the 2-category of small categories, the product of two categories $\mathscr{A}, \mathscr{B}$ is also their 2-product. Indeed write $\mathscr{A} \times \mathscr{B}$ for this product and $p_{\mathscr{A}}, p_{\mathscr{B}}$ for the two projection functors. Consider now a small category $\mathscr{C}$ and functors $F, F': \mathscr{C} \rightrightarrows \mathscr{A}$, $G, G': \mathscr{C} \rightrightarrows \mathscr{B}$, with the corresponding factorizations

$$\binom{F}{G}: \mathscr{C} \longrightarrow \mathscr{A} \times \mathscr{B}, \quad \binom{F'}{G'}: \mathscr{C} \longrightarrow \mathscr{A} \times \mathscr{B}.$$

The 2-dimensional property indicates that given natural transformations $\alpha: F \Rightarrow F'$, $\beta: G \Rightarrow G'$, there exists a unique natural transformation $\gamma: \binom{F}{G} \Rightarrow \binom{F'}{G'}$ such that $i_{p_{\mathscr{A}}} * \gamma = \alpha$, $i_{p_{\mathscr{B}}} * \gamma = \beta$. Indeed, it suffices to put $\gamma_C = \binom{\alpha_C}{\beta_C}$.

It should be noticed that in some 2-categories, the product of two objects can exist without being their 2-product (see exercise 7.10.4).

7.4.3.b   In the 2-category of small categories, consider the following diagram:

$$\mathscr{A} \overset{\overset{F}{\longrightarrow}}{\underset{G}{\longrightarrow}} {\Downarrow\alpha}\; \mathscr{B}.$$

Its 2-limit does exist. It is a subcategory of $\mathscr{A}$ whose objects A are characterized by the following two properties

(1) $FA = GA$;
(2) $\alpha_A: FA \longrightarrow GA$ is the identity morphism.

The morphisms to be considered between such objects are those equalized by F and G.

To conclude the present section, we introduce the notion of bilimit of a 2-functor, on which we shall not dwell very much. The idea is to have a notion of limit defined "up to equivalence" instead of "up to isomorphism".

Definition 7.4.4 *We keep the notation of the beginning of this section. The bilimit of a 2-functor $F\colon \mathscr{A} \longrightarrow \mathscr{B}$, if it exists, is a pair (L, π) where $L \in \mathscr{B}$ is an object of $\mathscr{B}$ and $\pi\colon \Delta_L \Rightarrow F$ is a 2-natural transformation such that the functor*

$$\mathscr{B}(B, L) \longrightarrow \text{2-Cone}(B, F)$$

of composition with π is an equivalence of categories, for each object $B \in \mathscr{B}$.

In more explicit terms, we have an arrow $\pi_A\colon L \longrightarrow FA$ for every $A \in \mathscr{A}$. Those arrows satisfy $Ff \circ \pi_A = \pi_{A'}$ for every $f\colon A \longrightarrow A'$ in $\mathscr{A}$, but also $F\alpha * i_{\pi_A} = i_{\pi_{A'}}$ for every $\alpha\colon f \Rightarrow g$ in $\mathscr{A}$. Now if some family $\sigma_A\colon B \longrightarrow FA$ has analogous properties, there exist a morphism $b\colon B \longrightarrow L$ and isomorphic 2-cells $\theta_A\colon \sigma_A \Rightarrow \pi_A \circ b$ for each $A \in \mathscr{A}$, such that $F\alpha * \theta_A = \theta_{A'}$ for each 2-cell $\alpha\colon f \Rightarrow g$ in $\mathscr{A}$. An additional property must hold. Choose another family $\sigma'_A\colon B \longrightarrow FA$, a corresponding family of isomorphic 2-cells $\theta'_A\colon \sigma'_A \longrightarrow \pi_A \circ b'$. Given a family $\Xi_A\colon \sigma_A \Rightarrow \sigma'_A$ of 2-cells such that $F\alpha * \Xi_A = \Xi_{A'}$ for every α as mentioned above, there exists a unique 2-cell $\beta\colon b \Rightarrow b'$ such that $\theta'_A \circ \Xi_A = (i_{\pi_A} * \beta) \circ \theta_A$.

Proposition 7.4.5 *Under the conditions of definition 7.4.4, two bilimits (L, π) and (B, σ) of the 2-functor F are necessarily weakly equivalent; this means the existence of factorizations $b\colon B \longrightarrow L$ and $b'\colon L \longrightarrow B$ as just described, together with isomorphic 2-cells $1_L \cong b \circ b'$, $1_B \cong b' \circ b$.*

Proof With the notation we have just used, we already have an arrow $b\colon B \longrightarrow L$ and an isomorphic modification $(\theta_A\colon \sigma_A \Rightarrow \pi_A \circ b)_{A \in \mathscr{A}}$. Permuting the roles of (L, π) and (B, σ), we obtain an arrow $b'\colon L \longrightarrow B$ and an isomorphic modification $(\theta'_A\colon \pi_A \Rightarrow \sigma_A \circ b')_{A \in \mathscr{A}}$.

Considering the equivalence of categories

$$\mathscr{B}(L, L) \longrightarrow \text{2-Cone}(L, F)$$

one observes that 1_L is mapped to π while $b \circ b'$ is mapped to the 2-natural transformation $(\pi_A \circ b \circ b')_{A \in \mathscr{A}}$. Those two "cones" are isomorphic via

$$\pi_A \xrightarrow{\theta'_A} \sigma_A \circ b' \xrightarrow{\theta_A * i_{b'}} \pi_A \circ b \circ b'.$$

Coming back along the equivalence, we find finally an isomorphic 2-cell $\mu\colon i_L \Rightarrow b \circ b'$ such that

$$i_{\pi_A} * \mu = (\theta_A * i_{b'}) \circ \theta'_A.$$

An analogous argument yields an isomorphic 2-cell $\nu: i_B \Rightarrow b' \circ b$ such that

$$i_{\sigma_A} * \nu = (\theta'_A * i_b) \circ \theta_A. \qquad \square$$

To prove that b and b', in the previous proof, actually constitute an equivalence in the sense of definition 7.1.2, it is necessary to check the two triangular conditions. For example, the first condition means

$$(i_b * \nu^{-1}) \circ (\mu * i_b) = i_b.$$

This is a compatibility condition between ν and μ, which have been constructed from two different equivalences given by definition 7.4.4. The best one can do is prove the existence of an isomorphic modification connecting the two sides of the equality. To get the equality, it would be necessary to strengthen definition 7.4.4 by requiring the existence of a coherent choice of the adjoints in the various equivalences; but this is most often impossible to verify in examples.

7.5 Lax functors and pseudo-functors

Just because many categorical constructions are defined "up to isomorphism", some constructions are functorial ... up to isomorphism! This is precisely the idea of what a pseudo-functor is. The notion of a lax functor is even weaker: it requires just functoriality up to arbitrary 2-cells instead of isomorphic ones.

Definition 7.5.1 *A lax functor* $F: \mathscr{A} \longrightarrow \mathscr{B}$ *between 2-categories* $\mathscr{A}, \mathscr{B}$ *consists of the following data:*

(1) for every object $A \in \mathscr{A}$, *an object* $FA \in \mathscr{B}$;

(2) for every pair of objects $A, B \in \mathscr{A}$, *a functor*

$$F_{AB}: \mathscr{A}(A, B) \longrightarrow \mathscr{B}(FA, FB);$$

(3) for every triple of objects $A, B, C \in \mathscr{A}$, *a natural transformation (see diagram 7.9)*

$$\gamma_{ABC}: c_{FA,FB,FC} \circ (F_{AB} \times F_{BC}) \Rightarrow F_{AC} \circ c_{ABC};$$

(4) for every object $A \in \mathscr{A}$, *a natural transformation (see diagram 7.10)*

$$\delta_A: u_{FA} \Rightarrow F_{AA} \circ u_A.$$

The natural transformations γ *and* δ *are required to satisfy the following coherence axioms.*

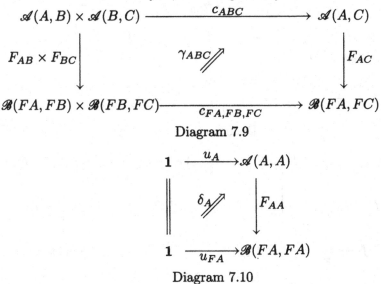

Diagram 7.9

Diagram 7.10

(1) Composition axiom: *for every triple of arrows*

$$A \xrightarrow{\ f\ } B \xrightarrow{\ g\ } C \xrightarrow{\ h\ } D$$

in $\mathscr{A}$, *the following equality between 2-cells holds (see diagram 7.11):*

$$\gamma_{g \circ f, h} \circ (i_{Fh} * \gamma_{f,g}) = \gamma_{f, h \circ g} \circ (\gamma_{g,h} * i_{Ff}),$$

where, for simplicity, we have written $\gamma_{f,g}$ *instead of* $(\gamma_{ABC})_{(f,g)}$.

(2) Unit axiom: *for every arrow* $f \colon A \longrightarrow B$ *in* $\mathscr{A}$, *the following equalities between 2-cells hold (see diagram 7.12):*

$$\gamma_{1_A, f} \circ (i_{Ff} * \delta_A) = i_{Ff}, \quad \gamma_{f, 1_B} \circ (\delta_A * i_{Ff}) = i_{Ff},$$

where, for simplicity, we have written δ_A *instead of* $(\delta_A)_*$.

When the natural transformations γ_{ABC} and δ_A are natural isomorphisms, F is called a *pseudo-functor*.

Definition 7.5.2 Consider two lax functors $F, G \colon \mathscr{A} \rightrightarrows \mathscr{B}$ between 2-categories $\mathscr{A}, \mathscr{B}$. A *lax-natural transformation* $\alpha \colon F \Rightarrow G$ consists in the following data:

(1) for every object $A \in \mathscr{A}$, a morphism $\alpha_A \colon FA \longrightarrow GA$;

(2) for every pair of objects $A, B \in \mathscr{A}$, a natural transformation

$$\tau_{AB} \colon \mathscr{B}(\alpha_A, 1) \circ G_{AB} \Rightarrow \mathscr{B}(1, \alpha_B) \circ F_{AB}$$

(see diagram 7.13), where $\mathscr{B}(\alpha_A, 1)$ and $\mathscr{B}(1, \alpha_B)$ are the functors obtained by fixing α_A or α_B in the bifunctors of composition.

These data are required to satisfy the following coherence axioms, where $\delta^F, \gamma^F, \delta^G, \gamma^G$ are the natural transformations of 7.5.1, respectively for

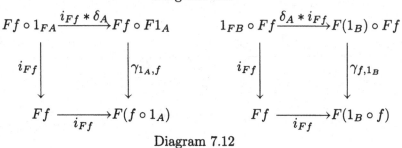

$$Fh \circ Fg \circ Ff \xrightarrow{i_{Fh} * \gamma_{f,g}} Fh \circ F(g \circ f)$$

$$\gamma_{g,h} * i_{Ff} \downarrow \qquad\qquad \downarrow \gamma_{g \circ f, h}$$

$$F(h \circ g) \circ Ff \xrightarrow{\gamma_{f,h \circ g}} F(h \circ g \circ f)$$

Diagram 7.11

$$Ff \circ 1_{FA} \xrightarrow{i_{Ff} * \delta_A} Ff \circ F1_A \qquad\qquad 1_{FB} \circ Ff \xrightarrow{\delta_A * i_{Ff}} F(1_B) \circ Ff$$

$$i_{Ff} \downarrow \qquad \downarrow \gamma_{1_A,f} \qquad\qquad i_{Ff} \downarrow \qquad \downarrow \gamma_{f,1_B}$$

$$Ff \xrightarrow{i_{Ff}} F(f \circ 1_A) \qquad\qquad Ff \xrightarrow{i_{Ff}} F(1_B \circ f)$$

Diagram 7.12

F and G. We keep the same abbreviated notation as in 7.5.1 and also write τ_f instead of $(\tau_{AB})_f$.

(1) For each object $A \in \mathscr{A}$, the following equality between 2-cells holds (see diagram 7.14):
$$\tau_{1_A} \circ \left(\delta_A^G * i_{\alpha_A}\right) \circ i_{\alpha_A} = \left(i_{\alpha_A} * \delta_A^F\right) \circ i_{\alpha_A}.$$

(2) For each pair of arrows
$$A \xrightarrow{f} B \xrightarrow{g} C$$
in $\mathscr{A}$, the following equality between 2-cells holds (see diagram 7.15):
$$(i_{\alpha_C} * \gamma_{f,g}) \circ (\tau_g * i_f) \circ (i_{Gg} * \tau_f) = \tau_{g \circ f} \circ (\gamma_{f,g} * i_{\alpha_A}).$$

When F, G are pseudo-functors and each τ_{AB} is a natural isomorphism, α is called a *pseudo-natural transformation*.

Definition 7.5.3 Consider two lax functors $F, G \colon \mathscr{A} \rightrightarrows \mathscr{B}$ between 2-categories $\mathscr{A}, \mathscr{B}$ and two lax-natural transformations $\alpha, \beta \colon F \Rightarrow G$. A *modification* $\Xi \colon \alpha \rightsquigarrow \beta$ is a family
$$\Xi_A \colon \alpha_A \Rightarrow \beta_A$$
of 2-cells of $\mathscr{B}$, for every object $A \in \mathscr{A}$. Such a family is required to satisfy the following property: for every pair of morphisms $f, g \colon A \rightrightarrows A'$ of $\mathscr{A}$ and every 2-cell $\alpha \colon f \Rightarrow g$, the equality
$$\Xi_{A'} * F\alpha = G\alpha * \Xi_A$$
holds in $\mathscr{B}$.

$$\mathscr{A}(A,B) \xrightarrow{\;F_{AB}\;} \mathscr{B}(FA, FB)$$

$$G_{AB} \Big\downarrow \qquad {}^{\tau_{AB}}\!\!\diagup\!\!\diagup \qquad \Big\downarrow \mathscr{B}(1, \alpha_B)$$

$$\mathscr{B}(GA, GB) \xrightarrow[\;\mathscr{B}(\alpha_A, 1)\;]{} \mathscr{B}(FA, GB)$$

Diagram 7.13

$$\alpha_A \xrightarrow{\;i_{\alpha_A}\;} 1_{GA} \circ \alpha_A \xrightarrow{\;\delta_A^G * i_{\alpha_A}\;} G(1_A) \circ \alpha_A$$

$$i_{\alpha_A} \Big\downarrow \qquad\qquad\qquad\qquad\qquad\qquad \Big\downarrow \tau_{1_A}$$

$$\alpha_A \circ 1_{FA} \xrightarrow[\;i_{\alpha_A} * \delta_A^F\;]{} \alpha_A \circ F(1_A)$$

Diagram 7.14

In particular, this definition applies when F, G are just pseudo-functors or even 2-functors (see 7.3.1).

The reader will observe that when all coherent isomorphisms in definitions 7.5.1, 7.5.2 are just identities, we recapture the notions of 2-functor and 2-natural transformations. We leave to the reader the straightforward proof of the following proposition:

Proposition 7.5.4 *There exists a 3-category structure constructed on the following data.*

- *Objects: the 2-categories.*

- *Arrows: the lax functors.*

- *2-cells: the lax-natural transformations.*

- *3-cells: the modifications.*

An analogous statement holds with "lax" replaced by "pseudo".   □

We refer to section 7.6 and chapter 8, volume 2, for examples of problems using pseudo-functors and pseudo-natural transformations.

Finally let us mention that in the case of lax functors and lax natural transformations, the notions can be in some manner dualized by reversing the direction of the 2-cells γ, δ, τ.

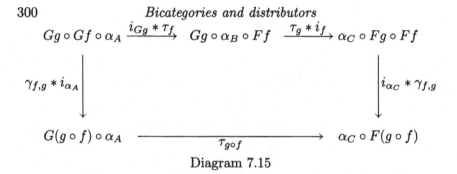

$$Gg \circ Gf \circ \alpha_A \xrightarrow{i_{Gg} * \tau_f} Gg \circ \alpha_B \circ Ff \xrightarrow{\tau_g * i_f} \alpha_C \circ Fg \circ Ff$$

$$\gamma_{f,g} * i_{\alpha_A} \downarrow \qquad\qquad\qquad\qquad \downarrow i_{\alpha_C} * \gamma_{f,g}$$

$$G(g \circ f) \circ \alpha_A \xrightarrow[\tau_{g \circ f}]{} \alpha_C \circ F(g \circ f)$$

Diagram 7.15

7.6 Lax limits and pseudo-limits

Pseudo-functors and lax functors emphasize the idea of replacing commutative diagrams by diagrams which commute up to isomorphism or even up to 2-cell. The classical notion of limit is precisely based on the idea of making commutative a family of triangles. The notions introduced in section 7.5 will allow us to introduce new corresponding notions of limits.

First of all, let us consider 2-categories $\mathscr{A}, \mathscr{B}$, with $\mathscr{A}$ small. For every object $B \in \mathscr{B}$, we consider as in section 7.4 the constant 2-functor on B written $\Delta_B \colon \mathscr{A} \longrightarrow \mathscr{B}$; in particular, this is a pseudo-functor and a lax functor where the coherent isomorphisms are just identities. Given a lax functor $F \colon \mathscr{A} \longrightarrow \mathscr{B}$, a lax-cone on F with vertex $B \in \mathscr{B}$ is a lax-natural transformation $\Delta_B \Rightarrow F$; we write $\mathsf{Lax\text{-}Cone}(B, F)$ for the category of these lax-cones and modifications between them. In the same way a pseudo-cone on F with vertex $B \in \mathscr{B}$ is a pseudo-natural transformation $\Delta_B \Rightarrow F$ and we write $\mathsf{Ps\text{-}Cone}(B, F)$ for the corresponding category of pseudo-cones and modifications between them.

Definition 7.6.1 *We use the notation we have just described. The lax limit of a lax functor $F \colon \mathscr{A} \longrightarrow \mathscr{B}$ between 2-categories $\mathscr{A}, \mathscr{B}$, if it exists, is a pair (L, π) where $L \in \mathscr{B}$ is an object of $\mathscr{B}$ and $\pi \colon \Delta_L \Rightarrow F$ is a lax-natural transformation such that the functor*

$$\mathscr{B}(B, L) \longrightarrow \mathsf{Lax\text{-}Cone}(B, F)$$

of composition with π is an isomorphism of categories, for each object $B \in \mathscr{B}$. Replacing "lax" by "pseudo" in the previous definition, we get the notion of pseudo-limit of a pseudo-functor.

Even when F is an actual 2-functor, the notions of lax limit, pseudo-limit and 2-limit of F produce results which are, in general, completely different. To emphasize this, let us write in more explicit terms the definition of the lax limit of F, in the case where F is a 2-functor.

The lax limit is given by a family $\pi_A\colon L \longrightarrow FA$ of arrows, for every object $A \in \mathscr{A}$, and a family $\tau_f\colon Ff \circ \pi_A \Rightarrow \pi_{A'}$ of 2-cells, for each arrow $f\colon A \longrightarrow A'$ in $\mathscr{A}$. These data are such that $\tau_{1_A} = i_{\pi_A}$ for each A and

$$\tau_g \circ (1_{Fg} * \tau_f) = \tau_{g \circ f}$$

for each pair of composable arrows $A \xrightarrow{f} A' \xrightarrow{g} A''$ in $\mathscr{A}$. Given other families $\sigma_A\colon B \longrightarrow FA$, $\theta_f\colon Ff \circ \sigma_A \Rightarrow \sigma_{A'}$ with analogous properties, there exists a unique arrow $b\colon B \longrightarrow L$ such that $\pi_A \circ b = \sigma_A$ for each $A \in \mathscr{A}$ and $\tau_f * 1_b = \theta_f$ for every $f\colon A \longrightarrow A'$. Now choose on the same object B two additional families $\sigma'_A\colon B \longrightarrow FA$, $\theta'_f\colon Ff \circ \sigma'_A \Rightarrow \sigma'_{A'}$ with analogous properties and the corresponding factorization $b'\colon B \longrightarrow L$; choose also a family of 2-cells $\Xi_A\colon \sigma_A \Rightarrow \sigma'_A$ such that for every pair $f, g\colon A \rightrightarrows A'$ of morphisms and every 2-cell $\alpha\colon f \Rightarrow g$, the relation

$$\theta'g \circ (F\alpha * \Xi_A) = \Xi_{A'} \circ \theta_f$$

holds; under these conditions, there must exist a unique 2-cell $\beta\colon b \Rightarrow b'$ such that for every $A \in \mathscr{A}$, $i_{\pi_A} * \beta = \Xi_A$.

As usual, the uniqueness of the factorization b in the previous discussion (or more precisely the corresponding discussion for a lax functor) implies immediately:

Proposition 7.6.2 *If (L, π) and (L, π') are two lax limits of the same lax functor $F\colon \mathscr{A} \longrightarrow \mathscr{B}$, they are isomorphic. A corresponding statement holds for the pseudo-limit of a pseudo-functor.* $\qquad\square$

Example 7.6.3

Let us choose for $\mathscr{P}$ the 2-category with three objects $0, 1, 2$, two non-trivial arrows $1 \longrightarrow 0$ and $2 \longrightarrow 0$, and no non-trivial 2-cells. A 2-functor $P\colon \mathscr{P} \longrightarrow \mathbf{Cat}$ to the 2-category of small categories is just the choice of two functors $F\colon \mathscr{A} \longrightarrow \mathscr{C}$, $G\colon \mathscr{B} \longrightarrow \mathscr{C}$. Even in this trivial case where no 2-cells are involved a priori, let us observe the striking differences between the notions of 2-limit, pseudo-limit and lax limit.

The 2-limit of P is just the usual pullback of F, G. The objects of the limit $\mathscr{L}$ are the pairs (A, B) with $A \in \mathscr{A}$, $B \in \mathscr{B}$ and $FA = GB$; an arrow $(f, g)\colon (A, B) \longrightarrow (A', B')$ is a pair $f\colon A \longrightarrow A'$, $g\colon B \longrightarrow B'$ of arrows in $\mathscr{A}$ and $\mathscr{B}$ respectively, such that $Ff = Gg$. The projections are the obvious ones. Thus the pullback of F, G coincides with their 2-pullback.

The lax limit object of the functor P is the category $\mathscr{L}$ whose objects are the quintuples (A, f, C, g, B), with $A \in \mathscr{A}$, $B \in \mathscr{B}$, $C \in \mathscr{C}$,

$f\colon FA \longrightarrow C$ and $g\colon GB \longrightarrow C$. An arrow

$$(A, f, C, g, B) \longrightarrow (A', f', C', g', B')$$

is a triple (a, c, b) with $a\colon A \longrightarrow A'$, $b\colon B \longrightarrow B'$, $c\colon C \longrightarrow C'$ and $c \circ f = f' \circ Fa$, $c \circ g = g' \circ Gb$. The projections are the obvious ones. The required natural transformations

$$F \circ \pi_{\mathscr{A}} \Rightarrow \pi_{\mathscr{C}} \Leftarrow G \circ \pi_{\mathscr{B}}$$

take values f, g on the object (A, f, C, g, B).

The pseudo-limit of the functor P is constructed in an analogous way, restricting one's attention to those quintuples (A, f, C, g, B) where f and g are isomorphisms.

In particular, it should be observed that none of the previous constructions gives the comma category (F, G), as defined in 1.6.1. The comma construction is in fact an example of a *weighted 2-limit* (see chapter 6, volume 2).

7.7 Bicategories

In sections 7.5, 7.6 we have indicated that since many constructions in category theory are just defined "up to isomorphism", it is sensible to consider constructions which are functorial "up to isomorphism" (or even up to 2-cell). But in some cases this leads to the consideration of category-like objects ... where the axioms for a category are just satisfied "up to isomorphism" (or even up to 2-cell).

This is the object of the notion of a bicategory. To be coherent with the terminology of the previous sections, it would be better to say "pseudo-category", but the term "bicategory" is now universal for the structure described in 7.7.1.

Definition 7.7.1 *A bicategory* $\mathscr{A}$ *is specified by the following data:*

(1) *a class* $|\mathscr{A}|$ *of "objects" (also called "0-cells");*

(2) *for each pair* A, B *of objects, a small category* $\mathscr{A}(A, B)$ *whose objects are called "arrows" (or "morphisms" or "1-cells") and whose morphisms are called "2-cells"; we write* $\alpha \circ \beta$ *for the composite of the 2-cells* α, β;

(3) *for each triple* A, B, C *of objects, a composition law given by a functor*

$$c_{ABC}\colon \mathscr{A}(A, B) \times \mathscr{A}(B, C) \longrightarrow \mathscr{A}(A, C);$$

$$\mathscr{A}(A,B) \times \mathscr{A}(B,C) \times \mathscr{A}(C,D) \xrightarrow{\quad 1 \times c_{BCD} \quad} \mathscr{A}(A,B) \times \mathscr{A}(B,D)$$

$$\Bigg\downarrow c_{ABC} \times 1 \qquad\qquad \alpha_{ABCD} \diagup\!\!\diagup \qquad\qquad \Bigg\downarrow c_{ABD}$$

$$\mathscr{A}(A,C) \times \mathscr{A}(C,D) \xrightarrow[\quad c_{ACD} \quad]{} \mathscr{A}(A,D)$$

Diagram 7.16

$$1 \times \mathscr{A}(A,B) \xleftarrow{\;\cong\;} \mathscr{A}(A,B) \xrightarrow{\;\cong\;} \mathscr{A}(A,B) \times 1$$

$$\Bigg\downarrow i_A \times 1 \quad \underset{\lambda_{AB}}{\xleftarrow{\hspace{1cm}}} \quad \Big\| \; \rho_{AB} \; \underset{\phantom{\rho_{AB}}}{\xrightarrow{\hspace{1cm}}} \quad \Bigg\downarrow 1 \times i_B$$

$$\mathscr{A}(A,A) \times \mathscr{A}(A,B) \xrightarrow[c_{AAB}]{} \mathscr{A}(A,B) \xleftarrow[c_{ABB}]{} \mathscr{A}(A,B) \times \mathscr{A}(B,B)$$

Diagram 7.17

given arrows $f\colon A \longrightarrow B$, $g\colon B \longrightarrow C$ of the bicategory $\mathscr{A}$, we write $g \circ f$ for their composite $c_{ABC}(f,g)$; given other arrows $f'\colon A \longrightarrow B$, $g'\colon B \longrightarrow C$ and 2-cells $\gamma\colon f \Rightarrow f'$, $\delta\colon g \Rightarrow g'$, we write $\delta * \gamma$ for their composite $c_{ABC}(\gamma,\delta)$;

(4) for each object $A \in \mathscr{A}$, an "identity arrow" $1_A\colon A \longrightarrow A$; we write i_A for the identity 2-cell on 1_A.

The associativity and identity axioms are now replaced by the existence of some isomorphisms, which are thus part of the data for a bicategory.

(1) Associativity isomorphisms: for each quadruple of objects A, B, C, D of $\mathscr{A}$, a natural isomorphism

$$\alpha_{ABCD}\colon c_{ACD} \circ (c_{ABC} \times 1) \Rightarrow c_{ABD} \circ (1 \times c_{BCD})$$

(see diagram 7.16).

(2) Unit isomorphisms: for each pair of objects $A, B \in \mathscr{A}$, two natural isomorphisms

$$\lambda_{AB}\colon 1 \Rightarrow c_{AAB} \circ (i_A \times 1), \quad \rho_{AB}\colon 1 \Rightarrow c_{ABB} \circ (1 \times i_B)$$

(see diagram 7.17).

Those various data are required to satisfy coherence conditions expressed by the following axioms.

(1) Associativity coherence: given arrows

$$A \xrightarrow{\;f\;} B \xrightarrow{\;g\;} C \xrightarrow{\;h\;} D \xrightarrow{\;k\;} E,$$

the following equality holds (see diagram 7.18):

$$((k \circ h) \circ g) \circ f \xrightarrow{\ \alpha_{g,h,k} \, * \, i_f\ } (k \circ (h \circ g)) \circ f \xrightarrow{\ \alpha_{f,h \circ g,k}\ } k \circ ((h \circ g) \circ f)$$

$$\Big\downarrow \alpha_{f,g,k \circ h} \qquad\qquad\qquad\qquad\qquad\qquad\qquad\qquad\qquad\qquad \Big\downarrow i_k * \alpha_{f,g,h}$$

$$(k \circ h) \circ (g \circ f) \xrightarrow[\qquad\qquad\qquad \alpha_{g \circ f,h,k} \qquad\qquad\qquad]{} k \circ (h \circ (g \circ f))$$

Diagram 7.18

$$(g \circ i_B) \circ f \xrightarrow{\ \alpha_{f,1_B,g}\ } g \circ (i_B \circ f)$$

$$\rho_g * i_f \searrow \qquad\qquad \swarrow i_g * \lambda_f$$

$$g \circ f$$

Diagram 7.19

$$(i_k * \alpha_{f,g,h}) \odot \alpha_{f,h \circ g,k} \odot (\alpha_{g,h,k} * i_f) = \alpha_{g \circ f,h,k} \odot \alpha_{f,g,k \circ h}.$$

where we have written $\alpha_{f,g,h}$ instead of $(\alpha_{ABCD})_{(f,g,h)}$, just for simplicity.

(2) Identity coherence: given arrows

$$A \xrightarrow{\ f\ } B \xrightarrow{\ g\ } , C$$

the following equality holds (see diagram 7.19):

$$(i_g * \lambda_f) \odot \alpha_{f,1_B,g} = \rho_g * i_f,$$

where, for simplicity again, we have written λ_f, ρ_g instead of $(\lambda_{AB})_f$, $(\rho_{BC})_g$.

Clearly, one could have considered an even more general notion where the associativity and identity isomorphisms α, λ, ρ are replaced by ordinary natural transformations, something one could have called a "lax category".

We leave to the reader the work of defining pseudo-functors and lax functors between bicategories and the corresponding notions of pseudo-limit, lax limit and even pseudo-bilimit or lax bilimit.

It is lengthy but straightforward to check that bicategories, lax functors and lax-natural transformations constitute a new bicategory and even, taking into account the modifications, something which could be called a tricategory. An analogous statement holds with "lax" replaced by "pseudo".

If all generalizations we have just mentioned are quite straightforward to define, the axioms nevertheless take a quite heavy form because of all

$$\begin{array}{ccccc}
1_B \circ f & \xrightarrow{\eta * i_f} & (f \circ g) \circ f & \xrightarrow{\alpha_{f,g,h}} & f \circ (g \circ f) \\
{\scriptstyle \lambda_f}\downarrow & & & & \downarrow{\scriptstyle i_f * \varepsilon} \\
f & & \xrightarrow[\rho_f]{\hspace{6em}} & & f \circ 1_A
\end{array}$$

$$\begin{array}{ccccc}
g \circ 1_B & \xrightarrow{i_g * \eta} & g \circ (f \circ g) & \xrightarrow{\alpha^{-1}_{g,f,g}} & (g \circ f) \circ g \\
{\scriptstyle \rho_g}\downarrow & & & & \downarrow{\scriptstyle \varepsilon * i_g} \\
g & & \xrightarrow[\lambda_g]{\hspace{6em}} & & 1_A \circ g
\end{array}$$

Diagram 7.20

the coherent isomorphisms it is necessary to introduce. As an example, let us make explicit the notion of an adjoint pair in a bicategory.

Definition 7.7.2 *Let $\mathscr{A}$ be a bicategory. Two arrows $f\colon A \longrightarrow B$ and $g\colon B \longrightarrow A$ are part of an adjoint pair when there exist 2-cells*

$$\eta\colon 1_B \Rightarrow f \circ g, \quad \varepsilon\colon g \circ f \Rightarrow 1_A$$

which satisfy the following equalities (see diagram 7.20):

$$(i_f * \varepsilon) \odot \alpha_{f,g,h} \odot (\eta * i_f) = \rho_f \odot \lambda_f,$$
$$(\varepsilon * i_g) \odot \alpha^{-1}_{g,f,g} \odot (i_g * \eta) = \lambda_g \odot \rho_g.$$

The reader will observe that, in this definition, we have used explicitly the fact that α is an isomorphism, not just a natural transformation.

Example 7.7.3

Let $\mathscr{C}$ be a small category with pullbacks. We shall construct the bicategory of spans of $\mathscr{C}$, closely related with the constructions of section 5.2 concerning the calculus of fractions. We use freely the axiom of choice.

The objects are those of $\mathscr{C}$ and a morphism $A \longrightarrow B$ is now a *span* on A, B, i.e. a pair of arrows $f\colon X \longrightarrow A$, $g\colon X \longrightarrow B$ of $\mathscr{C}$ (see diagram 7.21) with an arbitrary domain X. A 2-cell $\alpha\colon (f,g) \Rightarrow (f',g')$ between two spans on A, B is just a morphism $\alpha\colon X \longrightarrow X'$ such that $f' \circ \alpha = f$, $g' \circ \alpha = g$; see diagram 7.21. The composition of $\mathscr{C}$ immediately induces the structure of a category on the spans from A to B.

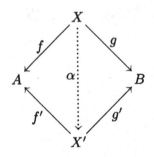

Diagram 7.21

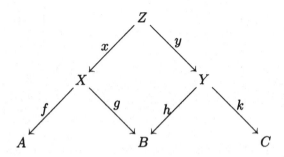

Diagram 7.22

Given two spans (f, g): $A \longrightarrow B$ and (h, k): $B \longrightarrow C$, we define their composite as

$$(h, k) \circ (f, g) = (f \circ x, k \circ y),$$

see diagram 7.22, where (x, y) is one arbitrarily specified pullback of (g, h). Given other spans (f', g'): $A \longrightarrow B$, (h', k'): $B \longrightarrow C$, we specify in the same way a pullback (x', y') of (g', h') and the corresponding composite

$$(h', k') \circ (f', g') = (f' \circ x', k' \circ y').$$

Now given 2-cells

$$\alpha \colon (f, g) \Rightarrow (f', g'), \quad \beta \colon (h, k) \Rightarrow (h', k'),$$

the equality

$$g' \circ \alpha \circ x = g \circ x = h \circ y = h' \circ \beta \circ y$$

in $\mathscr{C}$ implies the existence of a unique factorization γ: $Z \longrightarrow Z'$ through

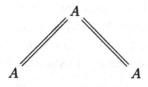

Diagram 7.23

the pullback (x', y'), with $x' \circ \gamma = \alpha \circ x$, $y' \circ \gamma = \beta \circ y$. From the relations

$$f' \circ x' \circ \gamma = f' \circ \alpha \circ x = f \circ x, \quad k' \circ y' \circ \gamma = k' \circ \beta \circ y = k \circ y,$$

we deduce that $\gamma \colon (f \circ x, k \circ y) \Rightarrow (f' \circ x', k' \circ y')$ is a 2-cell, which we choose as $\beta * \alpha$.

Since pullbacks are defined up to isomorphism and we made an arbitrary choice of them, we conclude that the associativity in the composition of spans holds up to isomorphism. This allows the definition of a natural isomorphism α as in 7.7.1. If one chooses to specify the identity always as the pullback of an identity arrow, one can choose the identity span $(1_A, 1_A)$ (see diagram 7.23) as identity on A and the identity natural transformations as isomorphisms λ, ρ in 7.7.1. We leave the verifications to the reader.

Example 7.7.4

Another canonical example of a bicategory can be found in the theory of bimodules. We choose as objects the rings with identity. An arrow from the ring R to the ring S is an (R, S)-bimodule M: this is an abelian group M provided with the structure of a left R-module and a right S-module, the axiom

$$r(ms) = (rm)s$$

being satisfied for all elements $r \in R$, $m \in M$, $s \in S$.

Given two (R, S)-bimodules M, N, we choose as 2-cells from M to N the (R, S)-linear mappings $f \colon M \longrightarrow N$, thus the group homomorphisms which are both R-linear on the left and S-linear on the right. This yields immediately a category $\mathsf{Bim}(R, S)$ of (R, S)-bimodules and their homomorphisms.

For a third ring T, the composition

$$\mathsf{Bim}(R, S) \times \mathsf{Bim}(S, T) \longrightarrow \mathsf{Bim}(R, T)$$

is just the "tensor product" functor. Indeed given an (R, S)-bimodule M and an (S, T)-bimodule N, the tensor product $M \otimes_S N$ over S produces an (R,T)-bimodule. This construction is well-known to be functorial. Now "the" tensor product of bimodules is associative in the sense that given rings R, S, T, V, an (R,S)-bimodule L, a (S,T)-bimodule M and a (T,V)-bimodule N, there exists a canonical isomorphism

$$L \otimes_S (M \otimes_T N) \cong (L \otimes_S M) \otimes_T N.$$

Moreover, given a ring R, this ring R itself can be seen as an (R,R)-bimodule and, up to an isomorphism, is an identity for tensoring over R.

From these considerations, the reader can check that we have effectively defined a bicategory of rings, bimodules and homomorphisms of bimodules.

7.8 Distributors

Roughly speaking, a distributor is to a functor what a relation is to a mapping.

More precisely, a set can be seen as a discrete category; the hom-sets are just the empty set (when the objects are different) and the singleton (when the objects are equal); in the definition of a category, the empty set and the singleton are replaced by arbitrary sets of morphisms.

The way one represents graphically a relation from a set A to a set B is to draw an arrow from $a \in A$ to $b \in B$ when those elements are in relation and no arrow when they are not. Once more with each pair $(a,b) \in A \times B$ we have associated a set of arrows which is just the singleton (when a, b are in relation) or the empty set (when a, b are not in relation). The correct generalization to the case of categories $\mathscr{A}, \mathscr{B}$ will be to define a "categorical relation" from $\mathscr{A}$ to $\mathscr{B}$ by associating an arbitrary set of "formal arrows" with each pair $(a,b) \in \mathscr{A} \times \mathscr{B}$ of objects. Clearly, such a "categorical relation" must be functorial with respect to the arrows of $\mathscr{A}$ and $\mathscr{B}$. Now intuitively, giving a relation from $\mathscr{A}$ to $\mathscr{B}$ must be equivalent to giving the opposite relation from $\mathscr{B}$ to $\mathscr{A}$ obtained by "reversing the direction of all formal arrows". To avoid inelegant contravariant behaviour in subsequent results, we in fact define a distributor from $\mathscr{A}$ to $\mathscr{B}$ as being a "categorical relation" from $\mathscr{B}$ to $\mathscr{A}$.

Definition 7.8.1 *By a distributor (also called profunctor or bimodule) from a category $\mathscr{A}$ to a category $\mathscr{B}$, we mean a bifunctor*

$$\phi \colon \mathscr{B}^* \times \mathscr{A} \longrightarrow \mathrm{Set}.$$

We shall write $\phi: \mathscr{A} \longrightarrow \mathscr{B}$ to indicate that ϕ is a distributor from $\mathscr{A}$ to $\mathscr{B}$. Given two distributors $\phi, \psi: \mathscr{A} \longrightarrow \mathscr{B}$, a morphism of distributors $\alpha: \phi \Rightarrow \psi$ is just a natural transformation $\alpha: \phi \Rightarrow \psi$.

In particular, when $\mathscr{A}$, $\mathscr{B}$ are small, the distributors from $\mathscr{A}$ to $\mathscr{B}$ and their morphisms organize themselves in a category $\mathrm{Dist}(\mathscr{A}, \mathscr{B})$, for the usual composition of natural transformations.

We are now going to introduce a composition law which will allow us to present the distributors as arrows of a bicategory. Let us recall that given two relations $R: A \longrightarrow B$, $S: B \longrightarrow C$ between sets, the relation $S \circ R: A \longrightarrow C$ is given by

$$(a, c) \in S \circ R \quad \text{iff} \quad \exists b \in B \; (a, b) \in R \text{ and } (b, c) \in S$$

Going back to the graphic representation of a relation, we view $(a, b) \in R$ as a formal arrow from a to b and in the same way for (b, c); (a, c) can thus be thought of as the formal composite of those formal arrows. In an analogous way, given two distributors $\phi: \mathscr{A} \longrightarrow \mathscr{B}$ and $\psi: \mathscr{B} \longrightarrow \mathscr{C}$, thought of as "categorical relations" from $\mathscr{B}$ to $\mathscr{A}$ and from $\mathscr{C}$ to $\mathscr{B}$, two elements

$$x \in \phi(B, A), \quad y \in \psi(C, B)$$

could be thought of as formal arrows

$$A \xleftarrow{\quad x \quad} B \xleftarrow{\quad y \quad} C$$

yielding a formal composite arrow $xy \in (\psi \circ \phi)(C, A)$. Now since $\mathscr{B}$ is a category, not just a set, we can have the situation

$$x \in \phi(B, A), \quad b \in \mathscr{B}(B', B), \quad y \in \psi(C, B'),$$

$$A \xleftarrow{\quad x \quad} B \xleftarrow{\quad b \quad} B' \xleftarrow{\quad y \quad} C,$$

and clearly we want a relation of the type $x(by) = (xb)y$ between the corresponding formal composites. It remains to define all those "formal composites". Observe that the formal composite $xb \in \phi(B', A)$ can be defined as the element $\phi(b, 1)(x)$; in the same way $by = \psi(1, b)(y)$.

Proposition 7.8.2 *Small categories, distributors and morphisms of distributors organize themselves in a bicategory.*

Proof We have already observed that given two small categories $\mathscr{A}, \mathscr{B}$, the distributors from $\mathscr{A}$ to $\mathscr{B}$ and their morphisms organize themselves in a category.

Now choose two distributors $\phi\colon \mathscr{A} \longrightarrow \mathscr{B}$ and $\psi\colon \mathscr{B} \longrightarrow \mathscr{C}$. We shall define their composite $\psi \circ \phi\colon \mathscr{A} \longrightarrow \mathscr{C}$. Given objects $A \in \mathscr{A}, C \in \mathscr{C}$ we put

$$(\psi \circ \phi)(C, A) = \frac{\coprod_{B \in \mathscr{B}} \psi(C, B) \times \phi(B, A)}{\approx}$$

where the equivalence relation is that generated by all pairs

$$(y, \phi(b, 1)(x)) \approx (\psi(1, b)(y), x)$$

for all $x \in \phi(B, A)$, $b \in \mathscr{B}(B', B)$, $y \in \psi(C, B')$. We must still define $\psi \circ \phi$ on the arrows. Choosing $a\colon A \longrightarrow A'$ in $\mathscr{A}$ and $c\colon C' \longrightarrow C$ in $\mathscr{C}$, $(\psi \circ \phi)(a, c)$ maps the equivalence class of the pair

$$(v, u) \in \psi(C, B) \times \phi(B, A)$$

to the equivalence class of the pair

$$\big(\psi(c, 1)(v), \phi(1, a)(u)\big) \in \psi(C', B) \times \phi(B, A').$$

This definition is compatible with the equivalence relations defining $\psi \circ \phi$; indeed, with the previous notation,

$$\begin{aligned}
(\psi \circ \phi)(a, c)\Big[\big(y, \phi(b, 1)(x)\big)\Big] &= \Big[\big(\psi(c, 1)(y), \phi(1, a)\phi(b, 1)(x)\big)\Big] \\
&= \Big[\big(\psi(c, 1)(y), \phi(b, a)(x)\big)\Big] \\
&= \Big[\big(\psi(c, 1)(y), \phi(b, 1)\phi(1, a)(x)\big)\Big] \\
&= \Big[\big(\psi(1, b)\psi(c, 1)(y), \phi(1, a)(x)\big)\Big] \\
&= \Big[\big(\psi(c, b)(y), \phi(1, a)(x)\big)\Big] \\
&= \Big[\big(\psi(c, 1)\psi(1, b)(y), \phi(1, a)(x)\big)\Big] \\
&= (\psi \circ \phi)(a, c)\Big[\big(\psi(1, b)(y), x\big)\Big].
\end{aligned}$$

This defines the mapping

$$(\psi \circ \phi)(c, a)\colon (\psi \circ \varphi)(C, A) \longrightarrow (\psi \circ \phi)(C', A').$$

It is straightforward to check that $\psi \circ \varphi$ is actually a bifunctor, thus a distributor $\mathscr{A} \longrightarrow \mathscr{C}$.

Now choose other distributors $\sigma\colon \mathscr{A} \longrightarrow \mathscr{B}$, $\tau\colon \mathscr{B} \longrightarrow \mathscr{C}$ and two morphisms of distributors $\alpha\colon \phi \Rightarrow \sigma$, $\beta\colon \psi \Rightarrow \tau$. We must define a morphism of distributors $\beta * \alpha$: $\psi \circ \phi \Rightarrow \tau \circ \sigma$. To define $(\beta * \alpha)_{CA}$ we map the equivalence class of the pair

$$(v, u) \in \psi(C, B) \times \phi(B, A)$$

to the equivalence class of the pair

$$(\beta_{CB}(v), \alpha_{BA}(u)) \in \tau(C, B) \times \sigma(B, A).$$

This definition is compatible with the equivalence relations defining $\psi \circ \phi$ and $\tau \circ \sigma$; indeed, with the previous notation,

$$
\begin{aligned}
(\beta * \alpha)_{CA}\Big[(y, \phi(b, 1)(x))\Big] &= \Big[(\beta_{CB'}(y), \alpha_{B'A}\phi(b, 1)(x))\Big] \\
&= \Big[(\beta_{CB'}(y), \sigma(b, 1)\alpha_{BA}(x))\Big] \\
&= \Big[(\tau(1, b)\beta_{CB'}(y), \alpha_{BA}(x))\Big] \\
&= \Big[(\beta_{CB}\psi(1, b)(y), \alpha_{BA}(x))\Big] \\
&= (\beta * \alpha)_{CA}\Big[(\psi(1, b)(y), x)\Big].
\end{aligned}
$$

This defines a mapping

$$(\beta * \alpha)_{CA} \colon (\psi \circ \varphi)(C, A) \longrightarrow (\tau \circ \sigma)(C, A).$$

It is straightforward to check that $\beta * \alpha$ is a natural transformation and, finally, that we have defined a bifunctor of composition

$$\mathsf{Dist}(\mathscr{A}, \mathscr{B}) \times \mathsf{Dist}(\mathscr{B}, \mathscr{C}) \longrightarrow \mathsf{Dist}(\mathscr{A}, \mathscr{C}).$$

The composite $\psi \circ \varphi$ of two distributors has been obtained via colimit processes in the category of sets. For these colimits to exist, observe it was necessary to assume $\mathscr{B}$ small. But those colimits are just defined up to an isomorphism. For this reason one can only prove the associativity law up to an isomorphism.

As far as identities are concerned, we have to specify a distinguished distributor $i_{\mathscr{A}} \colon \mathscr{A} \dashrightarrow \mathscr{A}$. This is just the hom-functor

$$\mathscr{A} \colon \mathscr{A}^* \times \mathscr{A} \longrightarrow \mathsf{Set}, \quad (A, A') \mapsto \mathscr{A}(A, A').$$

Indeed choose a distributor $\psi \colon \mathscr{A} \dashrightarrow \mathscr{C}$. When computing the colimit

$$(\psi \circ \mathscr{A})(C, A) = \frac{\coprod_{B \in \mathscr{A}} \psi(C, B) \times \mathscr{A}(B, A)}{\approx},$$

we observe that the equivalence relation is generated by the pairs

$$(y, x \circ b) \approx (\psi(1, b)(y), x)$$

where $x \in \mathcal{A}(B, A)$, $b \in \mathcal{A}(B', B)$ and $y \in \psi(C, B')$. In particular, given $(v, u) \in \psi(C, B) \times \mathcal{A}(B, A)$,

$$[(v, u)] = [(v, 1_A \circ u)] = \Big[\big(\psi(1_C, u)(v), 1_A \big) \Big],$$

so that we can write

$$(\psi \circ \mathcal{A})(C, A) \cong \frac{\psi(C, A)}{\approx}$$

where the equivalence relation is generated by the pairs

$$\psi(1_C, x \circ b)(y) \approx \psi(1_C, x)\psi(1_C, b)(y)$$

for all $x \in \mathcal{A}(B, A)$, $b \in \mathcal{A}(B', B)$ and $y \in \psi(C, B')$. But those pairs are pairs of equal elements, so that finally $(\psi \circ \mathcal{A})(C, A) \cong \psi(C, A)$. Once more we have found an isomorphism, not an identity. An analogous argument holds when composing with a distributor $\phi \colon \mathcal{D} \longrightarrow\!\!\!\circ\!\!\!\longrightarrow \mathcal{A}$.

It is now a lengthy but easy job to verify that all the axioms for a bicategory are satisfied. $\square$

Example 7.8.3

Consider a functor $F \colon \mathcal{A} \longrightarrow \mathcal{B}$. This yields two bifunctors

$$F_* \colon \mathcal{A}^* \times \mathcal{B} \longrightarrow \text{Set}, \quad (A, B) \mapsto \mathcal{B}(FA, B),$$
$$F^* \colon \mathcal{B}^* \times \mathcal{A} \longrightarrow \text{Set}, \quad (B, A) \mapsto \mathcal{B}(B, FA),$$

thus two distributors

$$F_* \colon \mathcal{B} \longrightarrow\!\!\!\circ\!\!\!\longrightarrow \mathcal{A}, \quad F^* \colon \mathcal{A} \longrightarrow\!\!\!\circ\!\!\!\longrightarrow \mathcal{B}.$$

Example 7.8.4

Consider now two functors $F, G \colon \mathcal{A} \rightrightarrows \mathcal{B}$ and a natural transformation $\alpha \colon F \Rightarrow G$. This yields two natural transformations defined by

$$(\alpha_*)_{AB} \colon \mathcal{B}(GA, B) \longrightarrow \mathcal{B}(FA, B), \quad (\alpha_*)_{AB} = \mathcal{B}(\alpha_A, 1_B),$$
$$(\alpha^*)_{BA} \colon \mathcal{B}(B, FA) \longrightarrow \mathcal{B}(B, GA), \quad (\alpha^*)_{BA} = \mathcal{B}(1_B, \alpha_A),$$

thus two morphisms of distributors

$$\alpha_* \colon G_* \Rightarrow F_*, \quad \alpha^* \colon F^* \Rightarrow G^*.$$

Let us write **Cat** for the 2-category of small categories, functors and natural transformations and **Dist** for the bicategory of small categories, distributors and natural transformations.

$$\mathcal{B}(FA, FA) \xrightarrow{\beta_{FA,A}} \mathcal{B}(FA, GA)$$

$$\mathcal{B}(f, 1_{FA}) \Big\downarrow \qquad\qquad \Big\downarrow \mathcal{B}(f, 1_{GA})$$

$$\mathcal{B}(B, FA) \xrightarrow[\beta_{B,A}]{} \mathcal{B}(B, GA)$$

Diagram 7.24

Proposition 7.8.5 *There exists an injective pseudo-functor*
$$\phi\colon \mathbf{Cat} \longrightarrow \mathbf{Dist}$$
with the property that for each pair of categories $\mathcal{A}, \mathcal{B}$
$$\phi_{\mathcal{A},\mathcal{B}}\colon \mathsf{Fun}(\mathcal{A}, \mathcal{B}) \longrightarrow \mathsf{Dist}(\phi\mathcal{A}, \phi\mathcal{B})$$
is full and faithful.

Proof Just define $\phi(\mathcal{A}) = \mathcal{A}$, $\phi(F) = F^*$ and $\phi(\alpha) = \alpha^*$. This clearly defines a pseudo-functor.

To prove the last assertion, choose two functors $F, G\colon \mathcal{A} \rightrightarrows \mathcal{B}$; we must prove that the mapping
$$\mathsf{Nat}(F, G) \longrightarrow \mathsf{Nat}(F^*, G^*), \quad \alpha \mapsto \alpha^*$$
is a bijection. First of all observe that given $A \in \mathcal{A}$
$$\alpha^*_{FA,A}\colon \mathcal{B}(FA, FA) \longrightarrow \mathcal{B}(FA, GA)$$
maps 1_{FA} to α_A; this proves the injectivity. Now starting from a natural transformation $\beta\colon F^* \Rightarrow G^*$, it suffices to consider for every $A \in \mathcal{A}$
$$\beta_{FA,A}\colon \mathcal{B}(FA, FA) \longrightarrow \mathcal{B}(FA, GA)$$
and put $\alpha_A = \beta_{FA,A}(1_{FA})$. The naturality of β immediately implies that of α. Moreover, considering the commutative diagram 7.24 where $f \in \mathcal{B}(B, FA)$, we conclude that
$$\beta_{B,A}(f) = \beta_{B,A} \circ \mathcal{B}(f, 1_{FA})(1_{FA}) = \mathcal{B}(f, 1_{GA}) \circ \beta_{FA,A}(1_{FA})$$
$$= \beta_{FA,A}(1_{FA}) \circ f = \alpha_A \circ f = \alpha^*_{B,A}(f),$$
proving the relation $\beta = \alpha^*$.

The pseudo-functor ϕ is certainly injective on the objects (it is the identity!); it remains to prove it is injective on the arrows. If two functors $F, G\colon \mathcal{A} \rightrightarrows \mathcal{B}$ are such that $F^* = G^*$, then given $A \in \mathcal{A}$, $\mathcal{B}(FA, FA) = \mathcal{B}(FA, GA)$ and we deduce that 1_{FA} is also a morphism $FA \longrightarrow GA$;

this yields $FA = GA$ (to be precise, we have to assume the disjointness of the various sets of morphisms in $\mathscr{B}$, which is no restriction at all). Finally choosing in $\mathscr{A}$ a morphism $a\colon A \longrightarrow A'$,

$$F(a) = F^*(1_{FA}, a)(1_{FA}) = G^*(1_{GA}, a)(1_{GA}) = G(a). \qquad \square$$

Proposition 7.8.5 justifies our choice of defining a distributor $\mathscr{A} \mathrel{-\!\!\!\circ\!\!\!\rightarrow} \mathscr{B}$ as something which appears intuitively as a categorical relation from $\mathscr{B}$ to $\mathscr{A}$. Making the opposite choice would have inverted the direction of the arrows in **Dist**, but not that of the 2-cells. Therefore ϕ would have been contravariant on the arrows and covariant on the 2-cells (or the converse if defining $\phi(F) = F_*$, $\phi(\alpha) = \alpha_*$): a situation which is not very elegant to handle, especially when one intends to view **Cat** as a 2-sub-category of **Dist**.

7.9 Cauchy completeness versus distributors

One of the reasons why distributors have been introduced is actually to produce a formal adjoint to every functor. Indeed, with the notation of 7.8.3

Proposition 7.9.1 *Consider a functor $F\colon \mathscr{A} \longrightarrow \mathscr{B}$ between small categories. The distributor $F_*\colon \mathscr{B} \mathrel{-\!\!\!\circ\!\!\!\rightarrow} \mathscr{A}$ is right adjoint to the distributor $F^*\colon \mathscr{A} \mathrel{-\!\!\!\circ\!\!\!\rightarrow} \mathscr{B}$ which, in view of 7.8.5, is the "functor F embedded in* **Dist**".

Proof Let us first compute $F_* \circ F^*$ and $F^* \circ F_*$. Given objects $A, A' \in \mathscr{A}$, $B, B' \in \mathscr{B}$,

$$(F_* \circ F^*)(A, A') = \frac{\coprod_{B \in \mathscr{B}} \mathscr{B}(FA, B) \times \mathscr{B}(B, FA')}{\approx},$$

$$(F^* \circ F_*)(B, B') = \frac{\coprod_{A \in \mathscr{A}} \mathscr{B}(B, FA) \times \mathscr{B}(FA, B')}{\approx}.$$

We must produce natural transformations

$$\eta\colon \mathscr{A} \Rightarrow F_* \circ F^*, \quad \varepsilon\colon F^* \circ F_* \Rightarrow \mathscr{B}$$

satisfying the conditions of 7.7.2.

First of all, in the definition of $(F_* \circ F^*)(A, A')$ observe that given

$$FA' \xleftarrow{\ x'\ } B \xleftarrow{\ x\ } FA =\!\!=\!\!= FA,$$

$x' \in \mathscr{B}(B, FA')$ and $x \in \mathscr{B}(FA, B)$, we get an equivalent pair

$$(x, x') \approx (1_{FA}, x' \circ x).$$

This implies the isomorphism

$$(F_* \circ F^*)(A, A') \cong \frac{\mathscr{B}(FA, FA')}{\approx}$$

where the equivalence is now induced by

$$y \circ (b \circ x) \approx (y \circ b) \circ x, \quad y \in \mathscr{B}(B', FA'), \quad b \in \mathscr{B}(B, B'), \quad x \in \mathscr{B}(FA, B).$$

Thus the equivalence on $\mathscr{B}(FA, FA')$ is generated by pairs of equal morphisms so that finally

$$(F_* \circ F^*)(A, A') \cong \mathscr{B}(FA, FA').$$

This lets us define $\eta_{AA'}$ as the mapping

$$\eta_{AA'} \colon \mathscr{A}(A, A') \longrightarrow \mathscr{B}(FA, FA'), \quad a \mapsto Fa.$$

Consider now the definition of $(F^* \circ F_*)(B, B')$. Given a pair of elements $(b, b') \in \mathscr{B}(B, FA) \times \mathscr{B}(FA, B')$, we get $b' \circ b \in \mathscr{B}(B, B')$. This construction is compatible with the equivalence relation involved in the definition of $(F^* \circ F_*)(B, B')$; indeed, given $b \colon B \longrightarrow FA$, $a \colon A \longrightarrow A'$ and $b' \colon FA' \longrightarrow B'$, we have an equivalent pair

$$\bigl(b, b' \circ F(a)\bigr) \approx \bigl(F(a) \circ b, b'\bigr)$$

and clearly $\bigl(b' \circ F(a)\bigr)b = b'\bigl(F(a) \circ b\bigr)$. This yields a mapping

$$\varepsilon_{BB'} \colon (F^* \circ F_*)(B, B') \longrightarrow \mathscr{B}(B, B'), \quad [(b, b')] \mapsto b' \circ b.$$

We leave to the reader the straightforward verifications that η, ε are natural and satisfy the conditions of 7.7.2. $\square$

Having in mind proposition 7.9.1, one might wonder if actual functors $F \colon \mathscr{A} \longrightarrow \mathscr{B}$ can be exactly characterized as those distributors $\mathscr{A} \longrightarrow\!\!\!\circ \mathscr{B}$ which have a right adjoint. The answer is "yes, if and only if $\mathscr{B}$ is Cauchy complete."

Let us write **1** for the final category. A distributor $\mathbf{1} \longrightarrow\!\!\!\circ \mathscr{A}$ is just a contravariant functor $\mathscr{A}^* \longrightarrow \mathsf{Set}$. Therefore we can view the covariant Yoneda embedding as a functor

$$Y \colon \mathscr{A} \longrightarrow \mathsf{Fun}(\mathscr{A}^*, \mathsf{Set}) \cong \mathsf{Dist}(\mathbf{1}, \mathscr{A}).$$

Proposition 7.9.2 *Consider a distributor* $\phi \colon \mathbf{1} \longrightarrow\!\!\!\circ \mathscr{A}$ *where* $\mathscr{A}$ *is a small category. The following conditions are equivalent:*

(1) *the distributor* ϕ *has a right adjoint* ψ;

(2) *via the isomorphism* $\mathsf{Fun}(\mathscr{A}^*, \mathsf{Set}) \cong \mathsf{Dist}(\mathbf{1}, \mathscr{A})$, ϕ *is isomorphic to a retract of a representable functor.*

Proof Suppose ψ is right adjoint to ϕ. We thus have two functors

$$\phi \colon \mathscr{A}^* \longrightarrow \mathsf{Set}, \ \ \psi \colon \mathscr{A} \longrightarrow \mathsf{Set}.$$

Since $\mathbf{1}$ has a single object and a single arrow (the identity), the two composites are given by

$$\phi \circ \psi \colon \mathscr{A}^* \times \mathscr{A} \longrightarrow \mathsf{Set},$$

$$(\phi \circ \psi)(A, B) = \phi A \times \psi B, \ \ (\psi \circ \phi)(*, *) = \frac{\coprod_A \psi A \times \phi A}{\approx},$$

where $\approx$ is the equivalence relation generated by

$$\big(x, (\phi f)(y)\big) \approx \big((\psi f)(x), y\big)$$

for every $f \colon A \longrightarrow A'$, $x \in \psi A$, $y \in \phi A'$ (see 7.8.1).

The two canonical morphisms of the adjunction, $\varepsilon \colon \phi \circ \psi \Rightarrow 1_{\mathscr{A}}$ and $\eta \colon 1_1 \Rightarrow \psi \circ \phi$, are thus

- a family of mappings $\varepsilon_{A,B} \colon \phi A \times \psi B \longrightarrow \mathscr{A}(A, B)$, natural in A and B,
- an element $\eta_* \in (\psi \circ \phi)(*, *)$.

Since $(\psi \circ \phi)(*, *)$ is defined as a quotient, choose an object $C \in \mathscr{A}$ and an element $(u, v) \in \psi C \times \phi C$ such that $\eta_* = \big[(u, v)\big]$.

In order to express the triangular identities for adjointness, let us compute

$$\phi \circ \psi \circ \phi \colon \mathscr{A}^* \longrightarrow \mathsf{Set}, \ \ \psi \circ \phi \circ \psi \colon \mathscr{A} \longrightarrow \mathsf{Set},$$

$$(\phi \circ \psi \circ \phi)(A) \cong \big(\phi \circ (\psi \circ \phi)\big)(A) \cong \phi A \times (\psi \circ \phi)(*, *)$$

$$\cong \big((\phi \circ \psi) \circ \phi\big)(A) \cong \frac{\coprod_B \phi(A) \times \psi(B) \times \phi(B)}{\approx_1},$$

$$(\psi \circ \phi \circ \psi)(B) \cong \big((\psi \circ \phi) \circ \psi\big)(B) \cong (\psi \circ \phi)(*, *) \times \phi(B)$$

$$\cong \big(\psi \circ (\phi \circ \psi)\big)(B) \cong \frac{\coprod_A \psi(A) \times \phi(A) \times \psi(B)}{\approx_2},$$

where $\approx_1$ is the equivalence relation generated by

$$\big(a, b, \phi(f)(c)\big) \approx \big(a, \psi(f)(b), c\big)$$

for $a \in \phi A$, $b \in \psi B$, $c \in \phi B'$, $f \colon B \longrightarrow B'$, while $\approx_2$ is the equivalence relation generated by

$$\big(r, \phi(f)(s), t\big) \approx \big(\psi(f)(r), s, t\big)$$

for $r \in \psi A$, $s \in \phi A'$, $t \in \psi B$, $f \colon A \longrightarrow A'$.

Consider now the first relation for adjunction in 7.7.2; up to canonical natural isomorphisms, we can write it $(\psi * \varepsilon) \circ (\eta * \psi) \cong 1_\psi$. Via the previous computations, this means just, for $x \in \psi B$,

$$
\begin{aligned}
x &= (\psi * \varepsilon)_B \circ (\eta * \psi)_B(x) \\
&= (\psi * \varepsilon)_B \big[(u, v, x) \big] \\
&= \psi \big(\varepsilon_{CB}(v, x) \big)(u).
\end{aligned}
$$

Next we consider the second relation for the adjunction, which reduces to $(\varepsilon * \phi) \circ (\phi * \eta) \cong 1_\phi$ up to canonical natural isomorphisms. Given $y \in \phi A$, this relation means

$$
\begin{aligned}
y &= (\varepsilon * \phi)_A \circ (\phi * \eta)_A(y) \\
&= (\varepsilon * \phi)_A \big[(y, u, v) \big] \\
&= \phi \big(\varepsilon_{AC}(y, u) \big)(v).
\end{aligned}
$$

Let us prove now that ϕ is a retract of the representable functor $\mathscr{A}(-, C)$, seen as a distributor $\Gamma \colon 1 \nrightarrow \mathscr{A}$. One easily defines two natural transformations

$$
\gamma \colon \phi \Rightarrow \mathscr{A}(-, C), \quad \gamma_A(y) = \varepsilon_{AC}(y, u),
$$
$$
\delta \colon \mathscr{A}(-, C) \Rightarrow \phi, \quad \delta_B(g) = \phi(g)(v),
$$

where $A, B \in \mathscr{C}$, $y \in \phi A$, $g \colon B \longrightarrow C$. Clearly, γ is natural since ε is and δ is natural just because ϕ is a functor. The second triangular equality for adjointness can be rewritten $\delta_A \circ \gamma_A(y) = y$, which proves that $\delta \circ \gamma = i_\phi$. Thus ϕ is indeed a retract of $\mathscr{A}(-, C)$.

Conversely suppose we are given

$$
\gamma \colon \phi \Rightarrow \mathscr{A}(-, C), \quad \delta \colon \mathscr{A}(-, C) \Rightarrow \phi
$$

such that $\delta \circ \gamma = i_\phi$. The composite

$$
\gamma \circ \delta \colon \mathscr{A}(-, C) \Rightarrow \mathscr{A}(-, C)
$$

is an idempotent on $\mathscr{A}(-, C)$. Since the Yoneda embedding is full and faithful (see 1.5.2), $\gamma \circ \delta = \mathscr{A}(-, e)$ for some idempotent $e \colon C \longrightarrow C$ in $\mathscr{A}$. Now consider the idempotent natural transformation

$$
\mathscr{A}(e, -) \colon \mathscr{A}(C, -) \Rightarrow \mathscr{A}(C, -).
$$

Since $\mathsf{Fun}(\mathscr{A}, \mathsf{Set})$ is complete (see 2.15.2), idempotents split (see 6.5.4) and we find a retract $\alpha, \beta \colon \psi \rightleftarrows \mathscr{A}(C, -)$ with the properties $\alpha \circ \beta =$

1_ψ and $\beta \circ \alpha = \mathscr{A}(e, -)$. Let us prove that ψ is right adjoint to ϕ, ϕ and ψ being considered as distributors.

Let us define $\varepsilon \colon \phi \circ \psi \Rightarrow 1_{\mathscr{A}}$ by

$$\varepsilon_{A,B} \colon (\phi \circ \psi)(A, B) = \phi A \times \psi B \xrightarrow{\hspace{2cm}} \mathscr{A}(A, B),$$

$$\varepsilon_{AB}(y, x) = \beta_B(x) \circ \gamma_A(y).$$

The naturality of $\varepsilon_{A,B}$ follows immediately from the naturalities of β and γ. Let us also define

$$u = \alpha_C(1_C) \in \psi C, \quad v = \delta_C(1_C) \in \phi C.$$

Putting $\eta_* = [(u, v)]$, we have defined $\eta \colon 1_1 \Rightarrow \psi \circ \phi$.

In order to check the two triangular equalities for adjunction, recall that

$$\gamma = \mathsf{Ker}\left(\mathscr{A}(-, 1_C), \mathscr{A}(-, e)\right), \quad \beta = \mathsf{Ker}\left(\mathscr{A}(1_C, -), \mathscr{A}(e, -)\right)$$

(see 6.5.4). Therefore given $z \in \phi(C)$ one has

$$\gamma_C \circ \phi(e)(z) = \mathscr{A}(C, e) \circ \gamma_C(z) = \mathscr{A}(C, 1_C) \circ \gamma_C(z) = \gamma_C(z).$$

As an equalizer, γ_C is injective (see 2.2.4.(3) and 2.15.3); this yields $\phi(e)(z) = z$ for every $z \in \phi(C)$. In an analogous way, one proves that $\psi(e)(z) = z$ for every $z \in \psi(C)$. In particular $\phi(e)(v) = v$ and $\psi(e)(u) = u$.

Now let us check the triangular identities for the adjunction (see 7.7.2). Both proofs are analogous; we develop the second one. Viewing ϕ, ψ as distributors, the second triangular identity reduces to

$$\left((\varepsilon * \phi) \circ (\phi * \eta)\right)_{(A,*)}(y) = y$$

for every $A \in \mathscr{A}$ and $y \in \phi(A, *)$. Going back to the description of $\phi \circ \psi \circ \phi$ at the beginning of the proof, y is mapped by $(\phi \circ \eta)_{(A,a)}$ to the equivalence class

$$[(y, u, v)] \in \phi(A) \times \psi(C) \times \phi(C)$$

and this class is itself mapped to the equivalence class

$$\left[(\beta_C(u) \circ \gamma_A(y), v)\right] \in (1_{\mathscr{A}} \circ \phi)(A, *).$$

By definition of the equivalence relation defining $(1_{\mathscr{A}} \circ \phi)(A, *)$, this is just the element

$$\phi\big(\beta_C(u) \circ \gamma_A(y)\big)(v) \in \phi(A) \equiv \phi(A, *).$$

Since ϕ is contravariant as a functor and $\phi(e)(v) = v$, one computes immediately that

$$\phi\big(\beta_C(u) \circ \gamma_A(y)\big)(v) = \phi\Big(\beta_C\big(\alpha_C(1_C)\big) \circ \gamma_A(y)\Big)(v)$$
$$= \phi\big(\mathscr{A}(e, C)(1_C) \circ \gamma_A(y)\big)(v)$$
$$= \phi\big(e \circ \gamma_A(y)\big)(v)$$
$$= \phi\big(\gamma_A(y)\big)\big(\phi(e)(v)\big)$$
$$= \phi\big(\gamma_A(y)\big)(v)$$
$$= \Big(\phi\big(\gamma_A(y)\big) \circ \delta_C\Big)(1_C)$$
$$= \Big(\delta_A \circ \mathscr{A}\big(\gamma_A(y), C\big)\Big)(1_C)$$
$$= \delta_A\big(\gamma_A(y)\big)$$
$$= y. \qquad\qquad \square$$

The relation between Cauchy completeness and the theory of distributors is expressed by the following theorem.

Theorem 7.9.3 *Given a small category $\mathscr{A}$, the following conditions are equivalent:*

(1) $\mathscr{A}$ is Cauchy complete;

(2) a distributor $\phi \colon 1 \longrightarrow \mathscr{A}$ has a right adjoint if and only if it is isomorphic to a functor;

(3) for every small category $\mathscr{B}$, a distributor $\theta \colon \mathscr{B} \longrightarrow \mathscr{A}$ has a right adjoint if and only if it is isomorphic to a functor.

Proof A functor $F \colon 1 \longrightarrow \mathscr{A}$ is just the choice of an object $F* \in \mathscr{A}$. Seen as a distributor it is

$$\mathscr{A}^* \cong \mathscr{A}^* \times 1 \longrightarrow \mathrm{Set}, \quad (A, *) \mapsto \mathscr{A}(A, F*);$$

thus it is the representable functor $\mathscr{A}(-, F*)$.

Suppose $\mathscr{A}$ is Cauchy complete. By 7.9.2 a distributor $\phi \colon 1 \longrightarrow \mathscr{A}$ has a right adjoint iff it is a retract of a representable functor, i.e. by 6.5.6 iff it is representable. This proves (1) $\Rightarrow$ (2).

(3) $\Rightarrow$ (2) is obvious. Let us prove (2) $\Rightarrow$ (3). If θ is a functor, we know it has a right adjoint distributor (see 7.9.1). Now suppose θ is just a distributor, with right adjoint τ. For every object $B \in \mathscr{B}$ consider the situation

$$1 \underset{\mathscr{B}(-, B)}{\overset{\mathscr{B}(B, -)}{\rightleftarrows}} \mathscr{B} \underset{\theta}{\overset{\tau}{\rightleftarrows}} \mathscr{A}$$

where $\mathcal{B}(B,-)$, seen as a distributor, is right adjoint to $\mathcal{B}(-,B)$, seen as another distributor (see 7.8.3). Composing the adjunctions we obtain that $\mathcal{B}(B,-) \circ \tau$ is right adjoint to $\theta \circ \mathcal{B}(-,B)$. Applying our assumption, the composite $\theta \circ \mathcal{B}(-,B)$ is a functor. Therefore we find an object $FB \in \mathcal{A}$ such that

$$\mathcal{A}(-,FB) \cong \theta \circ \mathcal{B}(-,B) \colon \mathcal{A}^* \longrightarrow \mathrm{Set}.$$

Let us explicitly compute $(\theta \circ \mathcal{B}(-,B))(A)$.

$$(\theta \circ \mathcal{B}(-,B))(A) = \frac{\coprod_{C \in \mathcal{B}} \theta(A,C) \times \mathcal{B}(C,B)}{\approx},$$

where $\approx$ is the equivalence relation generated by

$$(x, f \circ g) \approx (\theta(1_A, g)(x), f)$$

for $g \colon C \longrightarrow C'$, $x \in \theta(A,C)$ and $f \in \mathcal{B}(C',B)$. In particular, given elements $y \in \theta(A,C)$ and $h \in \mathcal{B}(C,B)$, one has

$$(x, 1_B \circ h) \approx (\theta(1_A, h)(x), 1_B),$$

so that every pair (y,h) is equivalent to a pair $(z, 1_B)$ with $z \in \theta(A,B)$. These considerations already yield a surjective mapping

$$\alpha \colon \theta(A,B) \longrightarrow (\theta \circ \mathcal{B}(-,B))(A), \quad z \mapsto [(z, 1_B)].$$

In order for them to yield another mapping

$$\beta \colon (\theta \circ (-,B))(A) \longrightarrow \theta(A,B), \quad [(x,h)] \mapsto \theta(1_A, h)(x)$$

we must prove that $(x,h) \approx (x',h')$ implies $\theta(1_A,h)(x) = \theta(1_A,h')(x')$, for $x' \in \theta(A,C')$ and $h' \in \mathcal{B}(C',B)$. It suffices to prove this for a couple of pairs generating the equivalence relation. Thus let us consider

$$(x, f \circ g) \approx (\theta(1_A, g)(x), f)$$

as above. We obviously have

$$\theta(1_A, f \circ g)(x) = \theta(1_A, f)(\theta(1_A, g)(x)),$$

which implies the existence of β. The relation $\beta \circ \alpha = 1$ is obvious and the relation $\alpha \circ \beta = 1$ holds by definition of the equivalence relation.

So we are at the point of having found an object $FB \in \mathcal{A}$ together with bijections

$$\mathcal{A}(A,FB) \cong (\theta \circ \mathcal{B}(-,B))(A) \cong \theta(A,B).$$

It remains to extend F to a functor, in such a way that these bijections become a natural isomorphism. A morphism $b: B \longrightarrow B'$ yields a natural transformation

$$\theta * \mathscr{B}(-,b): \theta \circ \mathscr{B}(-,B) \Rightarrow \theta \circ (-,B'),$$

thus a natural transformation $\mathscr{A}(-,FB) \Rightarrow \mathscr{A}(-,FB')$. Since the Yoneda embedding is full and faithful, this transformation has the form $\mathscr{A}(-,Fb)$ for some unique morphism $Fb: FB \longrightarrow FB'$ (see 1.5.2). This defines F on the arrows. The rest is now straightforward observations left to the reader.

It remains to prove (2) $\Rightarrow$ (1). Given a retract $R \xrightarrow{\longleftarrow} \mathscr{A}(-,A)$ of a representable functor, we consider the functor $R: \mathscr{A}^* \longrightarrow \mathsf{Set}$ which can be seen as a distributor $\phi: 1 \longrightarrow \mathscr{A}$. By 7.9.2, the distributor ϕ has a right adjoint distributor ψ. By assumption, this implies $\phi = F^*$ for some functor $F: 1 \longrightarrow \mathscr{A}$. In other words, $R(A) \cong \phi(A,*) \cong \mathscr{A}(A,F*)$ for some object $F* \in \mathscr{A}$, from which it follows immediately that R is the contravariant functor represented by $F*$. $\qquad\square$

Let us conclude this section with an interesting characterization of those categories with equivalent Cauchy completions.

Theorem 7.9.4 *Given two small categories $\mathscr{A}$ and $\mathscr{B}$, the following conditions are equivalent:*

(1) the categories of set-valued functors $\mathsf{Fun}(\mathscr{A}^*, \mathsf{Set})$ and $\mathsf{Fun}(\mathscr{B}^*, \mathsf{Set})$ are equivalent;

(2) the Cauchy completions of $\mathscr{A}$ and $\mathscr{B}$ are equivalent;

(3) there exist distributors $\phi: \mathscr{A} \longrightarrow \mathscr{B}$ and $\psi: \mathscr{B} \longrightarrow \mathscr{A}$ such that $\psi \circ \phi \cong 1_{\mathscr{A}}$ and $\phi \circ \psi \cong 1_{\mathscr{B}}$;

(4) $\mathscr{A}$ and $\mathscr{B}$ are equivalent in the bicategory Dist of small categories and distributors.

When these conditions are satisfied, $\mathscr{A}$ and $\mathscr{B}$ are called "Morita equivalent categories".

Proof The equivalence (1) $\Leftrightarrow$ (2) has already been proved in 6.5.11 and (3) is just a reformulation of (4).

Since the composite of two equivalences is obviously an equivalence, (2) $\Rightarrow$ (4) will be proved if we show that the inclusion $i: \mathscr{A} \hookrightarrow \overline{\mathscr{A}}$ of $\mathscr{A}$ in its Cauchy completion (see 6.5.9) is an equivalence in Dist. This inclusion yields the two adjoint distributors (see 7.9.1)

$$i^*: \overline{\mathscr{A}}^* \times \mathscr{A} \longrightarrow \mathsf{Set}, \quad (F,A) \mapsto \mathsf{Nat}\big(F, \mathscr{A}(-,A)\big),$$
$$i_*: \mathscr{A}^* \times \overline{\mathscr{A}} \longrightarrow \mathsf{Set}, \quad (A,F) \mapsto \mathsf{Nat}\big(\mathscr{A}(-,A),F\big).$$

As observed in the proof of 7.9.1, $(i_* \circ i^*)(A, B) \cong \overline{\mathscr{A}}(iA, iB)$, i.e. by definition of $\overline{\mathscr{A}}$ (see 6.5.9) and the Yoneda lemma (see 1.3.3),

$$(i_* \circ i^*)(A, B) \cong \overline{\mathscr{A}}\big(\mathscr{A}(-, A), \mathscr{A}(-, B)\big) \cong \mathscr{A}(A, B).$$

Going back again to the proof of 7.9.1, we deduce that the canonical natural transformation $\eta\colon \mathscr{A} \Rightarrow i_* \circ i^*$ is an isomorphism. Next we know, again by the proof of 7.9.1, that

$$(i^* \circ i_*)(F, G) \cong \frac{\coprod_{A \in \mathscr{A}} \mathsf{Nat}\big(F, \mathscr{A}(-, A)\big) \times \mathsf{Nat}\big(\mathscr{A}(-, A), G\big)}{\approx},$$

where the equivalence relation is generated by the pairs

$$\big(\alpha, \beta \circ \mathscr{A}(-, a)\big) \approx \big(\mathscr{A}(-, a) \circ \alpha, \beta\big)$$

for $a\colon A \longrightarrow B$ in $\mathscr{A}$, $\alpha\colon F \Rightarrow \mathscr{A}(-, A)$, $\beta\colon \mathscr{A}(-, B) \Rightarrow G$. The second natural transformation of the adjunction is given by

$$\varepsilon_{(F,G)}\colon (i^* \circ i_*)(F, G) \longrightarrow \overline{\mathscr{A}}(F, G), \quad [(\alpha, \beta)] \mapsto \beta \circ \alpha.$$

$\varepsilon_{(F,G)}$ is surjective since, by construction of the Cauchy completion as described in 6.5.9, G is a retract of a representable functor, i.e. we have $C \in \mathscr{A}$ and natural transformations $\tau\colon G \Rightarrow \mathscr{A}(-, C)$, $\sigma\colon \mathscr{A}(-, C) \Rightarrow G$ with $\sigma \circ \tau = 1_G$. Therefore every natural transformation $\gamma\colon F \Rightarrow G$ can indeed be written $\sigma \circ (\tau \circ \gamma)$, proving that γ is the image of $[(\tau \circ \gamma, \sigma)]$. $\varepsilon_{(F,G)}$ is also injective since given $\alpha\colon F \Rightarrow \mathscr{A}(-, A)$, $\beta\colon \mathscr{A}(-, A) \Rightarrow G$, with $\beta \circ \alpha = \gamma$, we get a natural transformation

$$\mathscr{A}(-, A) \Rightarrow G \Rightarrow \mathscr{A}(-, C)$$

and thus a morphism $f\colon A \longrightarrow C$ in $\mathscr{A}$ such that $\mathscr{A}(-, f) = \tau \circ \beta$ (see 1.5.2). We have then the equivalent pairs

$$(\alpha, \beta) = (\alpha, \sigma \circ \tau \circ \beta) = \big(\alpha, \sigma \circ \mathscr{A}(-, f)\big)$$
$$\approx \big(\mathscr{A}(-, f) \circ \alpha, \sigma\big) = (\tau \circ \beta \circ \alpha, \sigma) = (\tau \circ \gamma, \sigma),$$

proving the injectivity of $\varepsilon_{(F,G)}$.

It remains to prove (4) $\Rightarrow$ (2). By (2) $\Rightarrow$ (4), $\mathscr{A}$ is equivalent in **Dist** to its Cauchy completion $\overline{\mathscr{A}}$ and in the same way for $\mathscr{B}$ and $\overline{\mathscr{B}}$. Thus if $\mathscr{A}, \mathscr{B}$ are equivalent in **Dist**, $\overline{\mathscr{A}}$ and $\overline{\mathscr{B}}$ are equivalent in **Dist**. By 7.9.3 an equivalence $\phi\colon \overline{\mathscr{A}} \multimap \overline{\mathscr{B}}$, $\psi\colon \overline{\mathscr{B}} \multimap \overline{\mathscr{A}}$ is induced by functors $F\colon \overline{\mathscr{A}} \longrightarrow \overline{\mathscr{B}}$, $G\colon \overline{\mathscr{B}} \longrightarrow \overline{\mathscr{C}}$; the fullness and faithfulness condition in 7.8.5 implies that F, G constitute an equivalence in **Cat**. $\qquad\square$

And now that the whole story about Cauchy completeness has been told, it remains to justify the terminology. This is done in exercises 6.8.5 to 6.8.9 of volume 2, where it is shown that in a metric space (x, d), every Cauchy sequence has a limit if and only if a condition extending 7.9.3.(2) holds.

7.10 Exercises

7.10.1 In a 2-category, prove that when an arrow has an adjoint arrow, this adjoint arrow is defined uniquely up to isomorphism. Show that this result extends to the case of bicategories.

7.10.2 In the 2-category of groups, prove that the triangular conditions in definition 7.1.2 are redundant.

7.10.3 Prove the 2-Yoneda-lemma: given a 2-category $\mathscr{A}$, an object $A \in \mathscr{A}$ and a 2-functor $F: \mathscr{A} \longrightarrow \mathsf{Cat}$, there exists an isomorphism of categories

$$FA \simeq \text{2-Nat}\big(\mathscr{A}(A, -), F\big),$$

where $\mathscr{A}(A, -)$ is the 2-functor defined in 7.2.4 and the right-hand side is the category having the 2-natural transformations $\mathscr{A}(A, -) \Rightarrow F$ as objects and the modifications as arrows.

7.10.4 Consider a commutative monoid M. Prove that M can be seen as the category of 2-cells of a 2-category with just one object and one arrow, both composition laws on 2-cells being the multiplication on M. Prove that this 2-category admits products but not 2-products.

7.10.5 In the 2-category Cat, consider the diagram constituted of two functors $F, G: \mathscr{A} \rightrightarrows \mathscr{B}$. Compute its limit, its 2-limit, its pseudo-limit and its lax limit. Same question if a natural transformation $\alpha: F \Rightarrow G$ is added to this diagram.

7.10.6 Consider the 2-category **1** with a single object, a single arrow and a single 2-cell. Describe in explicit terms what a lax functor $\mathbf{1} \longrightarrow \mathsf{Cat}$ is. Compare with definition 4.1.1, volume 2.

7.10.7 Consider a bicategory $\mathscr{A}$ with a single object $*$. Thinking of the composition of arrows as a tensor product on the objects of the category $\mathscr{A}(*, *)$, compare with definition 6.1.1, volume 2.

7.10.8 Prove that in the bicategory of bimodules, both left and right Kan extensions always exist.

7.10.9 Consider two distributors $\phi\colon \mathscr{A} \longrightarrow\hspace{-0.7em}\circ\hspace{0.3em} \mathscr{B}$ and $\psi\colon \mathscr{B} \longrightarrow\hspace{-0.7em}\circ\hspace{0.3em} \mathscr{C}$ between small categories. The two bifunctors

$$\phi\colon \mathscr{B}^* \times \mathscr{A} \longrightarrow \mathsf{Set}, \quad \psi\colon \mathscr{C}^* \times \mathscr{B} \longrightarrow \mathsf{Set}$$

correspond to functors

$$\overline{\phi}\colon \mathscr{A} \longrightarrow \mathsf{Set}^{\mathscr{B}^*}, \quad \overline{\psi}\colon \mathscr{B} \longrightarrow \mathsf{Set}^{\mathscr{C}^*}.$$

Considering the covariant Yoneda embedding

$$Y\colon \mathscr{B} \longrightarrow \mathsf{Set}^{\mathscr{B}^*},$$

the left Kan extension $\widetilde{\psi}$ of $\overline{\psi}$ along Y exists (see 3.7.2). This yields the composite

$$\mathscr{A} \xrightarrow{\ \overline{\phi}\ } \mathsf{Set}^{\mathscr{B}^*} \xrightarrow{\ \widetilde{\psi}\ } \mathsf{Set}^{\mathscr{C}^*},$$

which corresponds to a bifunctor

$$\psi \otimes \phi\colon \mathscr{C}^* \times \mathscr{A} \longrightarrow \mathsf{Set}.$$

Prove that $\psi \otimes \phi$ is precisely the composite $\psi \circ \phi$ of the two distributors ϕ, ψ.

7.10.10 Prove that in the bicategory **Dist**, both left and right Kan extensions always exist.

7.10.11 Let $\phi\colon \mathscr{A} \longrightarrow\hspace{-0.7em}\circ\hspace{0.3em} \mathscr{B}$ be a distributor between small categories. Prove the existence of a category $\mathscr{C}$ and functors $F\colon \mathscr{A} \longrightarrow \mathscr{C}$, $G\colon \mathscr{B} \longrightarrow \mathscr{C}$ such that $\phi = G_* \circ F^*$. [Hint: to get $\mathscr{C}$, construct the disjoint union $\mathscr{A} \amalg \mathscr{B}$ and add the elements of $\phi(B, A)$ as arrows from B to A.]

7.10.12 Let $\phi\colon \mathscr{A} \longrightarrow\hspace{-0.7em}\circ\hspace{0.3em} \mathscr{B}$ be a distributor between small categories. Prove the existence of a category $\mathscr{D}$ and functors $F\colon \mathscr{D} \longrightarrow \mathscr{A}$, $G\colon \mathscr{D} \longrightarrow \mathscr{B}$ such that $\phi = G^* \circ F_*$. [Hint: consider the comma category constructed on the functors F, G of the previous exercise.]

8

Internal category theory

Up to now, the surrounding mathematical context in which we developed category theory was the category of sets: a small category has a set of objects and a set of arrows, together with some structure on those data. To conclude this first part, we would like to indicate that, for many purposes, the category of sets can be replaced by a rather arbitrary category with good properties (at least pullbacks).

8.1 Internal categories and functors

What we intend to generalize is the notion of a *small* category $\mathscr{A}$: such a category has a *set* $|\mathscr{A}|$ of objects and *sets* $\mathscr{A}(A, B)$ of arrows, for every pair A, B of objects. This last fact is equivalent to giving the disjoint union *set* $\coprod_{A,B} \mathscr{A}(A, B)$ of all arrows, together with the two mappings

$$d_0, d_1 \colon \coprod_{A,B} \mathscr{A}(A, B) \Longrightarrow |\mathscr{A}|$$

which map an arrow, respectively, to its source and its target.

Definition 8.1.1 *Let $\mathscr{C}$ be a category with pullbacks. By an internal category $\mathscr{A}$ in $\mathscr{C}$ we mean*

(1) an object $A_0 \in |\mathscr{C}|$, called the "object of objects",

(2) an object $A_1 \in |\mathscr{C}|$, called the "object of arrows",

(3) two morphisms $d_0, d_1 \colon A_1 \rightrightarrows A_0$ in $\mathscr{C}$, called respectively "source" and "target",

(4) an arrow $i \colon A_0 \longrightarrow A_1$ in $\mathscr{C}$, called "identity",

(5) an arrow $c \colon A_1 \times_{A_0} A_1 \longrightarrow A_1$ in $\mathscr{C}$, called "composition", where the pullback $(A_1 \times_{A_0} A_1, \pi_1, \pi_0)$ is that of d_0, d_1 (see diagram 8.1).

These data must satisfy the following axioms:

$$A_1 \times_{A_0} A_1 \xrightarrow{\pi_0} A_1$$

$$\pi_1 \downarrow \qquad\qquad \downarrow d_1$$

$$A_1 \xrightarrow{\quad d_0 \quad} A_0$$

Diagram 8.1

(1) $d_0 \circ i = 1_{A_0} = d_1 \circ i$;

(2) $d_1 \circ \pi_1 = d_1 \circ c, \quad d_0 \circ \pi_0 = d_0 \circ c$;

(3) $c \circ \begin{pmatrix} 1_{A_1} \\ i \circ d_0 \end{pmatrix} = 1_{A_1} = c \circ \begin{pmatrix} i \circ d_1 \\ 1_{A_1} \end{pmatrix}$;

(4) $c \circ (1_{A_1} \times_{A_0} c) = c \circ (c \times_{A_0} 1_{A_1})$

(consult the following comment as far as notation is concerned).

When $\mathscr{C}$ is the category of sets: A_0 is the set of objects; A_1 is the set of morphisms; d_0 maps an arrow to its source; d_1 maps an arrow to its target; i maps an object to the identity on this object; $A_1 \times_{A_0} A_1$ is the set of composable pairs (g, f) of arrows, i.e. the source of g equals the target of f; c is the composition mapping the pair (g, f) to $g \circ f$. Again for $\mathscr{C} = \mathsf{Set}$, the first axiom asserts that given an object $a \in A_0$, the identity on a is indeed a morphism from a to a. The second axiom indicates that a composite $g \circ f$ has for target the target of g and for source the source of f. The third axiom is that of identities; it makes good sense since the relations

$$d_1 \circ i \circ d_0 = d_0, \quad d_0 \circ i \circ d_1 = d_1$$

imply the existence of factorizations

$$\begin{pmatrix} 1 \\ i \circ d_0 \end{pmatrix} : A_1 \longrightarrow A_1 \times_{A_0} A_1, \quad \begin{pmatrix} i \circ d_1 \\ 1 \end{pmatrix} : A_1 \longrightarrow A_1 \times_{A_0} A_1,$$

through the pullback $A_1 \times_{A_0} A_1$. The fourth axiom expresses the associativity of the composition: the domain of the two composites is the "object of composable triples" $A_1 \times_{A_0} A_1 \times_{A_0} A_1$, i.e. the pullback of d_0 and $d_1 \circ \pi_1$ or, equivalently, the pullback of $d_0 \circ \pi_0$ and d_1; it is again routine to check the existence of the required factorizations

$$1 \times_{A_0} c \colon A_1 \times_{A_0} (A_1 \times_{A_0} A_1) \longrightarrow A_1 \times_{A_0} A_1,$$

$$c \times_{A_0} 1 \colon (A_1 \times_{A_0} A_1) \times_{A_0} A_1 \longrightarrow A_1 \times_{A_0} A_1$$

through the pullbacks.

Definition 8.1.2 *Let $\mathscr{C}$ be a category with pullbacks. Given two internal categories $\mathscr{A}$, $\mathscr{B}$, an internal functor $F\colon \mathscr{A} \longrightarrow \mathscr{B}$ is a pair of morphisms*

$$F_0\colon A_0 \longrightarrow B_0, \quad F_1\colon A_1 \longrightarrow B_1$$

which satisfies the following conditions:

(1) $d_0 \circ F_1 = F_0 \circ d_0, \quad d_1 \circ F_1 = F_0 \circ d_1$;
(2) $F_1 \circ i = i \circ F_0$;
(3) $F_1 \circ c = c \circ (F_1 \times_{F_0} F_1)$.

Again when $\mathscr{C}$ is the category of sets, F_0 maps an object a to the object $F(a)$ and F_1 maps an arrow $f\colon a \longrightarrow a'$ to the arrow $F(f)$. The first axiom indicates that $F(f)$ is an arrow from $F(a)$ to $F(a')$, while the second and the third axiom express respectively that F commutes with identities, and the composition law. For the sake of precision, let us make explicit the definition of $F_1 \times_{F_0} F_1$: it is the unique morphism

$$F_1 \times_{F_0} F_1\colon A_1 \times_{A_0} A_1 \longrightarrow B_1 \times_{B_0} B_1$$

such that $\pi_0 \circ (F_1 \times_{F_0} F_1) = F_1 \circ \pi_0, \quad \pi_1 \circ (F_1 \times_{F_0} F_1) = F_1 \circ \pi_1$.

Definition 8.1.3 *Let $\mathscr{C}$ be a category with pullbacks. Given two internal categories $\mathscr{A}$, $\mathscr{B}$ and two internal functors $F, G\colon \mathscr{A} \longrightarrow \mathscr{B}$, an internal natural transformation $\alpha\colon F \Rightarrow G$ is a morphism $\alpha\colon A_0 \longrightarrow B_1$ which satisfies the following conditions:*

(1) $d_0 \circ \alpha = F_0, \quad d_1 \circ \alpha = G_0$;
(2) $c \circ (\alpha \circ d_1, F) = c \circ (G, \alpha \circ d_0)$.

Again when $\mathscr{C} = \mathsf{Set}$, α maps an object $a \in A_0$ to the component α_a of the natural transformation; the first axiom indicates that α_a is a morphism from $F(a)$ to $G(a)$, while the second axiom is the usual naturality rule.

Proposition 8.1.4 *Let $\mathscr{C}$ be a category with pullbacks. Internal categories, internal functors and internal natural transformations organize themselves in a 2-category.*

Proof All the proofs are just diagram chasing arguments. Therefore we give the constructions and leave the verifications to the reader.

We consider internal functors $F\colon \mathscr{A} \longrightarrow \mathscr{B}$, and $G\colon \mathscr{B} \longrightarrow \mathscr{C}$, the composite $G \circ F\colon \mathscr{A} \longrightarrow \mathscr{C}$ is given by the pair of morphisms $G_0 \circ F_0$, $G_1 \circ F_1$.

Next, given internal functors $F, G, H\colon \mathscr{A} \longrightarrow \mathscr{B}$ and natural transformations $\alpha\colon F \Rightarrow G$ and $\beta\colon G \Rightarrow H$, the composite $\beta \circ \alpha\colon F \Rightarrow H$ is $c \circ (\beta, \alpha)$, where c is the composition of $\mathscr{B}$.

Finally, given internal functors $F, G\colon \mathscr{A} \rightrightarrows \mathscr{B}$, $F', G'\colon \mathscr{B} \rightrightarrows \mathscr{C}$ and internal natural transformations $\alpha\colon F \Rightarrow G$, $\alpha'\colon F' \Rightarrow G'$, the internal natural transformation $\alpha' * \alpha\colon F' \circ F \Rightarrow G' \circ G$ is defined by the composite

$$c \circ (G_1' \circ \alpha, \alpha' \circ F_0) = c \circ (\alpha' \circ G_0, F_1' \circ \alpha)$$

where c is the composition of $\mathscr{C}$.   □

Proposition 8.1.5 *Let $\mathscr{C}$ be a category with pullbacks. For every object $C \in \mathscr{C}$, the representable functor*

$$\mathscr{C}(C, -)\colon \mathscr{C} \longrightarrow \mathsf{Set}$$

maps internal categories, internal functors and internal natural transformations to small categories, functors and natural transformations respectively.

Proof From 2.9.4, $\mathscr{C}(C, -)$ preserves pullbacks.   □

Examples 8.1.6

8.1.6.a Consider a category $\mathscr{C}$ with pullbacks and an object $A \in \mathscr{C}$. One gets an internal "discrete" category by putting $A_0 = A = A_1$, $d_0 = d_1 = i = c = 1_A$.

8.1.6.b Consider a category $\mathscr{C}$ with finite limits, an internal category $\mathscr{B}$ and a morphism $b\colon \mathbf{1} \longrightarrow B_0$ (where $\mathbf{1}$ is the terminal object). Now b can be thought of as an "external object" of $\mathscr{B}$; see 8.1.5. Given another internal category $\mathscr{A}$, the data

$$A_0 \longrightarrow \mathbf{1} \xrightarrow{\ b\ } B_0, \quad A_1 \longrightarrow \mathbf{1} \xrightarrow{\ b\ } B_0 \xrightarrow{\ i\ } B_1$$

define an internal functor "constant on b".

8.1.6.c Consider a category $\mathscr{C}$ with pullbacks. Given an internal category $\mathscr{A}$, one gets a "dual" internal category $\mathscr{A}^*$ by permuting the roles of d_0, d_1 and twisting the composition morphism accordingly.

8.2 Internal base-valued functors

The category of sets is not small, thus cannot be seen as a category internal to Set. Nevertheless, given a small category $\mathscr{A}$, the functors $\mathscr{A} \longrightarrow \mathsf{Set}$ play a key role in category theory. Therefore, given a category $\mathscr{A}$ internal to $\mathscr{C}$, we want to internalize the notion of a functor $P\colon \mathscr{A} \longrightarrow \mathscr{C}$. To do this, we must define first the "family of objects $\bigl(P(a) \in \mathscr{C}\bigr)_{a \in A_0}$"; from 1.2.7.a, this should be an arrow $p\colon P \longrightarrow A_0$ in

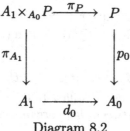

Diagram 8.2

$\mathscr{C}$, where P is thought of as the disjoint union $P = \coprod_{a \in A_0} P(a)$ and p is thought of as the mapping sending $x \in P(a)$ to the index a. Clearly, one must also define the action of the "arrows" of $\mathscr{A}$ on the "elements of P".

Definition 8.2.1 *Let $\mathscr{C}$ be a category with pullbacks and let $\mathscr{A}$ be an internal category. By an internal $\mathscr{C}$-valued functor $P \colon \mathscr{A} \longrightarrow \mathscr{C}$ we mean*

(1) an object $P \in |\mathscr{C}|$ together with a morphism $p_0 \colon P \longrightarrow A_0$ of $\mathscr{C}$,
(2) an arrow $p_1 \colon A_1 \times_{A_0} P \longrightarrow P$ of $\mathscr{C}$, where $(A_1 \times_{A_0} P, \pi_{A_1}, \pi_P)$ is the pullback of d_0, p_0 (see diagram 8.2).

These data are required to satisfy the following axioms:

(1) $p_0 \circ p_1 = d_1 \circ \pi_{A_1}$;
(2) $p_1 \circ (i \circ p_0, 1_P) = 1_P$;
(3) $p_1 \circ (1_{A_1} \times_{A_0} p_1) = p_1 \circ (c \times_{A_0} 1_P)$.

When $\mathscr{C} = \mathbf{Set}$ and $P \colon \mathscr{A} \longrightarrow \mathbf{Set}$ is a functor: $P = \coprod_{a \in A_0} P(a)$; $p_0(x) = a$ if $x \in P(a)$; $p_1(f, x) = P(f)(x)$ if $f \colon a \longrightarrow b$ and $x \in P(a)$. The first axiom indicates that $P(f)(x) \in P(b)$, while the second and the third axioms are the usual compatibility rules with the identities and the composition. (Observe that the third axiom expresses the equality between two arrows defined on $A_1 \times_{A_0} A_1 \times_{A_0} P$.)

In the situation of 8.2.1, an internal $\mathscr{C}$-valued functor $P \colon \mathscr{A}^* \longrightarrow \mathscr{C}$ on the dual internal category (see 8.1.6.c) is also called an *internal presheaf*, extending the terminology of 3.2.2, volume 3.

Definition 8.2.2 *Let $\mathscr{C}$ be a category with pullbacks, $\mathscr{A}$ an internal category and $P, Q \colon \mathscr{A} \rightrightarrows \mathscr{C}$ two internal $\mathscr{C}$-valued functors, written explicitly as $P = (P, p_0, p_1)$, $Q = (Q, q_0, q_1)$. By an internal natural transformation $\alpha \colon P \Rightarrow Q$ we mean an arrow $\alpha \colon P \longrightarrow Q$ such that*

(1) $q_0 \circ \alpha = p_0$,
(2) $\alpha \circ p_1 = q_1 \circ (1_{A_1} \times_{A_0} \alpha)$.

In the case $\mathscr{C} = \mathbf{Set}$, α maps $x \in \boldsymbol{P}(a)$ $(a \in A_0)$ on some element $\alpha(x)$ which, by axiom (1), lies in $\boldsymbol{Q}(a)$. Axiom (2) is the usual naturality condition.

Proposition 8.2.3 *Let $\mathscr{C}$ be a category with pullbacks and $\mathscr{A}$ an internal category. The internal $\mathscr{C}$-valued functors $\mathscr{A} \longrightarrow \mathscr{C}$ and internal natural transformations between them organize themselves in a category.*

Proof Composition of internal natural transformations is just composition of arrows in $\mathscr{C}$. $\qquad\qquad\qquad\qquad\qquad\qquad\qquad\qquad\qquad\qquad\square$

Example 8.2.4

Let $\mathscr{C}$ be a category with finite limits and $\mathscr{A}$ an internal category. Given an object $X \in \mathscr{C}$, one defines a "constant" $\mathscr{C}$-valued internal functor $\boldsymbol{P} \colon \mathscr{A} \longrightarrow \mathscr{C}$ by putting

- $P = X \times A_0$,
- $p_0 \colon P \longrightarrow A_0$ is the second projection,
- $p_1 \colon A_1 \times_{A_0} P \longrightarrow P$ is the second projection.

Now given an arrow $f \colon X \longrightarrow Y$ of $\mathscr{C}$ and the corresponding $\mathscr{C}$-valued internal constant functor on Y, written $\boldsymbol{Q} \colon \mathscr{A} \longrightarrow \mathscr{C}$,

$$f \times 1 \colon X \times A_0 \longrightarrow Y \times A_0$$

defines an internal "constant" natural transformation $\boldsymbol{P} \Rightarrow \boldsymbol{Q}$. When $\mathscr{C} = \mathbf{Set}$, we obviously recapture the notions of constant functor and constant natural transformation (see 1.2.8.e and 1.3.6.d).

Proposition 8.2.5 *Let $\mathscr{C}$ be a finitely complete category. We consider a fixed object $A \in \mathscr{C}$ and the corresponding discrete internal category $\mathscr{A}$ (see 8.1.6.a). The category of internal $\mathscr{C}$-valued functors on $\mathscr{A}$ is equivalent to the category $\mathscr{C}/A$ of arrows over A.*

Proof Given an arrow $p \colon P \longrightarrow A$, the pullback $A_1 \times_{A_0} P$ is just P again, since $d_0 \colon A_1 \longrightarrow A_0$ is the identity arrow. Therefore putting $p_0 = p$, $p_1 = 1_P$ yields an internal $\mathscr{C}$-valued functor.

Choose now $\boldsymbol{P} = (P, p_0, p_1)$ an internal $\mathscr{C}$-valued functor as in 8.2.1. Since $d_0 \colon A_1 \longrightarrow A_0$ is the identity arrow, $A_1 \times_{A_0} P$ can be identified with P and thus p_1 can be seen as a morphism $p_1 \colon P \longrightarrow P$. The first axiom in 8.1.1 becomes $p_0 \circ p_1 = p_0$. Since the projection $A_1 \times_{A_0} P \longrightarrow P$ is now the identity, the arrow $(i \circ p_0, 1_P)$ in the second axiom is in particular such that

$$(i \circ p_0, 1_P) = \pi_2 \circ (i \circ p_0, 1_P) = 1_P,$$

so that $p_1 = 1_P$. The last axiom is now redundant. Thus P is just the triple $(P, p_0, 1_P)$.

Now given two $\mathscr{C}$-valued internal functors

$$P = (P, p_0, 1_P), \quad Q = (Q, q_0, 1_Q)$$

on $\mathscr{A}$, the second axiom in 8.2.2 becomes a tautology so that an internal natural transformation $\alpha \colon P \Rightarrow Q$ is just a morphism $\alpha \colon P \longrightarrow Q$ such that $q_0 \circ \alpha = p_0$. □

Let us conclude this section with a straightforward but very important observation:

Proposition 8.2.6 *Let $\mathscr{C}$ be a category with pullbacks, $\mathscr{A}$ an internal category and $P \colon \mathscr{A} \longrightarrow \mathscr{C}$ an internal $\mathscr{C}$-valued functor. One gets a new internal category $\mathscr{P}$ by defining*

- $P_0 = P$,
- $P_1 = A_1 \times_{A_0} P$, *the pullback of d_0, p_0,*
- $\overline{d_0} \colon P_1 \longrightarrow P_0$, *the second projection of $A_1 \times_{A_0} P$,*
- $\overline{d_1} \colon P_1 \longrightarrow P_0$, *the arrow p_1,*
- $\overline{i} \colon P_0 \longrightarrow P_1$, *the arrow $(i \circ p_0, 1_P)$,*
- $\overline{c} \colon P_1 \times_{P_0} P_1 \longrightarrow P_1$, *the arrow $\bigl(c \circ (\pi_1, \pi_3), \pi_4\bigr)$,*

where $\pi_1, \pi_2, \pi_3, \pi_4$ are the four projections of

$$(A_1 \times_{A_0} P) \times_P (A_1 \times_{A_0} P) \cong P_1 \times_{P_0} P_1.$$

This new internal category $\mathscr{P}$ is called the "internal category of elements" of P. □

When $\mathscr{C} = \mathsf{Set}$, the elements of P_0 are indeed those of $\coprod_{a \in A_0} P(a)$, thus the pairs (x, a) where $x \in P(a)$. Now if $f \colon (x, a) \longrightarrow (y, b)$ is an arrow in the category of elements of P (see 1.6.4), $f \in A_1$, b is determined as $d_1(f)$ and y is determined as $P(f)(x)$. Therefore giving an arrow in $\mathsf{Elts}(P)$ is equivalent to giving an object (x, a) and an arrow in $\mathscr{A}$ with domain a; this proves that $A_1 \times_{A_0} P$ is indeed the set of arrows of $\mathsf{Elts}(P)$. The rest is obvious.

Going back to the definition of a flat functor $P \colon \mathscr{A} \longrightarrow \mathsf{Set}$ (see 6.3.1), it is thus sensible to say that the internal $\mathscr{C}$-valued functor $P \colon \mathscr{A} \longrightarrow \mathscr{C}$ of 8.2.6 is *flat* when the corresponding internal category $\mathscr{P}$ is cofiltered, i.e. when the dual internal category $\mathscr{P}^*$ (see 8.1.6.c) is filtered. This last notion can be easily defined.

Definition 8.2.7 *Let $\mathscr{C}$ be a finitely complete category and $\mathscr{A}$ an internal category. Using the notation of 8.1.1 we say that $\mathscr{A}$ is filtered when*

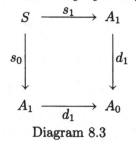

Diagram 8.3

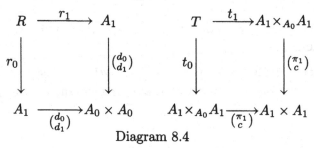

Diagram 8.4

(1) the unique morphism $A_0 \longrightarrow 1$ is a strong epimorphism,

(2) the morphism $\begin{pmatrix} d_0 \circ s_0 \\ d_0 \circ s_1 \end{pmatrix} : S \longrightarrow A_0 \times A_0$ is a strong epimorphism, where (S, s_0, s_1) is the kernel pair of d_1 (see 2.5.4 and diagram 8.3),

(3) the morphism $t: T \longrightarrow R$ is a strong epimorphism, where (R, r_0, r_1), (T, t_0, t_1) are the kernel pairs of, respectively, $\binom{d_0}{d_1}$ and $\binom{\pi_1}{c}$ (see 2.5.4 and diagram 8.4) while t is the unique morphism such that $r_0 \circ t = \pi_0 \circ t_0$, $r_1 \circ t = \pi_0 \circ t_1$.

First of all observe that this definition makes sense since, in condition (3), writing p_0, p_1 for the two projections of the product $A_0 \times A_0$, and $\overline{p_0}, \overline{p_1}$ for those of the product $A_1 \times A_1$, the first projections satisfy the following relations

$$p_0 \circ \binom{d_0}{d_1} \circ \pi_0 \circ t_0 = d_0 \circ \pi_0 \circ t_0$$
$$= d_0 \circ c \circ t_0$$
$$= d_0 \circ \overline{p_1} \circ \binom{\pi_1}{c} \circ t_0$$
$$= d_0 \circ \overline{p_1} \circ \binom{\pi_1}{c} \circ t_1$$
$$= d_0 \circ c \circ t_1$$
$$= d_0 \circ \pi_0 \circ t_1$$
$$= p_0 \circ \binom{d_0}{d_1} \circ \pi_0 \circ t_1,$$

and analgously for the second projections

$$p_1 \circ \left(\begin{smallmatrix}d_0\\d_1\end{smallmatrix}\right) \circ \pi_0 \circ t_0 = d_1 \circ \pi_0 \circ t_0$$
$$= d_0 \circ \pi_1 \circ t_0$$
$$= d_0 \circ \overline{p_0} \circ \left(\begin{smallmatrix}\pi_1\\c\end{smallmatrix}\right) \circ t_0$$
$$= d_0 \circ \overline{p_0} \circ \left(\begin{smallmatrix}\pi_1\\c\end{smallmatrix}\right) \circ t_1$$
$$= d_0 \circ \pi_1 \circ t_1$$
$$= d_1 \circ \pi_0 \circ t_1$$
$$= p_1 \circ \left(\begin{smallmatrix}d_0\\d_1\end{smallmatrix}\right) \circ \pi_0 \circ t_1.$$

Thus $\left(\begin{smallmatrix}d_0\\d_1\end{smallmatrix}\right) \circ \pi_0 \circ t_0 = \left(\begin{smallmatrix}d_0\\d_1\end{smallmatrix}\right) \circ \pi_0 \circ t_1$ and t is correctly defined.

In the case of the category $\mathscr{C} = \mathsf{Set}$ of sets, where strong epimorphisms are just surjections (see 1.8.5.a), the three axioms become, with shorthand notation:

(1) $\exists A \in A_0$,
(2) $S = \{(f,g) \,|\, f\colon A \longrightarrow C, g\colon B \longrightarrow C\}$,
 $\forall A, B \ \ \exists(A \overset{f}{\longrightarrow} C, B \overset{g}{\longrightarrow} C)$,
(3) $R = \{(f,g) \mid f\colon A \longrightarrow B, g\colon A \longrightarrow B\}$,
 $T = \{(f,g,h) \,|\, f\colon A \longrightarrow B, g\colon A \longrightarrow B, h\colon B \longrightarrow C, h \circ f = h \circ g\}$,
 $\forall f, g\colon A \rightrightarrows B \ \ \exists h\colon B \longrightarrow C \ \ h \circ f = h \circ g$.

This is precisely definition 2.13.1.

Definition 8.2.8 *Let $\mathscr{C}$ be a finitely complete category and $\mathscr{A}$ an internal category. Consider an internal $\mathscr{C}$-valued functor $\boldsymbol{P}\colon \mathscr{A} \longrightarrow \mathscr{C}$ and its internal category $\mathscr{P}$ of elements. $\boldsymbol{P}$ is flat when the internal dual category $\mathscr{P}^*$ is filtered.*

8.3 Internal limits and colimits

Consider a finitely complete category $\mathscr{C}$ and an internal category $\mathscr{A}$. Let us write $\mathscr{C}^{\mathscr{A}}$ for the category of $\mathscr{C}$-valued internal functors on $\mathscr{A}$ and internal natural transformations between them (see 8.2.3). We also write (see 3.2.3)

$$\Delta_{\mathscr{A}}\colon \mathscr{C} \longrightarrow \mathscr{C}^{\mathscr{A}}$$

to denote the functor which maps an object $X \in \mathscr{C}$ to the $\mathscr{C}$-valued internal constant functor on X and an arrow $f\colon X \longrightarrow Y$ on the constant internal natural transformation on f; see 8.2.4. With the considerations of 3.2.3 in mind, we define:

$$A_1 \times_{A_0} P \xrightarrow{\ 1 \times \mu\ } A_1 \times_{A_0} (M \times A_0)$$

$$
\begin{array}{ccc}
& p_1 \Big\downarrow & \Big\downarrow \pi_2 \\
\end{array}
$$

$$P \xrightarrow[\ \mu\ ]{} M \times A_0$$

Diagram 8.5

Definition 8.3.1 Let $\mathscr{C}$ be a finitely complete category, $\mathscr{A}$ an internal category in $\mathscr{C}$ and $P: \mathscr{A} \longrightarrow \mathscr{C}$ a $\mathscr{C}$-valued internal functor. With the previous notation,

- by an internal limit of P, we mean a coreflection of P along the functor $\Delta_{\mathscr{A}}$,
- by an internal colimit of P, we mean a reflection of P along the functor $\Delta_{\mathscr{A}}$.

$\mathscr{C}$ is said to be internally (co)complete when the internal (co)limit exists for every $\mathscr{A}$ and every P.

A condition for internal cocompleteness is fairly easy to obtain.

Proposition 8.3.2 *A category with finite limits and coequalizers is internally cocomplete.*

Proof With the notation of 8.3.1, 8.1.1 and 8.2.1, we consider the coequalizer l of π_2, p_1,

$$A_1 \times_{A_0} P \underset{p_1}{\overset{\pi_2}{\rightrightarrows}} P \xrightarrow{\ l\ } L,$$

where the pullback $A_1 \times_{A_0} P$ is that of d_0, p_0 and π_2 is the second projection. This yields an object $L \in \mathscr{C}$ and an obvious internal natural transformation $\lambda: P \Rightarrow \Delta_{\mathscr{A}}(L)$ determined by the arrow $\binom{l}{p_0}: P \longrightarrow L \times A_0$. We shall prove that (L, λ) is the coreflection of P along $\Delta_{\mathscr{A}}$.

Choose another object $M \in \mathscr{C}$ together with an internal natural transformation $\mu: P \Rightarrow \Delta_{\mathscr{A}}(M)$. This yields an arrow

$$\mu: P \longrightarrow M \times A_0$$

which, by the naturality of μ, makes diagram 8.5 commute. This at once implies

$$\mu \circ p_1 = \pi_2 \circ (1 \times \mu) = \mu \circ \pi_2,$$

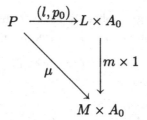

Diagram 8.6

from which we get the existence of a unique factorization $m: L \longrightarrow M$ such that $m \circ l = \mu$. But by the naturality of μ, $\pi_2 \circ \mu = p_0$. Therefore the relation $m \circ l$ is equivalent to the commutativity of diagram 8.6, i.e. to the relation $\Delta_{\mathscr{A}}(m) \circ \lambda = \mu$. □

Let us now consider the more difficult case of internal limits. We shall study cartesian closed categories in chapter 6 of volume 2. A finitely complete category $\mathscr{C}$ is cartesian closed when, for every object $A \in \mathscr{C}$, the functor "product with A",

$$- \times A: \mathscr{C} \longrightarrow \mathscr{C}, \quad X \mapsto X \times A,$$

has a right adjoint, written

$$(-)^A: \mathscr{C} \longrightarrow \mathscr{C}, \quad X \mapsto X^A.$$

In 3.1.6.b we have already observed that the category Set of sets is cartesian closed; in that case, the functor $(-)^A$ is just $\mathsf{Set}(A, -)$ or, in other words, the functor "raising to the power A". This last fact admits an obvious generalization. First of all we need a definition.

Definition 8.3.3 *Let $\mathscr{C}$ be a finitely complete category.*

- *By an internal product in $\mathscr{C}$ we mean the internal limit of a $\mathscr{C}$-valued internal functor $P: \mathscr{A} \longrightarrow \mathscr{C}$, where $\mathscr{A}$ is a discrete internal category in the sense of 8.1.6.a.*
- *Let X, A be objects of $\mathscr{C}$. By the internal power $X^A \in \mathscr{C}$ we mean, if it exists, the internal limit of the constant $\mathscr{C}$-valued functor on X (see 8.2.4) defined on the discrete internal category on A (see 8.1.6.a).*

In the case $\mathscr{C} = \mathsf{Set}$, these definitions exactly describe the usual notions of product and power. Observe that given a $\mathscr{C}$-valued functor (P, p_0, p_1) on the discrete internal category $\mathscr{A}$, the corresponding product in the case $\mathscr{C} = \mathsf{Set}$ is $\prod_{a \in A_0} p_0^{-1}(a)$, since P is the functor mapping $a \in A_0$ to $p_0^{-1}(a)$.

It is now interesting to observe that in the definition of a cartesian closed category, the object X^A is indeed an internal power.

Proposition 8.3.4 *Let $\mathscr{C}$ be a finitely complete category. The following conditions are equivalent:*

(1) $\mathscr{C}$ is cartesian closed;
(2) for every pair $X, A \in \mathscr{C}$ of objects, the internal power X^A exists.

Proof   The constant $\mathscr{C}$-valued internal functor on X defined on the discrete internal category on A is given by

$$P = X \times A, \quad p_0 = \pi_A \colon X \times A \longrightarrow A, \quad p_1 = 1_{X \times A}.$$

The internal limit of such a functor, if it exists, is a pair (L, λ) universal for the properties $L \in \mathscr{C}$ and $\lambda \colon L \times A \longrightarrow X \times A$, with $\pi_A \circ \lambda = \pi_A$ (see 8.2.5). This is equivalent to a pair (L, λ_X) universal for the properties $L \in \mathscr{C}$ and $\lambda_X \colon L \times A \longrightarrow X$. This is exactly the definition of the coreflection of X along the functor $- \times A \colon \mathscr{C} \longrightarrow \mathscr{C}$.   □

Let us now consider a finitely complete category $\mathscr{C}$ and the discrete internal category $\mathscr{A}$ on some object $A \in \mathscr{C}$ (see 8.1.6.a). By 8.2.5, we know that the category of $\mathscr{C}$-valued internal functors on $\mathscr{A}$ is just $\mathscr{C}/A$. The corresponding functor

$$\Delta_{\mathscr{A}} \colon \mathscr{C} \longrightarrow \mathscr{C}^{\mathscr{A}}$$

considered at the beginning of this section is thus (see 8.2.4)

$$\Delta_A \colon \mathscr{C} \longrightarrow \mathscr{C}/A, \quad X \mapsto \left(X \times A \xrightarrow{\ \pi_C\ } A \right).$$

The existence theorem for internal limits is then contained in the following equivalences:

Proposition 8.3.5 *Let $\mathscr{C}$ be a finitely complete category. The following conditions are equivalent:*

(1) $\mathscr{C}$ is cartesian closed;
(2) $\mathscr{C}$ admits all internal powers;
(3) each functor $\Delta_A \colon \mathscr{C} \longrightarrow \mathscr{C}/A$ has a right adjoint $\prod_A$;
(4) $\mathscr{C}$ admits all internal products;
(5) $\mathscr{C}$ is internally complete.

Proof   We have proved the equivalence (1) $\Leftrightarrow$ (2) in 8.3.4. The equivalence (3) $\Leftrightarrow$ (4) is attested by 8.2.5 and the previous considerations about Δ_A. Obviously (5) $\Rightarrow$ (4) and (4) $\Rightarrow$ (2). So it remains to prove (1) $\Rightarrow$ (5). As a lemma, we first prove (1) $\Rightarrow$ (3).

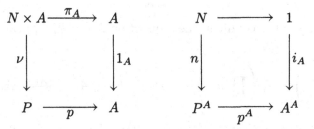

$$\begin{array}{ccc}
M & \longrightarrow & 1 \\
m\downarrow & & \downarrow i_A \\
P^A & \xrightarrow{\;p^A\;} & A^A
\end{array}
\qquad
\begin{array}{ccc}
M \times A & \xrightarrow{\;\pi_A\;} & A \\
\mu\downarrow & & \downarrow 1_A \\
P & \xrightarrow{\;p\;} & A
\end{array}$$

Diagram 8.7

$$\begin{array}{ccc}
N \times A & \xrightarrow{\;\pi_A\;} & A \\
\nu\downarrow & & \downarrow 1_A \\
P & \xrightarrow{\;p\;} & A
\end{array}
\qquad
\begin{array}{ccc}
N & \longrightarrow & 1 \\
n\downarrow & & \downarrow i_A \\
P^A & \xrightarrow{\;p^A\;} & A^A
\end{array}$$

Diagram 8.8

We suppose $\mathscr{C}$ cartesian closed and consider an object $p\colon P \longrightarrow A$ of $\mathscr{C}/A$. We compute the left-hand side pullback in diagram 8.7, where $i_A\colon \mathbf{1} \longrightarrow A^A$ is the morphism corresponding to $1_A\colon A \longrightarrow A$ by cartesian adjointness. Observe that when $\mathscr{C} = \mathsf{Set}$,

$$\begin{aligned}
M &= \left\{ (x_a)_{a\in A} \,\middle|\, x_a \in P;\ \big(p(x_a)\big)_{a\in A} = (a)_{a\in A} \right\} \\
&= \left\{ (x_a)_{a\in A} \,\middle|\, x_a \in p^{-1}(a) \right\} \\
&= \prod_{a\in A} p^{-1}(a).
\end{aligned}$$

Applying the cartesian adjunction to the previous pullback, we get the commutative right-hand square in diagram 8.7 and therefore a morphism $\mu\colon (M \times A, \pi_A) \longrightarrow (P,p)$ in $\mathscr{C}/A$. We shall prove that (M, μ) is the coreflection of (P,p) along Δ_A.

Given $N \in \mathscr{C}$ and $\nu\colon (N \times A, \pi_A) \longrightarrow (P,p)$, the commutativity of the first square in diagram 8.8 implies, by cartesian adjunction, the commutativity of the second square. This implies the existence of a unique factorization $l\colon N \longrightarrow M$ through the pullback defining M. The relation $m \circ l = n$ is equivalent to $\nu \circ (l \times 1_A) = \mu$, concluding the proof that (M, μ) is the coreflection of (P,p) along Δ_A.

Now we prove (1) $\Rightarrow$ (5). We consider an internal category $\mathscr{A}$ and a $\mathscr{C}$-valued internal functor $\boldsymbol{P}\colon \mathscr{A} \longrightarrow \mathscr{C}$. Rephrasing the construction of

2.8.1, it suffices to define the internal limit of P via the equalizer

$$L \xrightarrow{\quad l \quad} \prod_{A_0}\left(P \xrightarrow{\quad p_0 \quad} A_0\right) \underset{\beta}{\overset{\alpha}{\Longrightarrow}} \prod_{A_1}\left(P \times_{A_0} A_1 \xrightarrow{\quad \pi_{A_1} \quad} A_1\right)$$

where the pullback $P \times_{A_0} A_1$ is that of p_0, d_1 and $\prod_{A_0}, \prod_{A_1}$ are the right adjoints to $\Delta_{A_0}, \Delta_{A_1}$. It remains to define α and β. The composite

$$u: \prod_{A_0}(p_0) \times A_1 \xrightarrow{\quad 1 \times d_1 \quad} \prod_{A_0}(p_0) \times A_0 \xrightarrow{\quad \eta_{(P,p_0)} \quad} P$$

where $\eta_{(P,p_0)}$ is the counit of the adjunction $\Delta_{A_0} \dashv \prod_{A_0}$, gives rise to a morphism

$$\binom{u}{1_{A_1}}: \sigma_{A_1}\left(\prod_{A_0}(p_0)\right) \longrightarrow \left(P \times_{A_0} A_1 \xrightarrow{\quad \pi_{A_1} \quad} A_1\right),$$

which corresponds to α via the adjunction $\Delta_{A_1} \dashv \prod_{A_1}$. On the other hand the composite

$$v: \prod_{A_0}(p_0) \times A_1 \xrightarrow{\quad 1 \times d_0 \quad} \prod_{A_0}(p_0) \times A_0 \xrightarrow{\quad \eta_{(P,p_0)} \quad} P$$

gives rise to a morphism

$$\binom{1_{A_1}}{v}: \prod_{A_0}(p_0) \times A_1 \longrightarrow A_1 \times_{A_0} P,$$

where the pullback $A_1 \times_{A_0} P$ is that of d_0, p_0. The composite

$$w: \prod_{A_0}(p_0) \times A_1 \xrightarrow{\quad \binom{1}{v} \quad} A_1 \times_{A_0} P \xrightarrow{\quad p_1 \quad} P$$

now gives rise to a morphism

$$\binom{w}{1_{A_1}}: \sigma_{A_1}\left(\prod_{A_0}(p_0)\right) \longrightarrow \left(P \times_{A_0} A_1 \xrightarrow{\quad \pi_{A_1} \quad} A_1\right),$$

which corresponds to β via the adjunction $\Delta_{A_1} \dashv \prod_{A_1}$. Finally l corresponds via the adjunction $\Delta_{A_0} \dashv \prod_{A_0}$ to a morphism

$$\lambda: (L \times A_0, \pi_{A_0}) \longrightarrow (P, p_0)$$

in $\mathscr{C}/A_0$. We leave to the reader the diagram chasing argument verifying that $(L, \lambda: L \times A_0 \longrightarrow P)$ defines the internal limit of P.   □

8.4 Exercises

8.4.1 Let $\mathscr{C}$ be a finitely complete category. Prove that the category of internal categories and internal functors in $\mathscr{C}$ is itself finitely complete.

8.4.2 Let $\mathscr{C}$ be a finitely complete category. An internal monoid in $\mathscr{C}$ is an internal category $\mathscr{A}$ whose "object of objects" A_0 is the terminal object. Prove that the category of internal monoids in the category Gr of groups is equivalent to the category of abelian groups.

8.4.3 Let $\mathscr{C}$ be a category with finite limits and coequalizers. Given two internal categories $\mathscr{A}, \mathscr{B}$, an internal distributor $\mathscr{A} \multimap \mathscr{B}$ is a $\mathscr{C}$-valued internal functor $\mathscr{B}^* \times \mathscr{A} \longrightarrow \mathscr{C}$ (see 8.4.1). A morphism of internal distributors is just an internal natural transformation between them. Considering 8.2.6 and the construction in the proof of 7.8.2, define the composite of two internal distributors.

8.4.4 Let $\mathscr{C}$ be a category with pullbacks. Given an internal category $\mathscr{A}$, construct the internal category of arrows of $\mathscr{A}$ (see 1.2.7.c).

8.4.5 Let $\mathscr{C}$ be a finitely complete cartesian closed category. Prove that internal categories and internal functors in $\mathscr{C}$ again constitute a cartesian closed category.

8.4.6 Let $\mathscr{C}$ be a finitely complete category. $\mathscr{C}$ is said to be *locally cartesian closed* when for each object $A \in \mathscr{C}$, the category $\mathscr{C}/A$ is cartesian closed. Prove that $\mathscr{C}$ is locally cartesian closed iff for every arrow $f \colon A \longrightarrow B$ in $\mathscr{C}$, the functor

$$f^* \colon \mathscr{C}/B \longrightarrow \mathscr{C}/A$$

obtained by pulling back along f has a right adjoint $\prod_f$.

8.4.7 In the non-abelian cohomology of groups, one chooses as a system of coefficients a device called a *crossed module*. This is a quadruple (H, Π, ρ, Φ) where H and Π are groups and

$$\rho \colon H \longrightarrow \Pi, \quad \Phi \colon \Pi \longrightarrow \mathsf{Aut}(H)$$

are group homomorphisms, with $\mathsf{Aut}(H)$ the group of automorphisms of H; the following axioms are required from these data:

(1) $\forall h, k \in H \quad \Phi\big(\rho(h)\big)(k) = hkh^{-1}$;
(2) $\forall h \in H \quad \forall f \in \Pi \quad \rho\big(\Phi(f)(h)\big) = f\rho(h)f^{-1}$.

A morphism $(H, \Pi, \rho, \Phi) \longrightarrow (H', \Pi', \rho', \Phi')$ of crossed modules is a pair

$$(\eta \colon H \longrightarrow H', \quad \theta \colon \Pi \longrightarrow \Pi')$$

of group homomorphisms such that

(1) $\rho' \circ \eta = \theta \circ \rho$,

(2) $\forall h \in H \quad \forall f \in \Pi \quad \eta\big(\Phi(f)(h)\big) = \Phi'\big(\theta(f)\big)\big(\eta(h)\big)$.

Prove that the category of crossed modules is equivalent to the category of internal categories and internal functors in the category of groups.

Bibliography

J. Adamek, H. Herrlich, G. Strecker, *Abstract and concrete categories*, John Wiley, 1990

M. Barr, C. Wells, *Category theory for computing science*, Prentice Hall, 1990

J. Bell, M. Machover, *A course in mathematical logic*, North Holland, 1977

J. Bénabou, Critères de représentabilité des foncteurs, *Comptes Rendus de l'Académie des Sciences de Paris* **260**, 1965, 752–755

J. Bénabou, *Introduction to bicategories*, Springer LNM **40**, 1967, 1–77

J. Bénabou, *Les distributeurs*, Université Catholique de Louvain, Institut de Mathématique Pure et Appliquée, rapport **33**, 1973

A. Blass, The interaction between category theory and set theory, *AMS series in Contemporary Mathematics* **30**, 1984, 5–29

F. Borceux, B. Veit, On the left exactness of orthogonal reflections, *Journal of Pure and Applied algebra* **49**, 1987, 33–42

N. Bourbaki, *Eléments de mathématique* **XXII**, *Théorie des ensembles, livre I, Structures, chapitre 4*, Hermann, 1957

I. Bucur, A. Deleanu, *Introduction to the theory of categories and functors*, John Wiley, 1968

B. Eckmann, P. Hilton, Commuting limits with colimits, *Journal of Algebra* **11**, 1969, 116–144

C. Ehresmann, *Catégories et structures*, Dunod, 1965

C. Ehresmann, Sur l'existence de structures libres et de foncteurs adjoints, *Cahiers de Topologie et Géométrie Différentielle* **IX**, 1967, 1–146

S. Eilenberg, S. Mac Lane, Natural isomorphisms in group theory, *Proceedings of the National Academy of Sciences USA* **28**, 1942, 537–543

S. Eilenberg, S. Mac Lane, General theory of natural equivalences, *Transactions of the AMS* **58**, 1945, 231–294

S. Eilenberg, J.C. Moore, *Foundations of relative homological algebra*, Memoirs of the AMS, **55**, 1965

C. Faith, *Algebra II: Ring theory*, Springer, 1976

P. Freyd, *Abelian categories*, Harper and Row, 1964

P. Freyd, G.M. Kelly, Categories of continuous functors I, *Journal of Pure and Applied Algebra* **2**, 1972, 169–191

P. Freyd, A. Scedrov, *Categories; allegories*, North Holland, 1990

P. **Gabriel**, M. **Zisman**, *Calculus of fractions and homotopy theory*, Springer, 1967

P. **Gabriel**, F. **Ulmer**, *Lokal präsentierbare Kategorien*, Springer LNM **221**, 1971

J. **Gray**, *Formal category theory I: Adjointness for 2-categories*, Springer LNM **391**, 1974

A. **Grothendieck**, J.L. **Verdier**, *Préfaisceaux*, Springer LNM **269**, 1970, 1–218

H. **Herrlich**, *Topologische Reflexionen und Coreflexionen*, Springer LNM **78**, 1968

J. **Isbell**, Structure of categories, *Bulletin of the AMS* **72**, 1966, 619–655

J. **Isbell**, Small subcategories and completeness, *Mathematical System Theory* **2**, 1968, 27–50

D. **Kan**, Adjoint functors, *Transactions of the AMS* **87**, 1958, 294–329

J.L. **Kelley**, *General topology*, Van Nostrand, 1955, and Springer, 1975

G.M. **Kelly**, Monomorphisms, epimorphisms and pullbacks, *Journal of the Australian Mathematical Society* **9**, 1969, 124–142

G.M. **Kelly**, R. **Street**, *Review of the elements of 2-categories*, Springer LNM **420**, 1974, 75–103

G.M. **Kelly**, A unified treatment of transfinite constructions for free algebras, free monoids, colimits, associated sheaves, and so on, *Bulletin of the Australian Mathematical Society* **22**, 1980, 1–83

J.F. **Kennison**, On limit preserving functors, *Illinois Journal of Mathematics* **12**, 1968, 616–619

A.G. **Kuroš**, *Theory of groups*, volumes I, II, Chelsea, 1956

J. **Lambeck**, *Completion of categories*, Springer LNM 24, 1966

J. **Lambeck**, Non-commutative localizations, *Bulletin of the AMS* **79**, 1973, 857–872

F.W. **Lawvere**, S. **Schanuel**, *Conceptual mathematics: a first introduction to categories*, Buffalo Workshop Press, 1991

S. **Mac Lane**, *Categories for the working mathematician*, Springer, 1971

B. **Mitchell**, *Theory of categories*, Academic Press, 1965

B. **Mitchell**, The dominion of Isbell, *Transactions of the AMS* **167**, 1972, 319–331

M. **Naimark**, *Normed rings*, Noordhof, 1959

R. **Paré**, On absolute colimits, *Journal of Algebra* **19**, 1971, 80–95

R. **Paré**, Connected components and colimits, *Journal of Pure and Applied Algebra* **3**, 1973, 21–42

B. **Pareigis**, *Categories and functors*, Academic Press, 1970

H. **Schubert**, *Categories*, Springer, 1970

E. **Spanier**, *Algebraic topology*, McGraw-Hill, 1966

L. **Steen**, J. **Seebach**, *Counterexamples in topology*, Holt, Rinehart and Winston 1970, Springer 1978

R. **Street**, Two constructions on lax-functors, *Cahiers de Topologie et Géométrie Différentielle* **XIII**, 1972, 217–264

F. **Ulmer**, Properties of dense and relative adjoint functors, *Journal of Algebra* **8**, 1968, 77–95

F. **Ulmer**, Representable functors with values in arbitrary categories, *Journal of Algebra* **8**, 1968, 96–129

N. **Yoneda**, On the homology theory of modules, *Journal of the Faculty of Sciences of Tokyo* **I-7**, 1954, 193–227

Index